AF388970

Grid-Connected Renewable Energy Sources

Grid-Connected Renewable Energy Sources

Editor

Jesus C. Hernández

MDPI • Basel • Beijing • Wuhan • Barcelona • Belgrade • Manchester • Tokyo • Cluj • Tianjin

Editor
Jesus C. Hernández
University of Jaen
Spain

Editorial Office
MDPI
St. Alban-Anlage 66
4052 Basel, Switzerland

This is a reprint of articles from the Special Issue published online in the open access journal *Electronics* (ISSN 2079-9292) (available at: https://www.mdpi.com/journal/electronics/special_issues/grid_renewable_energy).

For citation purposes, cite each article independently as indicated on the article page online and as indicated below:

LastName, A.A.; LastName, B.B.; LastName, C.C. Article Title. *Journal Name* **Year**, *Volume Number*, Page Range.

ISBN 978-3-0365-0998-3 (Hbk)
ISBN 978-3-0365-0999-0 (PDF)

Contents

About the Editor

Jesus C. Hernández was born in Jaén, Spain. He received his M.Sc. and Ph.D. degrees from the University of Jaén, in 1994 and 2003, respectively. Since 1995, he has been an Associate Professor in the Department of Electrical Engineering, University of Jaén. His current research interests are smart grids, smart meters, renewable energy, and power electronics.

Electronics

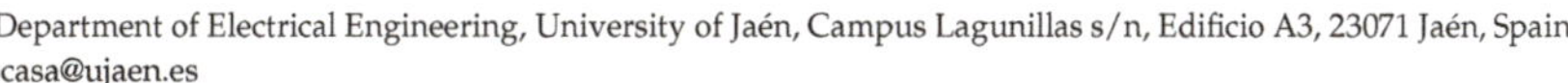

Editorial

Grid-Connected Renewable Energy Sources

Jesus C. Hernández

Department of Electrical Engineering, University of Jaén, Campus Lagunillas s/n, Edificio A3, 23071 Jaén, Spain; jcasa@ujaen.es

Abstract: The use of renewable energy sources (RESs) is a need of global society. This editorial, and its associated Special Issue "Grid-Connected Renewable Energy Sources", offer a compilation of some of the recent advances in the analysis of current power systems composed after the high penetration of distributed generation (DG) with different RESs. The focus is on both new control configurations and novel methodologies for the optimal placement and sizing of DG. The eleven accepted papers certainly provide a good contribution to control deployments and methodologies for the allocation and sizing of DG.

Keywords: renewable energy conversion; power conditioning devices; renewable energy policies; power quality; computations methods; control strategies; electric vehicle charging; energy management systems; ancillary services; monitoring; prognostic and diagnostic

Citation: Hernández, J.C. Grid-Connected Renewable Energy Sources. *Electronics* **2021**, *10*, 588. https://doi.org/10.3390/electronics 10050588

Received: 8 February 2021
Accepted: 1 March 2021
Published: 3 March 2021

1. Introduction

A significant share of the electricity presently produced worldwide is generated by centralized systems based on conventional fossil fuel plants or nuclear power. Nonetheless, energy systems across the globe are undergoing a significant transformation as society transitions towards the widespread use of clean and sustainable energy sources. Thus, renewable distributed generation (DG) can play a major role in the future world's energy generation. As a result, the architecture of energy generation is rapidly shifting from centralized to decentralized power plants. Instead of depending on only one energy source, a wide range of types can be used. This will eventually lead to the extensive inclusion of power electronics based on non-synchronous or renewable DG. The grid-interactive power converters involved will significantly improve the flexibility, controllability, and efficiency of conventional power systems. Smart control strategies can thus enable energy management capabilities as well as the provision of ancillary services to the grid from renewable DG. Nevertheless, maintaining a reliable and safe power system poses significant challenges. The optimization of the allocation and sizing of renewable DG is also an important task in this context.

2. A Short Review of the Contributions in This Issue

To cover the above-mentioned promising areas of research and development, this Special Issue collects the latest research on relevant topics, and more importantly, addresses current issues related to more sustainable, safer, and more resilient power systems. This Special Issue received fifteen submissions, of which eleven were accepted for publication. Various topics are addressed in these manuscripts, mainly on energy storage and photovoltaic and wind power technologies. The contents of these papers are summarized hereafter.

In the paper "Intermittent Renewable Energy Sources: The Role of Energy Storage in the European Power System of 2040" [1], the authors address the challenges of variable renewable energy integration in Europe in terms of the power capacity and energy capacity of stationary storage technologies.

Two papers discuss the issues related to the time framework. In article [2], the enhanced time delay compensator approach manages varying time delays inherent to communications schemes in power systems. This introduces the perspective of network latency instead of dead time. The potential provision of advanced energy services from the small grid-connected renewable DG is described in article [3]. This work shows that the smart meter roll-out in household-prosumers offers an easy access to granular meter measurements for future advanced energy applications. The development and calibration of a smart meter prototype is adjusted as required in the provision of advanced energy services.

The grid integration of renewable DG is increasingly pursued all over the world due to several technical, economical, and environmental benefits. Consequently, three articles work on the optimal placement and sizing of DG in distribution networks. The work done by S. Katyara et al. [4] exploits genetic algorithms for the proper placement of a new DG; meanwhile, the energy management is designed using a fuzzy inference system. A hybrid master–slave optimization procedure is proposed in article [5]. In the master stage, the discrete version of the sine–cosine algorithm determines the optimal location of the DG. In the slave stage, the problem of the optimal sizing of the DG is solved through the implementation of the second-order cone programming equivalent model to obtain solutions for the resulting optimal power flow problem. Still on the topic of DG, the comparison between AC and DC distribution networks to provide electricity to rural and urban areas from the point of view of grid energy losses and greenhouse gas emissions impact is analyzed in [6]. Results confirm that power distribution with DC technology is more efficient than its AC counterpart.

Electrical system performance can be enhanced to maximize economic benefits by incorporating an appropriate electric energy control scheme. Accordingly, this Special Issues includes four papers focused on converter control. The research in [7] introduces an LC impedance source bi-directional DC–DC converter by redesigning after rearranging the reduced number of components of a switched boost bi-directional DC–DC converter. This novel design decreases the stress on the capacitor voltage compared to existing topologies in literature. The second paper, [8], proposes a proportional-integral passivity-based controller to regulate the amplitude and frequency of the three-phase output voltage in a DC–AC converter with an LC filter. The third paper, [9], authored by Md.R. Hazari et al., presents a novel control scheme for a battery-based energy storage system (ESS) in coordination with an SCIG-based wind turbine generator (WTG), which improves the low voltage ride through capability. A closely related work of [9] is [10], which automatically identifies the frequency stabilization by WTG and ESS. The work models a control scheme that shares their releasable and absorbable energies between both sources.

Another range of topics is addressed in this Special Issue. Thus, a microcontroller-based PV source emulator is presented in [11]. This modeling and design is based on a completely new technique, which consists in subtracting an adequate amount of current from a fixed direct current source so as to reproduce the desired I–V characteristic.

3. Future

While the potential of grid-connected renewable DG has been extensively recognized by the research community, several significant obstacles still remain, and therefore, research and technology are essentials tools for attaining a new energy paradigm, which is going towards the responsible and careful use of the environment's resources. In the future, it can be expected that more friendly and pollution-free energy sources will be required in large amounts for sustainable societies. In this circumstance, for the optimal planning of DG in electrical distribution networks, appropriate converter control strategies and approaches should be ready.

Acknowledgments: First of all, the Guest Editor would like to thank all the authors for their great contributions in this Special Issue. We are grateful to all the peer reviewers who helped evaluate the submitted manuscripts and made beneficial suggestions for their improvement. We would also like to acknowledge the editorial board of *Electronics*, who graciously invited me to guest edit this Special

Issue. Finally, we are grateful to the *Electronics* Editorial Office staff; they worked tirelessly to ensure the peer-review schedule and timely publication.

Conflicts of Interest: The author declares no conflict of interest.

References

1. Zsiborács, H.; Baranyai, N.H.; Vincze, A.; Zentkó, L.; Birkner, Z.; Máté, K.; Pintér, G. Intermittent renewable energy sources: The role of energy storage in the european power system of 2040. *Electronics* **2019**, *8*, 729. [CrossRef]
2. Molina-Cabrera, A.; Ríos, M.A.; Besanger, Y.; Hadjsaid, N.; Montoya, O.D. Latencies in power systems: A database-based time-delay compensation for memory controllers. *Electronics* **2021**, *10*, 208. [CrossRef]
3. Sanchez-Sutil, F.; Cano-Ortega, A.; Hernandez, J.C.; Rus-Casas, C. Development and calibration of an open source, low-cost power smart meter prototype for PV household-prosumers. *Electronics* **2019**, *8*, 878. [CrossRef]
4. Katyara, S.; Shaikh, M.F.; Shaikh, S.; Khand, Z.H.; Staszewski, L.; Bhan, V.; Majeed, A.; Shah, M.A.; Zbigniew, L. Leveraging a genetic algorithm for the optimal placement of distributed generation and the need for energy management strategies using a fuzzy inference system. *Electronics* **2021**, *10*, 172. [CrossRef]
5. Montoya, O.D.; Molina-Cabrera, A.; Chamorro, H.R.; Alvarado-Barrios, L.; Rivas-Trujillo, E. A Hybrid approach based on SOCP and the discrete version of the SCA for optimal placement and sizing DGs in AC distribution networks. *Electronics* **2021**, *10*, 26. [CrossRef]
6. Montoya, O.D.; Serra, F.M.; De Angelo, C.H. On the efficiency in electrical networks with AC and DC operation technologies: A comparative study at the distribution stage. *Electronics* **2020**, *9*, 1352. [CrossRef]
7. Raveendhra, D.; Dhaouadi, R.; Rehman, H.; Mukhopadhyay, S. LC impedance source bi-directional converter with reduced capacitor voltages. *Electronics* **2020**, *9*, 1062. [CrossRef]
8. Serra, F.M.; Fernández, L.M.; Montoya, O.D.; Gil-González, W.; Hernández, J.C. Nonlinear voltage control for three-phase DC-AC converters in hybrid systems: An application of the PI-PBC method. *Electronics* **2020**, *9*, 847. [CrossRef]
9. Hazari, M.R.; Jahan, E.; Mannan, M.A.; Tamura, J. Coordinated Control scheme of battery storage system to augment LVRT capability of SCIG-based wind turbines and frequency regulation of hybrid power system. *Electronics* **2020**, *9*, 239. [CrossRef]
10. Kang, M.; Yoon, G.; Hong, S.; Park, J.; Kim, J.; Baek, J. Coordinated frequency stabilization of wind turbine generators and energy storage in microgrids with high wind power penetration. *Electronics* **2019**, *8*, 1390. [CrossRef]
11. Merenda, M.; Iero, D.; Carotenuto, R.; Della Corte, F.G. Simple and low-cost photovoltaic module emulator. *Electronics* **2019**, *8*, 1445. [CrossRef]

electronics

Article

Latencies in Power Systems: A Database-Based Time-Delay Compensation for Memory Controllers

Alexander Molina-Cabrera [1,2], Mario A. Ríos [1], Yvon Besanger [3,*], Nouredine Hadjsaid [3] and Oscar Danilo Montoya [4,5]

[1] Department of Electrical and Electronic Engineering, School of Engineering, Universidad de los Andes, Carrera 1 No. 18A-12, Bogotá D.C. 111711, Colombia; almo@utp.edu.co (A.M.-C.); mrios@uniandes.edu.co (M.A.R.)
[2] Department of Electric Power Engineering, Universidad Tecnológica de Pereira, Pereira 660003, Colombia
[3] Department of Electrical Power Engineering, Universite Grenoble Alpes, CNRS, Grenoble INP, G2Elab, 21 Avenue des Martyrs, 38000 Grenoble, France; nouredine.hadj-said@grenoble-inp.fr
[4] Facultad de Ingeniería, Universidad Distrital Francisco José de Caldas, Bogotá D.C. 11021, Colombia; odmontoyag@udistrital.edu.co
[5] Laboratorio Inteligente de Energía, Universidad Tecnológica de Bolívar, Cartagena 131001, Colombia
* Correspondence: Yvon.Besanger@g2elab.grenoble-inp.fr

Abstract: Time-delay is inherent to communications schemes in power systems, and in a closed loop strategy the presence of latencies increases inter-area oscillations and security problems in tie-lines. Recently, Wide Area Measurement Systems (WAMS) have been introduced to improve observability and overcome slow-rate communications from traditional Supervisory Control and Data Acquisition (SCADA). However, there is a need for tackling time-delays in control strategies based in WAMS. For this purpose, this paper proposes an Enhanced Time Delay Compensator (ETDC) approach which manages varying time delays introducing the perspective of network latency instead dead time; also, ETDC takes advantage of real signals and measurements transmission procedure in WAMS building a closed-loop memory control for power systems. The strength of the proposal was tested satisfactorily in a widely studied benchmark model in which inter-area oscillations were excited properly.

Keywords: power systems analysis; interconnected power systems; latencies; time-delay effects; wide area monitoring systems

Citation: Molina-Cabrera, A.; Ríos, M.A.; Besanger, Y.; Hadjsaid, N.; Montoya, O.D. Latencies in Power Systems: A Database-Based Time-Delay Compensation for Memory Controllers. *Electronics* **2021**, *10*, 208. https://doi.org/10.3390/electronics10020208

Received: 27 November 2020
Accepted: 13 January 2021
Published: 18 January 2021

Publisher's Note: MDPI stays neutral with regard to jurisdictional claims in published maps and institutional affiliations.

1. Introduction

Wide Area Measurement Systems (WAMS) bring information to the control center in modern power systems to improve observability for achieving stability and security [1,2]. WAMS are integrated by Phasor Measurement Units (PMU) and a sophisticated communication infrastructure [3–7]. This communication infrastructure is based on several standards, interoperability of devices, language, and agents involved in the procedure. Also, this infrastructure involves protocols such as TCP/IP and UDP/IP to provide redundancy, guarantee information integrity, solve traffic problems, and tackle failures of some links [8–11]. For observability purposes, WAMS is better than traditional supervisory control and data acquisition (SCADA). Unfortunately, the measurements managed by WAMS reach the control center with time-delay due to the size of large power systems monitored as well as procedures such as filtering, digitalization, time stamping, and labeling [8,12]. The time-delay is problematic in closed loop control for power systems; i.e., time-lapse in the backward channel is an important issue that emerges with undesirable effects on the performance of transferred power due to inter-area oscillations and frequency oscillations, among others [13–17].

There are many authors who are committed to tackling the time-delay problem in power systems, but the main problem remains unsolved. In the most common perspective, the time compensators were used considering time-lapse as a dead-time phenomenon.

In this direction, one of the first time-delay strategies for compensation was the Smith Predictor (SP), primarily used in chemical process [18–20]. In [19], Chaudhuri et al. implemented a unified SP to design a damping control, but its success dependeds strongly on the exactitude of the model. Then, they proposed a unified Smith phasorial time-delay compensator which runs fast with few calculations; nonetheless, it only works well for small values of delay [21]. Moktari developed a time compensator based on fuzzy logic, which works well for higher values of time delays close to tens of milliseconds; however, it fails in cases of disturbances associated with tripping lines [22]. Another more sophisticated perspective considers the complexities of actual WAMS. For instance, [23] obtains time-delay values from isolated arriving signals using time-stamp from the data package. The authors of [10] employ the knowledge of the WAMS only to simulate the communication procedure in a Hardware In the Loop (HIL) test but leave these data out of the compensation strategy. The first perspective presented above fails because it considers time-delays as a dead-time phenomenon, as in a chemical process; in the second perspective, the complexity of WAMS is considered, but no capitalization of the valuable information available from the communication process is carried out. Generally, recent investigations suffer from misconceptions in modeling and simulating scenarios of time-delay performance [24–26]. Following with the literature revision, the authors in [27] obtain worthy results using buffers and a wide-area power oscillation damper (WPOD) to compensate for delays and packet dropouts; nevertheless, the implementation is based on a straightforward model for single-input single-output applied in a Double-Fed Induction Generator (DFIG). Another drawback in the proposal: it takes a long time to stabilize signals (more than 20 seconds) with dangerous power explorations (more than one hundred percent and negative values). Consequently, the main gap in the literature to be filled is the absence of a definitive proposal to face power system oscillations increased by time-delays, considering a more realistic performance of WAMS with high values for network latencies [10,28–31].

There are previous contributions to the aforementioned gap: our early works include an adapted Model Predictive Control (MPC) capable of dealing with the nonlinear large-scale nature of delayed power system (power systems with delays in WAMS) to maintain stability; then, we introduced time-delay compensation suitable for tackling fixed and varying values of latencies [32–34]. This paper contributes in time compensation strategies for reaching a memory closed loop control in delayed power systems; furthermore, this work details the WAMS' performance to offer the background needed for the proposal (to run more realistic simulations). The strategy is named Enhanced Time-Delay Compensation (ETDC): it features a Kalman-based time-compensator and a time-organized database of measurements. The ETDC introduces the concept of the Most Updated Available (MUA) information, which is the key value to feed the MPC. The inclusion of historic data in the control closed loop leads to a memory controller to face latencies.

The paper is organized as follows: Section 2 summarizes the general problem of latencies in communications, including some details of their behavior, typical values, and shape. Subsequently, a typical performance of time-delays is illustrated, which will be considered for simulations. Section 3 describes the model of power systems with delays in the backward channel to offer a better picture of the control problem from a math perspective, thereby allowing us to hypothesize the possible solution. Then, in Section 3.2, the Enhanced Time-Delay Compensation is introduced, and some statements are made to study the convenience of the solution. Finally, the results of the simulation (Section 4), conclusions, and further works (Section 5) are presented.

2. Latencies in Wams Communications Infrastructure: Reaching a More Realistic Model

2.1. The General Delayed Communication Infrastructure

At present, energy management systems (EMS) have been improved with the introduction of Phasor Measurement Units (PMUs) and their supporting infrastructure. PMUs are fundamental elements of the modern Wide Area Measurement Systems (WAMS), whose capabilities allow the development of Wide Area Monitoring and Control Systems (WAMCS) [3,34–36].

Secondly, signals from several PMUs installed in different locations are gathered with Phasor Data Concentrators (PDCs) to run an additional routine of synchronization; then, each PDC sends the new data packet to the super PDC (SPDC) or directly to the control center. However, these signals do not arrive at the same time due to the inherent latency of communication channels. For this reason, before running the synchronization, PDC manages the absence of simultaneity of the arrival of signals with the assistance of either TCP/IP or UDC/IP protocol. Despite this compensatory mechanism, the total latency increases [8,9].

The measurements are taken with electrical sensors in substations and main buses; then, they are synchronized with PMUs and sent to the control center. The resulting information packet complies with C.37.117.7 and IEC 61850 standards [37]. This procedure includes not only metering but also filtering, processing, digitalization, and time-stamp labelling. Obviously, this procedure add delays to the signal and it is considered the first component of the latency in WAMS.

In cases of a wide area power system, the utilities gather the information of PMUs and other regional PDCs through a Super PDC (SPDCs). Then the SPDCs send the signals to the control center. The time spent by the SPDCs in this process increases the time-delay [8,9,38]. Finally, at the control center, all the signals are gathered to allow control in Wide Area Monitoring Protection and Control (WAMPaC) [6,9].

Another important component of latencies is the signal flying time to travel through the medium and routers at each link. Their stochastic behaviour contributes to the total time delay [35].

In addition to the aforementioned physical infrastructure, the TCP/IP protocol, as well as C.37.117.7 and IEC 61850 standards, are introduced in power system communications to provide flexible, reliable, and standardized communications [8,37]. Standards C.37.117.7 and IEC 61850 provide values of satellite-synchronized time-stamp for each measurement in the WAMS. This time stamp is valuable in the proposal because it allows to determine each signal latency value, which is subsequently included in the compensation scheme. Now, the TCP/IP protocol is responsible for the information interchange among the agents in the communication network. Based on the protocol, the sender always guarantees the reception of the data packets in the final destination. Despite this guaranteed reception, some authors have focused their efforts on the need to face a non-existent loss of data packets [39].

The aforementioned WAMS description shows the capabilities of the power systems' communication infrastructure. In WAMS, the well-structured packets transferred with TCP/IP should be organized in databases due to their useful information (e.g., time stamps), with the purpose of improving power system control and time compensators.

2.2. Behavior and Modeling of the Random Time Delays in the Pmu Communication Infrastructure

The communication infrastructure can be understood as a net of devices interconnected through communication links. They collect and process information in some topological nodes. Two major delay components are considered to describe the latency. Firstly, τ_v denotes the total time delay produced by the devices mentioned in the previous section (PMU, PDC, SPDC). Secondly, τ_{link} is the total additional latency due to the links of the communication process; τ_{link} is related to the weather conditions and the medium and is generally greater than τ_v [35]. The sum of these values is given by:

$$\tau_d = \tau_v + \tau_{\text{link}}, \tag{1}$$

Table 1 shows the typical corresponding ranges of τ_d for different communication links in power systems.

Table 1. Ranges of Latencies in Communications [40].

Communication Link	Associated Delay τ_d (ms)
Fiber-Optic cables	100–150
Microwave Links	100–150
Power line carrier (PLC)	150–50
Telephone lines	200–300
Satellite link	500–700

The development of tools to deal with delayed systems requires clarity of the network latencies. The time-delay is stochastic and unpredictable; however, it is possible to model time delays with a probability density function as in [8]. The authors in [8] gathered empirical data, and then they made a goodness-of-fit test. They found that Gaussian shape properly models the time delay behavior of τ_d with a formal math representation given by:

$$\tau_d = \mathcal{N}(\mu_G, \sigma_G), \tag{2}$$

The advantages of the τ_d representation in (2) are a better description of the varying time-delay and the possibility of running more realistic simulations. Figure 1 illustrates an example: the histogram of events called D_G with a mean value $\mu_G = 300$ ms and standard deviation $\sigma_G = 60$ ms. The maximum value for the time delay in this set is close to 550 ms. Typical values μ_G and σ_G presented in [8,40] were adapted for this research to include the effects of the latencies in the simulation of the power systems' behavior.

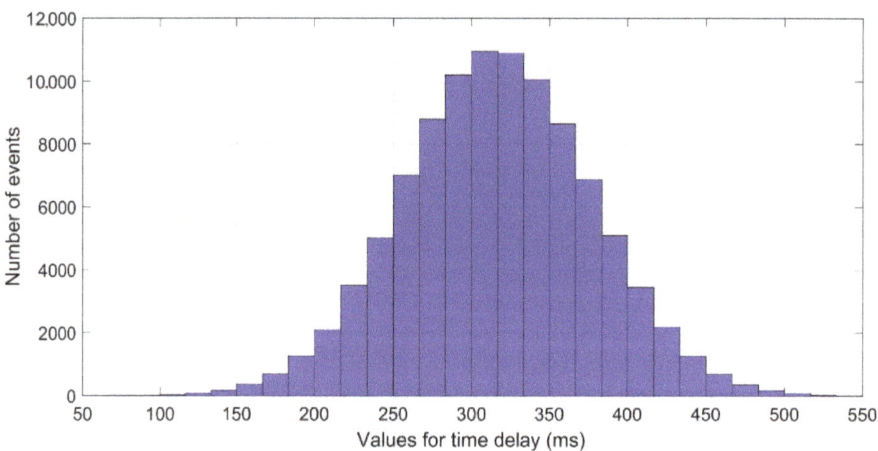

Figure 1. Typical behavior of latencies in a Phasor Measurement Unit (PMU)-based communication system.

3. The Enhanced Time-Delay Compensator for Time-Delayed Power Systems Control: Modeling the Problem to Propose a Solution

In this section, we derive the math model of the closed loop considering the power system with its nonlinearities because the proposed control must act over the actual nonlinear power system. The math model includes the behavior of network latencies in the feedback channel and the control law based on the estimated delayed states. Based on this model, it was possible to determine the complexities involved in the whole problem to hypothesize a solution. Then, in Section 3.2, the structure of the ETDC is described, taking into consideration the WAMS' description in Section 2 and the math model in Section 3.1. Following this, ETDC consistency and stability are studied trough some statements at the end of Section 3.2.

3.1. Time-Delayed Power Systems Modeling

The present paper proposes a formal model to include the nonlinearities of power systems and the variability of time-delays, which is made with the purpose to provide a more appropriate representation of the WAMS communications infrastructure. The formulation for modeling the dynamic behavior of nonlinear power systems is as described by (3)–(7):

$$\dot{x}(t) = f(x(t), u(t)), \tag{3}$$

In this case, $u(t)$ denotes the feedback control law. Basically, Equation (3) describes a memoryless closed loop strategy and if the control law is $u(t) = \gamma(x(t))$, then it turns into:

$$\dot{x}(t) = f(x(t), \gamma(x(t))), \tag{4}$$

Equation (4) represents the ideal case with a proper control law and accessibility to actualized states. Hence, considering the latency in communications in the feedback channel affecting $u(t)$, the nonlinear problem formulation turns into the autonomous model:

$$\dot{x}(t) = (x(t), \gamma(x(t - \tau_d))), \tag{5}$$

Now, by including the expression for τ_d denoted by (2):

$$\dot{x}(t) = (x(t), \gamma(x(t - \mathcal{N}(\mu_G, \sigma_G)))), \tag{6}$$

Equation (6) gathers two emerging difficulties to be tackled in real applications on power systems. The first is the power system's nonlinear nature (including parameter-changing, uncertainties and its large number of variables); the second is associated with time delays in the closed loop control strategy. In this regard, although the nonlinearity was successfully addressed in [32,33], the dead time misconception is yet to be faced.

Finally, the computation of the control law in Equation (4) requires the values of $x(t)$. The typical way to obtain the states in such a complex system is by using state estimators like Kalman filters. Basically, estimators derive estimated values of states, $\hat{x}(t)$, from output signals $y(t)$. Hence, Equation (6) turns into:

$$\dot{x}(t) = (x(t), \gamma(\hat{x}(t - \mathcal{N}(\mu_G, \sigma_G)))), \tag{7}$$

Note the following: Equation (7) establishes the complexity of the problem, which is highly nonlinear, and the controller must use estimated states $\hat{x}(t)$ instead of actual states $x(t)$. In addition, it is included the stochastic behaviour of the time-delay trough $\mathcal{N}(.)$. From this point, a valuable strategy must keep values of (7) as close as possible from (4) to achieve a good performance.

3.2. The Enhanced Time-Delay Compensator

This section shows an improved time compensator named Enhanced Time-Delay Compensator (ETDC), which is more suitable for practical power systems and represents a major improvement when compared to previous research in two main aspects: (a) it manages varying values of latencies under a new paradigm that re-evaluates the dead time misconception and (b) incorporates real WAMS operational elements.

As mentioned in the introduction, time-delay compensation is a strategy included in the closed loop controllers to obtain actualized states for the feedback control law in which signals reach the controller with a retard [18–20,41,42]. The main objective of a time compensator is to reduce the error $e(t_i) = \hat{x}(t_i) - x(t_i)$ at a given time t_i, where $\hat{x}(t_i)$ is calculated by the time compensator and $x(t_i)$ represents the actual and unknown states. In the compensation strategy, the vector $\hat{x}(t_i)$ is calculated from the delayed values $x(t_i - \tau_d)$ [18–20]. Most compensators work satisfactorily under two main conditions: the precise model of the system and the knowledge of the constant time-delay value.

However, it is almost impossible to have an exact model of the system. Consequently, the implementation of those compensators in real systems has been thwarted: instabilities emerge even with small errors in the model [18,41,43]. The challenge at this regard is identifying how to compensate delayed signals without adding instabilities: here ETDC plays an important role. In Figure 2, the traditional scheme of compensations is illustrated.

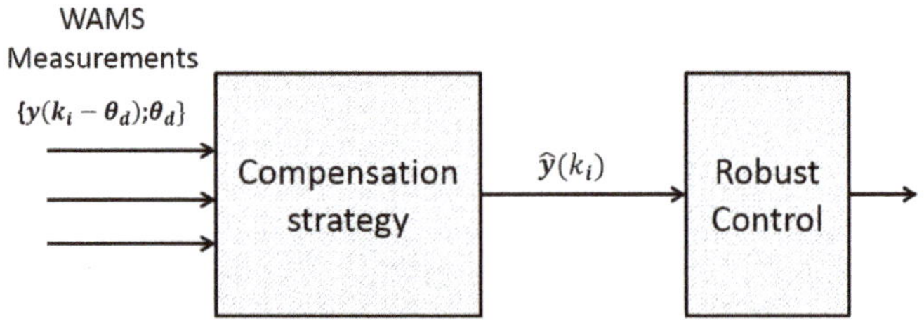

Figure 2. The Enhanced Time-Delay Compensator scheme.

The proposed ETDC is composed by two main components to be presented: the Sliding Prediction Block, and the added database. By this way, in the closed loop control the ETDC compensates the network latencies; next, signals are delivered to the control strategy. Then, control signals are sent to the power system.

The first component of the ETDC is the Sliding Prediction Block. The authors have been developing tools for damping oscillations in power systems with delays in their communication [33]. As a previous contribution, a time-delay compensator called Sliding Prediction Block (SPB) has been developed. It performs properly with the Model Predictive Control (MPC) strategy adapted to power systems. The SPB is as follows: the classic Kalman filter is fed by the couple $y(t_i)$-$u(t_i)$ to obtain $\hat{x}(t_i)$. The novelty here is: SPB is fed by the delayed states $\hat{x}(t_i - \tau_d)$ and the known historical control sequence $U(t_i) = [u(t_i - \tau_d) \dots u(t_i - \tau_m) \dots u(t_i)]$. The values of $\hat{x}(t_i - \tau_d)$ are obtained by a previous Kalman filter fed by $y(t_i - \tau_d)$-$u(t_i - \tau_d)$. Another Kalman filter stage is used recursively to obtain $\hat{x}(t_i)$ from $\hat{x}(t_i - \tau_d)$. Then, the states are brought to the control strategy.

The second main component of the ETDC is the database which allows keep old measurements. Discrete time is considered for the description. The database takes arriving signals $\{\hat{x}(k_i - \theta_d); \theta_d\}$ and lists them according to the value of the time delay (here, θ_d is the corresponding time-delay for the delayed signal $\hat{x}(k_i - \theta_d)$). The less-delayed data packet is allocated at the top of the list and denoted by $\{\hat{x}(k_i - \theta_{dm}); \theta_{dm}\}$. This packet will become the Most Updated Available state, $\hat{x}_{MUA}$, and it allows the building of a memory control strategy for delayed power systems. Then, $\hat{x}_{MUA}$ is delivered to the SPB for the time-compensation. The listing procedure is possible owing to the processing of signals during PMU measurements in compliance with IEEE C37.118 data formatting. Basically, from a specific signal, the PMU takes measurements and organizes them into data packets. Within the information included in the data packet, the time stamp is crucial for both the listing and time compensation.

The Algorithm 1 illustrates the ETDC with the two components. As shown, the simplicity of the procedure allows fast calculations and easy-implementation; also, it is higly scalable. In brief, the strength of the ETDC lies in his simplicity, with very good results.

Algorithm 1: Enhanced Time-delay Compensator.

Data: Read the information of the power system.

1 **Require:** Delayed states $\hat{x}(k_i - \theta_d)$; Time Delay θ_d; Buffered Control Signal $U(k_i)$;

2 **Ensure:** Estimated states $\hat{x}(k_i)$;

3 **Initialize:** iter $= 1$;

4 θ_d; $x_{DEL} \leftarrow \hat{x}(k_i - \theta_d)$;

5 **if** *iter* $= 1$ **then**

6      $x_{MUA} \leftarrow x_{DEL}$;

7      $\theta_{dm} \leftarrow \theta_d$;

8   iter $=$ iter $+ 1$;

9 **else**

10   **BEGIN** Database sorting and Listing;

11   **read database** (θ_{dm}, x_{MUA});

12   **if** $\theta_{dm} + 1 \geq \theta_d$ **then**

13          $x_{MUA} \leftarrow x_{DEL}$;

14          $\theta_{dm} \leftarrow \theta_d$;

15   **else**

16          $x_{MUA} \leftarrow x_{MUA}$;

17          $\theta_{dm} \leftarrow \theta_{dm} + 1$;

18   **end**

19   **END** Database sorting and Listing **BEGIN** Sliding Prediction Procedure;

20   **for** $j = k_i - \theta_d$ **to** $j = k_i - 1$ **do**

21          $x(j+1) = A(j) + Bu(j)$;

22          $\hat{x}(k_i) \leftarrow x(j+1)$;

23   **end**

24   **END** Sliding Prediction Procedure

25 **end**

26 iter $=$ iter $+ 1$.;

     Result: Return $\hat{x}(k_i)$

Now, once the signals are compensated by the ETDC, the signals are brought to the robust control strategy. In the case of this work, it was used Model Predictive Control (MPC). As illustrated below, all the strategy is coherently implemented considering the functioning of the MPC. According to Figure 3, the outputs of the Power System are measured, then, data packages are sent to the control center; and they arrive with network latencies. The output and control signals are used by the Kalman filter to obtain the states. Here, ETDC acts to compensate the time-delay in order to obtain an estimation of the current states. This work's main contribution is providing a very good estimation that allows a good performance of the closed control loop strategy. The MPC receives the estimated states to create the control sequence.

Model Predictive Control is responsible for the control task. The MPC strategy creates a time evolution of the states in a horizon of prediction using as initial condition the values of $\hat{x}(k_i)$ and the state-space model of the power system [44,45]. There, the evolution of the states are dependent from the control signals $U(k)$. Thereby, an optimal control problem is built considering an objective function and several constraints. The objective function includes minimization of efforts in control signals and the error in reference, the variable of interest here is the control signal. Physical limits and other considerations are included in the set of constrains. In this way, we derive a convex problem to be solved by any optimization technique [45–47]. The solution is a sequence of control signals in time, the first of which is applied to the power system. This procedure is made recursively at each sampling time using values of $\hat{x}(k_i)$ by the ETDC as the initial condition [44,45].

The whole compensation scheme proposed here, including database and SPB, is illustrated in Figure 3.

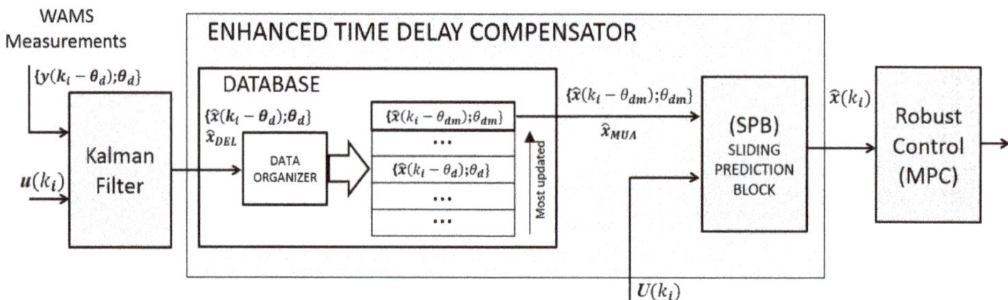

Figure 3. The Enhanced Time-Delay Compensator scheme.

Due to the ETDC, values of time-delay for the compensation strategy has quantitative reductions that can be established by comparison of the datasets from $\hat{x}_{\text{DEL}}$ and $\hat{x}_{\text{MUA}}$ (see signals in Figure 3). As an example, we took the Gaussian test-data called D_G of the Section 2 (Figure 1) and built with all the signals $\hat{x}_{\text{DEL}}$ (in Figure 3). Those signals were processed by the proposed storage block of the ETDC; so a new set of data D_W was obtained (corresponding to the set of signals $\hat{x}_{\text{MUA}}$ in Figure 3). The resulting histogram of events for the D_W dataset had Weibull shape with lower means values than the original sets. The dataset D_W (obtained with the MUA processing) has mean value $\mu_W = 254$ ms being almost 50 ms smaller than the mean value $\mu_G = 300$ ms for the dataset D_G (without the MUA processing). The data dispersion of the same set of data is also reduced and the maximum value for the latencies after the MUA processing is less than 400 ms, as Figure 4 shows. That is, while a traditional compensator is fed by signals with time-delays around 550 ms (histogram without MUA processing), with the same dataset, the ETDC will feed the SPB with time-delays under 400 ms (histogram of latencies with MUA processing) improving the performance of the compensator.

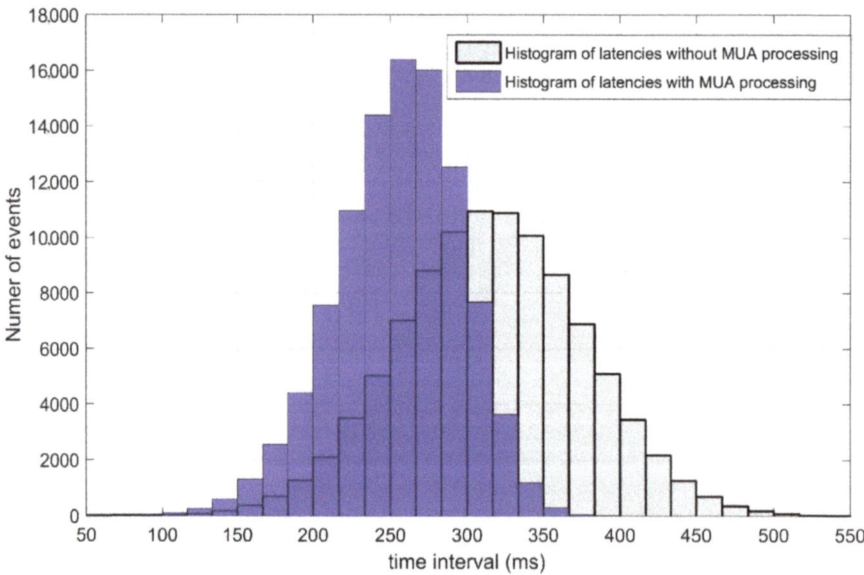

Figure 4. Time Delay shape after using enhanced time delay compensator (ETDC) compared with Sliding Prediction Block (SPB).

The use of the $\hat{x}_{\text{MUA}}$ information not only provokes changes in the shape of the data, but also enhances the performance of the SPB. Time-compensator routines are related with the time delay that needs to be compensated for; hence, the computational burden is lowered because $\hat{x}_{\text{MUA}}$ are less delayed. Additionally, the error in prediction is improved due to the reduction of the horizon time.

In order to offer information about the convenience of the solution based on the database added to the SPB compensation strategy, some statements are made by the authors.

Firstly, for the compensation procedure in discrete time k, a possible statement is: let Ω be an invariant set for $x(k)$ and $\hat{x}(k)$, let X be the time evolution of real states for the autonomous system $x(k+1) = f(x(k), \gamma(x(t)))$, and $\hat{X}$; the resulting trajectory of the time compensator with a representation $\hat{x}(k+1) = F(\hat{x}(k))$; both X and $\hat{X}$ exist in the interval of time $[k_i, k_i + T_{he}]$ and have $x_0 = x(k_i) \in \Omega$ as the initial condition. T_{he} represents the horizon of evolution. Additionally, the relationship between $f(\cdot)$ and $F(\cdot)$ includes the error $E(\cdot)$:

$$f(\cdot) = F(\cdot) + E(\cdot), \tag{8}$$

Given a small scalar $\epsilon > 0$, and with associated value $\delta > 0$, which defines a set of functions β:

$$\beta = \{F(\hat{x}(k)) \mid \|F(\hat{x}(k)) - f(x(k), \gamma(x(k)))\| < \delta\}, \ \forall k \in [k_i, k_i + T_{he}], \tag{9}$$

The states trajectory derived by the compensator satisfies the following:

$$\|\hat{X} - X\| \leq \epsilon, \ \forall k \in [k_i, k_i + T_{he}], \tag{10}$$

As such, with limited $E(\cdot)$, it corresponds to an appropriate representation $F(\cdot)$ of the real system $f(\cdot)$. Then, $\hat{X}$ and X are close trajectories remaining in the invariant set Ω.

Secondly, regarding the error in the compensation, a statement could be formulated: let $\zeta = \|\hat{x}(k_i + T_{he}) - x(k_i + T_{he})\|$ be the error between $\hat{x}(k_i + T_{he})$ and $x(k_i + T_{he})$ at the end of the time interval $[k_i, k_i + T_{he}]$. Let $[k_i, k_i + T_{db}]$ be a new interval for the evolution of $x(k)$ and $\hat{x}(k)$. Given a small scalar value for $\zeta > 0$, and with the same associated value of $\delta > 0$ for the same compact set of functions β (see Equation (9)), the final values obtained by the compensator satisfy:

$$\|\hat{x}(k_i + T_{db}) - x(k_i + T_{db})\| < \zeta, \ \forall T_{db} < T_{he}, \tag{11}$$

This means that although the trajectories $\hat{X}$ and X are close in the time interval $[k_i, k_i + T_{he}]$, there is a small value $\zeta > 0$, due to the error $E(\cdot)$ in the model representation. In addition, for shorter horizons of evolution T_{db}, the difference between the values $\hat{x}(k_i + T_{db})$ and $x(k_i + T_{db})$ at the end of the interval is limited by ζ, according to Equation (11). In practical terms: the shorter the horizons of evolution to be compensated, the lower the error in the compensated signal.

The previous statements support an important achievement owing to the ETDC reducing the value of network latencies; the compensated signals are closer to the real ones, hence they accomplish the reduction of the error.

Finally, the block diagram of Figure 5 includes the proposed ETDC into the power system control of the control center. The closed loop is built with the communication system feeding the control center, which, in turn, acts over the nonlinear power system (Equations (3)–(6)). The state estimator is also illustrated; and since the state estimator receives delayed measurements values, $y(t - \tau_d)$, it obtains delayed values of the states $\hat{x}(t - \tau_d)$. With the incorporation of the database, the strategy leads to a kind of nonlinear memory controller [48,49].

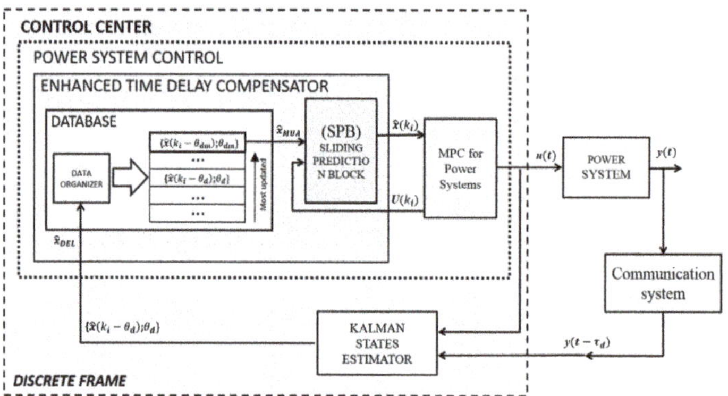

Figure 5. ETDC + Model Predictive Control (MPC) scheme for time delayed Power Systems.

4. Application Test

4.1. Test System and Scenarios

Kundur's benchmark system was used to validate the approach [1]. Despite its small size, this test system performances well in real inter-area oscillations due to time-delays in a single channel; in this system, we can create a scenario with oscillations specifically provoked by time-delays in WAMS. The IEEE 14 bus system and NETS 39-Bus system could be used to validate multiple channel time-delay and to control multiple sources of oscillations in furtherworks. Kundur's test system has two coherent generation areas with four machines (Figure 6), each one with its corresponding governor and Automatic Voltage Regulator (AVR). Two tie-lines guarantee power interchange between both areas; in case of a tie-line tripping, the other one preserves the connectivity. Using the time-delay model from Section 2, the simulations for the communications of the monitoring loop were run with latencies varying from 100 ms to 500 ms [8]. During the monitoring process, a single PMU collected and transmitted data to the control center.

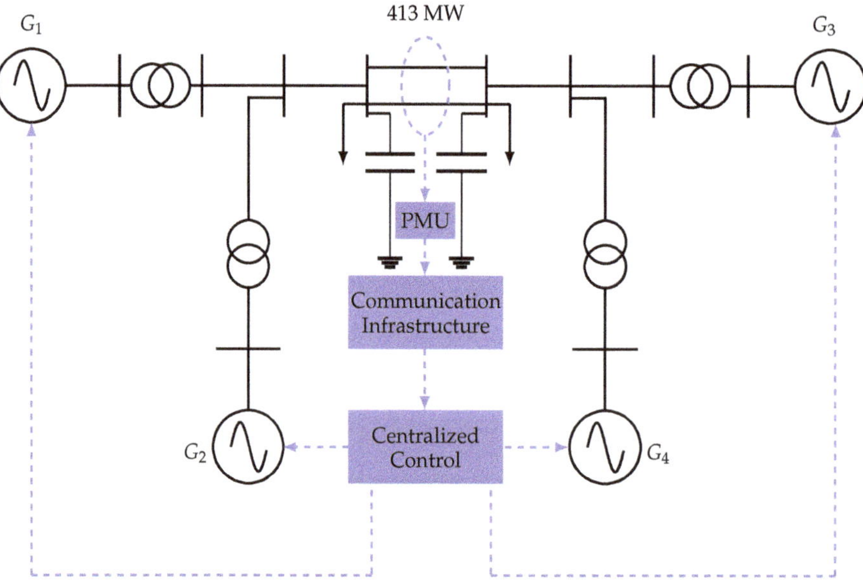

Figure 6. Kundur's test system with the control strategy.

As illustrated in Figure 7, the block diagram of the power system to be managed is a Multiple Input Single Output (MISO) representation, in that we have a power system with four inputs (supplementary signals sent by the control scheme to the four generators) and one output measured by the WAMS (inter-area power flow).

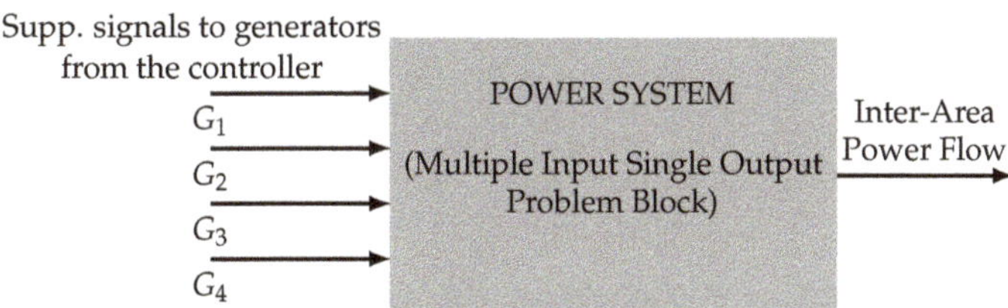

Figure 7. Multiple Input Single Output (MISO) representation of the power system.

In order to control the test system, the centralized scheme was employed with the proposed approach described in Section 3.2. As illustrated in Figure 6, the centralized controller receives the measure from the inter-area power flow, then it sends four supplementary control signals to G1, G2, G3, G4.

Two strong disturbances were simulated for the power system in a steady state. The first one consisted of a three-phase fault with a tie-line tripping; the inter-area oscillation modes took place in the test system. Then, the steady state was reached, and an additional level of higher stress was provoked with an abrupt change of power reference in the non-tripped tie line. In Figure 8, the inter-area oscillation modes are shown, excited due to the three-phase fault; the figure shows power flow response in low frequency oscillations with and without Power System Stabilizer (PSS).

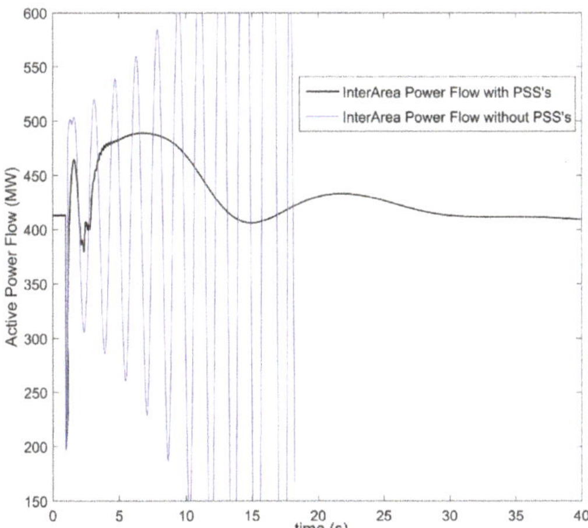

Figure 8. Inter-area oscillations in power flow after three-phase fault.

4.2. Performance Comparison between Spb + Mpc and Etdc + Mpc

Simulations were run with the same test system controlled by two different strategies: (a) the SPB + MPC and (b) the proposal of this paper ETDC + MPC. Both faced three different sequential conditions of operation: (1) initial steady state with a transferred power of 413 MW, (2) a three-phase failure at $t = 1$ s, (transient condition I) and (3) change of power reference with $\Delta P = +25$ MW at $t = 10$ s (transient condition II), once the system returns to steady state.

In the case of SPB + MPC, the compensation scheme considers the arriving signal with its corresponding time delay to obtain the current states without using databases. Hence, it works as a memoryless scheme of compensation and control.

Using SPB + MPC (case a), and due to the failure with line tripping, the active power flow reached a dangerous overshoot of 13.8% at $t = 2.5$ s, with real value of 471 MW (Figure 9). Subsequently, SPB + MPC stabilized the power flow close to the initial pre-fault value in a time close to $t = 6.2$ s. With respect to the power system behavior following the change of reference ($\Delta P = +25$ MW), the power flow reached a steady state with a new reference of 438 MW after undergoing a second overshoot of 2.66% (calculated with the new reference).

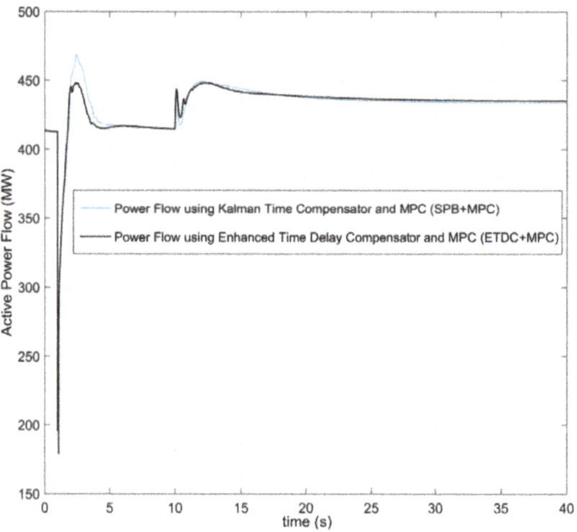

Figure 9. Comparative performance using SPB + MPC and the enhanced ETDC + MPC.

Secondly, using ETDC + MPC (case b), the overshoot was 8.4% at $t = 2.5$ s, with a real value of 447.6 MW (Figure 9). After the first overshoot, the ETDC + MPC reached a steady state $t = 3.8$ s. Then, once the reference was changed, the overshoot reached a value of 2.5% followed by the settling time at 15 s. Table 2 illustrates the values.

Table 2. Overshoot and Settling Times in the two Transient Conditions Introduced

Transient Condition	Overshoot (%)		Settling Time (s)	
	SPB + MPC	EDTC + MPC	SPB + MPC	ETDC + MPC
I	13.8	8.4	6.2	3.8
II	2.66	2.5	17.4	15.2

Next, five different tests were performed with different changes of power reference to add more stress to the controller. The aforementioned test conditions (1)–(3) (Section 4.2) are kept for the sake of comparison. In all the cases, the overshoot after failure was less abrupt (8.9% variation close to 450 MW); then, the active power reached a steady state value close to the initial power reference (see Figure 10). The error after some seconds was less than 4 MW with a downward tendency as in the previous test. At time $t = 10$ s, Figure 10 depicts the behavior of the power flow in the face of reference changes. The five changes in the references and their respective errors are reported in Table 3. The approach can even manage changes in references with $\Delta P = 30$ MW.

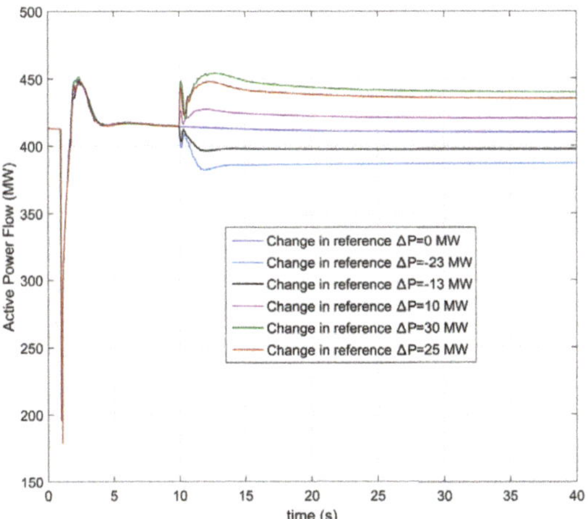

Figure 10. Power flow controlled by ETDC + MPC.

Table 3. Errors obtained changing references.

New Reference and ΔP (MW)	Associated Error (MW)
390 ($\Delta P = -23$)	2.8
400 ($\Delta P = -13$)	2.6
413 ($\Delta P = 0$)	2.5
423 ($\Delta P = 10$)	2.5
438 ($\Delta P = 25$)	3.0
443 ($\Delta P = 30$)	3.2

5. Conclusions and Future Works

The communication infrastructure in power systems based on PMUs, PDCs, SPDCs, protocols, and standards create a complex but useful monitoring system. Thus, WAMS, WAC, WAMC, and finally WAMPaC can be supported by that infrastructure.

The communication infrastructure has an inherent delay due to both the devices and the links, and this issue produces instability problems in the closed loop control strategy. The model of the total latency is not deterministic but stochastic, and the shape of the time delays in typical power communication systems is Gaussian.

Use of the database derived from the MUA concept yields a delayed signal preprocessing to reduce the maximum time delay and mean values. The resulting shape of time delays after using MUA is Weibull. This implies less effort for the time compensation strategies, and, especially, the reduction of latencies leads to better convergence and performance of both the time compensator and the MPC. Improvements achieved are backed up by the results.

The database introduced complies with IEC C.37.117.7, IEC 61850 and TCP/IP, which is the underlying path of the proposed memory controller. Hence, delays were faced as network latencies instead of dead time.

The MPC with the time compensation scheme increases the transfer capabilities in tie lines on the test system; but with the enhanced time delay compensator (ETDC + MPC), it is possible to reduce overshoots and dangerous power excursions. In fact, the achieved reduction of overshoot (almost 39%) implies less stress over the thermal limits and less risk of isolating due to the activation of protection relays.

Further works should examine the performance of the tool considering larger power systems, with several channels (each one with its own stochastic time delay behavior). It is also important to consider the time delay in the control signals during the sending procedure from the control center to generators.

Author Contributions: Conceptualization, A.M.-C., M.A.R., Y.B., N.H., and O.D.M.; methodology, A.M.-C., M.A.R., Y.B., N.H., and O.D.M.; formal analysis, A.M.-C., M.A.R., Y.B., N.H., and O.D.M.; investigation, A.M.-C., M.A.R., Y.B., N.H., and O.D.M.; resources, A.M.-C., M.A.R., Y.B., N.H., and O.D.M.; writing—original draft preparation, A.M.-C., M.A.R., Y.B., N.H., and O.D.M. All authors have read and agreed to the published version of the manuscript.

Funding: The third author was supported by the Austrian Science Fund FWF (Grant M 2967).

Acknowledgments: The first author want to thank Vicerrectoria de Investigación, Innovación y Extensión from Universidad Tecnológica de Pereira for the support provided in this investigation. The fifth author want to thank to the Centro de Investigación y Desarrollo Científico de la Universidad Distrital Francisco José de Caldas under grant 1643-12-2020 associated with the project: "Desarrollo de una metodología de optimización para la gestión óptima de recursos energéticos distribuidos en redes de distribución de energía eléctrica."

Conflicts of Interest: The authors declare no conflicts of interest.

Abbreviations

Acronyms and variables

MPC	Model Predictive Control
PDC	Phasor Data Concentrator
SPDC	Super Phasor Data Concentrator
SPB	Sliding Prediction Block
PMU	Phasor Measurement Unit
ETDC	Enhanced Time Delay Compensator
τ_d	Time delay
$u(t)$	Control signal
$x(t)$	states of the power system
$x(t - \tau_d)$	delayed states
$\hat{x}_r(t - \tau_d)$	delayed estimated states
WAMS	Wide Area Monitoring System
WAMC	Wide Area Monitoring and Control
WAMPaC	Wide Area Monitoring Protection and Control
PSS	Power System Stabilizer

References

1. Kundur, P.; Balu, N.J.; Lauby, M.G. *Power System Stability and Control*; McGraw-Hill: New York, NY, USA, 1994.
2. Machowski, J.; Lubosny, Z.; Bialek, J.W.; BumbyJan, J.R. *Power System Dynamics: Stability and Control*; Wiley: Hoboken, NJ, USA, 2020.
3. Mittelstadt, W.A.; Krause, P.E.; Wilson, R.E.; Overholt, P.N.; Sobajic, D.J.; Hauer, J.F.; Rizy, D.T. The DOE Wide Area Measurement System (WAMS) Project: Demonstration of dynamic information technology for the future power system. In Proceedings of the Joint Conference on Fault and Disturbance Analysis: Precise Measurement in Power Systems, Washington DC, USA, 9–11 April 1996.
4. Ivanescu, D.; Hadjsaid, N.; Snyder, A.; Dion, J.M.; Dugard, L. Robust Stabilizing Control for an Interconnected Power System: Time Delay Approach. In Proceedings of the Fourteenth International Symposium of Mathematical Theory of Networks and Systems, Perpignan, France, 19–23 June 2000.
5. Ivanescu, D.; Snyder, A.F.; Dion, J.M.; Dugard, L.; Georges, D.; Hadjsaid, N. Control of an Interconnected Power System: A Time Delay Approach. *IFAC Proc. Vol.* **2001**, *34*, 449–454. [CrossRef]
6. Younis, M.R.; Iravani, R. Wide-area damping control for inter-area oscillations: A comprehensive review. In Proceedings of the 2013 IEEE Electrical Power & Energy Conference, Halifax, NS, Canada, 21–23 August 2013. [CrossRef]
7. Aboul-Ela, M.; Sallam, A.; McCalley, J.; Fouad, A. Damping controller design for power system oscillations using global signals. *IEEE Trans. Power Syst.* **1996**, *11*, 767–773. [CrossRef]
8. Zhu, K.; Chenine, M.; Nordström, L.; Holmström, S.; Ericsson, G. An Empirical Study of Synchrophasor Communication Delay in a Utility TCP/IP Network. *Int. J. Emerging Electr. Power Syst.* **2013**, *14*, 341–350. [CrossRef]
9. Li, Y.; Yang, D.; Liu, F.; Cao, Y.; Rehtanz, C. *Interconnected Power Systems*; Springer: Berlin/Heidelberg, Germany, 2016. [CrossRef]
10. Li, Y.; Zhou, Y.; Liu, F.; Cao, Y.; Rehtanz, C. Design and Implementation of Delay-Dependent Wide-Area Damping Control for Stability Enhancement of Power Systems. *IEEE Trans. Smart Grid* **2017**, *8*, 1831–1842. [CrossRef]
11. Got Latency? Available online: https://selinc.com/solutions/synchrophasors/report/115256/ (accessed on 1 November 2012).
12. Molina-Cabrera, A. Inter-area Oscillations in Time Delayed Power Systems: A Kalman Time Compensator and a Model Predictive Control Approach. Ph.D. Thesis, Universidad de los Andes, Bogotá D.C, Colombia, May 2018.

13. Milano, F.; Anghel, M. Impact of Time Delays on Power System Stability. *IEEE Trans. Circuits Syst. I Regul. Pap.* **2012**, *59*, 889–900. [CrossRef]
14. Taleb, M.; Zribi, M.; Rayan, M. On the Control of Time Delay Power Systems. *Int. J. Innov. Inf. Control.* **2013**, *9*, 769–792.
15. Bokharaie, V.; Sipahi, R.; Milano, F. Small-signal stability analysis of delayed power system stabilizers. In Proceedings of the 2014 Power Systems Computation Conference, Wroclaw, Poland, 18–22 August 2014. [CrossRef]
16. Snyder, A.; Ivanescu, D.; HadjSaid, N.; Georges, D.; Margotin, T. Delayed-input wide-area stability control with synchronized phasor measurements and linear matrix inequalities. In Proceedings of the 2000 Power Engineering Society Summer Meeting (Cat. No.00CH37134), Seattle, WA, USA, 16–20 July 2000. [CrossRef]
17. Wu, H.; Tsakalis, K.; Heydt, G. Evaluation of Time Delay Effects to Wide-Area Power System Stabilizer Design. *IEEE Trans. Power Syst.* **2004**, *19*, 1935–1941. [CrossRef]
18. Normey-Rico, J.E.; Camacho, E.F. Dead-time compensators: A survey. *Control Eng. Pract.* **2008**, *16*, 407–428. [CrossRef]
19. Majumder, R.; Chaudhuri, B.; Pal, B.; Zhong, Q.C. A unified Smith predictor approach for power system damping control design using remote signals. *IEEE Trans. Control Syst. Technol.* **2005**, *13*, 1063–1068. [CrossRef]
20. Molina-Cabrera, A.; Gomez, O.; Rios, M.A. Smith predictor based backstepping control for damping power system oscillations. In Proceedings of the 2014 IEEE PES Transmission & Distribution Conference and Exposition—Latin America (PES T&D-LA), Medellin, Colombia, 10–13 September 2014. [CrossRef]
21. Chaudhuri, N.R.; Ray, S.; Majumder, R.; Chaudhuri, B. A New Approach to Continuous Latency Compensation With Adaptive Phasor Power Oscillation Damping Controller (POD). *IEEE Trans. Power Syst.* **2010**, *25*, 939–946. [CrossRef]
22. Mokhtari, M.; Aminifar, F.; Nazarpour, D.; Golshannavaz, S. Wide-area power oscillation damping with a fuzzy controller compensating the continuous communication delays. *IEEE Trans. Power Syst.* **2013**, *28*, 1997–2005. [CrossRef]
23. Yao, W.; Jiang, L.; Wen, J.Y.; Cheng, S.J.; Wu, Q.H. Networked predictive control based wide-area supplementary damping controller of SVC with communication delays compensation. In Proceedings of the 2013 IEEE Power & Energy Society General Meeting, Vancouver, BC, Canada, 21–25 July 2013. [CrossRef]
24. Esquivel, P.; Romero, G.; Ornelas-Tellez, F.; Reyes, E.; Castañeda, C.E.; Morfin, O. Statistical inference of multivariable modal stability margins of time-delay perturbed power systems. *Electr. Power Syst. Res.* **2020**, *181*, 106186. [CrossRef]
25. Nie, Y.; Zhang, P.; Cai, G.; Zhao, Y.; Xu, M. Unified Smith predictor compensation and optimal damping control for time-delay power system. *Int. J. Electr. Power Energy Syst.* **2020**, *117*, 1–11. [CrossRef]
26. Nie, Y.; Zhang, P.; Cai, G.; Zhao, Y.; Xu, M. Fixed Low-Order Wide-Area Damping Controller Considering Time Delays and Power System Operation Uncertainties. *IEEE Trans. Power Syst.* **2020**, *35*, 3918–3926. [CrossRef]
27. Nan, J.; Yao, W.; Wen, J.; Peng, Y.; Fang, J.; Ai, X.; Wen, J. Wide-area power oscillation damper for DFIG-based wind farm with communication delay and packet dropout compensation. *Int. J. Electr. Power Energy Syst.* **2021**, *124*, 1–11. [CrossRef]
28. Ye, H.; Liu, Y. Design of model predictive controllers for adaptive damping of inter-area oscillations. *Int. J. Electr. Power Energy Syst.* **2013**, *45*, 509–518. [CrossRef]
29. Shiroei, M.; Ranjbar, A. Supervisory predictive control of power system load frequency control. *Int. J. Electr. Power Energy Syst.* **2014**, *61*, 70–80. [CrossRef]
30. Ma, M.; Chen, H.; Liu, X.; Allgöwer, F. Distributed model predictive load frequency control of multi-area interconnected power system. *Int. J. Electr. Power Energy Syst.* **2014**, *62*, 289–298. [CrossRef]
31. Li, Y.; Rehtanz, C.; Yang, D.; Rüberg, S.; Häger, U. Robust high-voltage direct current stabilising control using wide-area measurement and taking transmission time delay into consideration. *IET Gener. Transm. Distrib.* **2011**, *5*, 289–297. [CrossRef]
32. Molina-Cabrera, A.; Rios, M.A.; Velasquez, M.A. Model Predictive Control for non-linear delayed power systems. In Proceedings of the 2015 IEEE Eindhoven PowerTech, Eindhoven, The Netherlands, 29 June–2 July 2015. [CrossRef]
33. Molina-Cabrera, A.; Rios, M.A. A Kalman Latency Compensation Strategy for Model Predictive Control to Damp Inter-Area Oscillations in Delayed Power Systems. *Int. Rev. Electr. Eng.* **2016**, *11*, 296. [CrossRef]
34. Molina-Cabrera, A.; Rios, M.A.; Besanger, Y.; HadjSaid, N. A latencies tolerant model predictive control approach to damp Inter-area oscillations in delayed power systems. *Int. J. Electr. Power Energy Syst.* **2018**, *98*, 199–208. [CrossRef]
35. Johnson, A.; Wen, J.; Wang, J.; Liu, E.; Hu, Y. Integrated system architecture and technology roadmap toward WAMPAC. In Proceedings of the ISGT 2011, Anaheim, CA, USA, 17–19 January 2011. [CrossRef]
36. Ashton, P.M.; Taylor, G.A.; Irving, M.R.; Carter, A.M.; Bradley, M.E. Prospective Wide Area Monitoring of the Great Britain Transmission System using Phasor Measurement Units. In Proceedings of the 2012 IEEE Power and Energy Society General Meeting, San Diego, CA, USA, 22–26 July 2012. [CrossRef]
37. Hossain, E.; Han, Z.; Vincent, H. *Smart Grid Communications and Networking*; Cambridge University Press: Cambridge, UK, 2009. [CrossRef]
38. IEEE Standard for Synchrophasor Measurements for Power Systems. Available online: https://ieeexplore.ieee.org/document/61 11219 (accessed on 28 December 2011).
39. Liu, W.; Luo, H.; Li, S.; Gao, D. Investigation and Modeling of Communication Delays in Wide Area Measurement System. In Proceedings of the 2012 Asia-Pacific Power and Energy Engineering Conference, Shanghai, China, 27–29 March 2012. [CrossRef]
40. Mohagheghi, S.; Venayagamoorthy, G.K.; Harley, R.G. Optimal Wide Area Controller and State Predictor for a Power System. *IEEE Trans. Power Syst.* **2007**, *22*, 693–705. [CrossRef]

41. Albertos, P.; Garcia, P.; Sanz, R. Some contributions to the design of dead-time compensators. In Proceedings of the 2016 14th International Conference on Control, Automation, Robotics and Vision (ICARCV), Phuket, Thailand, 13–15 November 2016. [CrossRef]
42. Naduvathuparambil, B.; Valenti, M.; Feliachi, A. Communication delays in wide area measurement systems. In Proceedings of the Thirty-Fourth Southeastern Symposium on System Theory (Cat. No.02EX540), Huntsville, AL, USA, 19 March 2002. [CrossRef]
43. García, P.; Albertos, P. Dead-time-compensator for unstable MIMO systems with multiple time delays. *J. Process Control* **2010**, *20*, 877–884. [CrossRef]
44. James, B.R.; David, Q.M.; Moritz, M.D. *Model Predictive Control Theory And Design*; Nob Hill Publishing: London, UK, 2009.
45. Molina-Cabrera, O.D.; Gil-González, W.; Molina-Cabrera, A. Second-Order Cone Approximation for Voltage Stability Analysis in Direct Current Networks. *Symmetry* **2020**, *12*, 1587. [CrossRef]
46. Gil-Gonzalez, W.; Serra, F.; Dominguez, J.; Campillo, J.; Montoya, O.D. Predictive Power Control for Electric Vehicle Charging Applications. In Proceedings of the 2020 IEEE ANDESCON, Quito, Ecuador, 13–16 October 2020. [CrossRef]
47. Grisales-Noreña, W.; Molina-Cabrera, A.; Montoya, O.D.; Grisales-Noreña, L.F. An MI-SDP model for optimal location and sizing of distributed generators in DC grids that guarantees the optimal global solution. *Appl. Sci.* **2020**, *10*, 7681. [CrossRef]
48. Rawlings, J.B.; Mayne, D.Q.; Diehl, M.M. *Model Predictive Control: Theory, Computation, and Design*; Nob Hill Publishing: London, UK, 2013.
49. Boukas, E.K.; Liu, Z.K. *Deterministic and Stochastic Time-Delay Systems*; Birkhäuser: Boston, MA, USA, 2002. [CrossRef]

Article

Leveraging a Genetic Algorithm for the Optimal Placement of Distributed Generation and the Need for Energy Management Strategies Using a Fuzzy Inference System

Sunny Katyara [1,*,†], **Muhammad Fawad Shaikh** [1,†], **Shoaib Shaikh** [1], **Zahid Hussain Khand** [1], **Lukasz Staszewski** [2], **Veer Bhan** [1], **Abdul Majeed** [1], **Madad Ali Shah** [1,*] and **Leonowicz Zbigniew** [2,*]

[1] Department of Electrical Engineering, Sukkur IBA University, Sukkur 65200, Pakistan; muhammadfawad@iba-suk.edu.pk (M.F.S.); shoaibahmed@iba-suk.edu.pk (S.S.); registrar@iba-suk.edu.pk (Z.H.K.); veer.bhan@iba-suk.edu.pk (V.B.); abdulmajeed@iba-suk.edu.pk (A.M.)
[2] Department of Electrical Power Engineering, Wroclaw University of Science and Technology, 50-370 Wroclaw, Poland; lukasz.staszewski@pwr.edu.pl
* Correspondence: sunny.katyara@iba-suk.edu.pk (S.K.); madad@iba-suk.edu.pk (M.A.S.); Zbigniew.Leonowicz@pwr.edu.pl (L.Z.); Tel.: +48-7132-0262-6 (L.Z.)
† These authors contributed equally to this work.

Citation: Katyara, S.; Shaikh, M.F.; Shaikh, S.; Khand, Z.H.; Staszewski, L.; Bhan, V.; Majeed, A.; Shah, M.A.; Zbigniew, L. Leveraging a Genetic Algorithm for the Optimal Placement of Distributed Generation and the Need for Energy Management Strategies Using a Fuzzy Inference System. *Electronics* **2021**, *10*, 172. https://doi.org/10.3390/electronics10020172

Received: 14 December 2020
Accepted: 11 January 2021
Published: 14 January 2021

Publisher's Note: MDPI stays neutral with regard to jurisdictional claims in published maps and institutional affiliations.

Abstract: With the rising load demand and power losses, the equipment in the utility network often operates close to its marginal limits, creating a dire need for the installation of new Distributed Generators (DGs). Their proper placement is one of the prerequisites for fully achieving the benefits; otherwise, this may result in the worsening of their performance. This could even lead to further deterioration if an effective Energy Management System (EMS) is not installed. Firstly, addressing these issues, this research exploits a Genetic Algorithm (GA) for the proper placement of new DGs in a distribution system. This approach is based on the system losses, voltage profiles, and phase angle jump variations. Secondly, the energy management models are designed using a fuzzy inference system. The models are then analyzed under heavy loading and fault conditions. This research is conducted on a six bus radial test system in a simulated environment together with a real-time Power Hardware-In-the-Loop (PHIL) setup. It is concluded that the optimal placement of a 3.33 MVA synchronous DG is near the load center, and the robustness of the proposed EMS is proven by mitigating the distinct contingencies within the approximately 2.5 cycles of the operating period.

Keywords: DG placement; evolutionary algorithms; energy management; fuzzy controller

1. Introduction

Notwithstanding the modernization of the existing power system, its operation, control, monitoring, protection, and communication are still challenging aspects to address. Further, the improper placement of the equipment responsible for the performance, safety, and security improvement (i.e., capacitor banks, Distributed Generators (DGs), protective relays, etc.) may result in increased power losses, power quality worsening, and ineffective coordination of the protection devices [1,2]. Amongst all the factors, the level of threat varies, but surely, the non-optimal location and sizing of DGs pose some of the most serious threats, not only to the utility, but also to the end-users. Apart from the fact that the DGs mostly rely on non-conventional energy sources and are environmentally friendly, their exponential growth may still lead to voltage instability, false tripping, power losses, short-circuit level increase, reverse power flow, and faster equipment degradation [3,4].

Prior to the installation of new DGs, many technical data, as well as the environmental and regulatory characteristics of the distribution system need to be analyzed. Amongst the many parameters, the key ones deciding the effectiveness of the DGs placed in the already existing networks are the voltage profiles, system losses, and power flows [5]. However, investigations into the behavior of the power system under normal and fault conditions require thorough load flow studies and contingency analysis. These affect the power system operation differently, and if the installed protection schemes do not function in a timely manner, the fault situation may result in either islanding or even complete blackout [6,7]. Therefore, to alleviate the adverse effects of different contingencies with the existence of DGs in the network, designing a proper Energy Management System (EMS) is highly required.

The EMS is an optimal control strategy, designed to regulate the energy flow in the network based on the network characteristics, as shown in Figure 1, wherein the sources are the group of existing energy providers in the network, storage represents the energy preserved during the off-load periods, loads are scheduled as pre-defined from the load curves and also dynamically integrated on run-time, and optimal control indicates the series of actions taken to ensure the optimal energy flow and regulation. The EMS provides promising functionalities to regulate the power system instabilities and contingencies within the previously defined constraints [8,9]. Furthermore, the EMS can manage the load profile of the power system following the available energy sources and thus minimizes unnecessary energy usage, extra operational costs, and the prevailing safety issues. An effective EMS makes a significant difference in the power system operation and efficiency, after the proper placement of DGs, by mitigating the undesirable situations, i.e., network unbalance conditions, system faults, islanding, etc. [10,11].

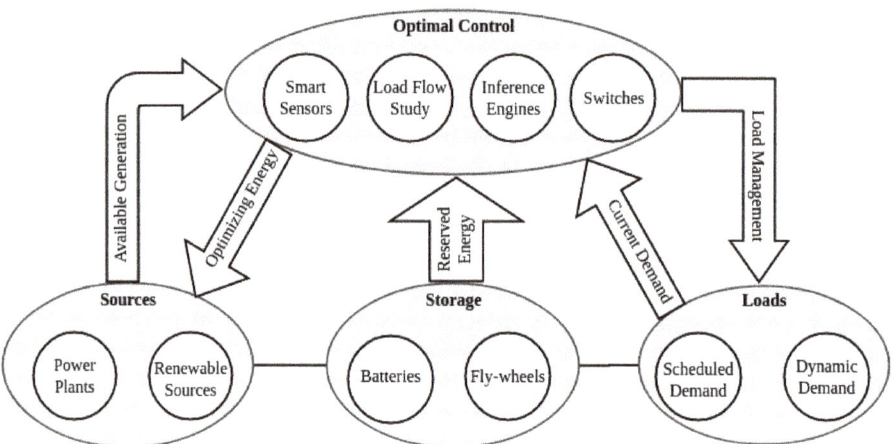

Figure 1. General sketch of the energy management system for the optimal allocation of resources in accordance with the network conditions.

Proper placement of DGs and the use of different energy management strategies in the distribution system have been well studied in the literature [12–15], but the close relationship between the two optimization problems, i.e., optimal DG placement and energy management, has not been investigated under different network conditions. In this regard, an analytical approach was proposed in [16] for the proper placement of DGs in a radial network with the aim of minimizing the total power loss without exploiting admittance matrices. The method is computationally promising and robust for a single DG, but its accuracy and performance are compromised for multiple DGs in more complex utility networks. Similarly, a loss sensitivity factor together with the meta-heuristic technique was discussed in [17] for optimal sizing and siting of new DGs sequentially. This suggested approach, however, uses a multi-dimensional cost function for the search algorithm, but is slow and computationally expensive for a large network. Therefore, an efficient and modular algorithm for large power systems was discussed in [18]. The suggested idea exploits quadratic programming to identify and select the optimal position and size of the DG based on the active loss minimization, power balance compensation, and voltage limits' satisfaction.

The algorithm uses a passive optimization and controller, which limit its application to networks with the least penetration of DGs and a lack of an effective EMS. To address the issue of energy management in micro-grids with environmental uncertainties related to properly installed DGs, a framework was presented in [19] that uses a fuzzy prediction model as an inference to predict the dynamic characteristics and uncertainties of available energy sources. The method, however, exploits the notion of fuzzy logic to deal with the probabilistic nature of network uncertainties, but still lacks an active controller to make decisions in the transient conditions. Hence, active controller schemes were presented in [20,21] that make use of a fuzzy inference system for battery management and fault classification. However, their application is limited to contingency detection and regulation, and this motivates us to extend the benefits of the fuzzy controller for power system energy management as well.

To the best of our knowledge, there is no single framework available that initially applies a path search algorithm to optimally place the new DGs using the key parameters of the network highlighted earlier and then under different network conditions, exploiting the supervised learning approach to realize the energy management strategies to lessen the adverse effects during distinct contingency scenarios. Overall, the key contributions of this work are (1) using a GA for the proper placement of new DGs using three critical system parameters, i.e., power loss, voltage profile, and phase angle jump, (2) designing the energy management system using a fuzzy controller with 12 inference rules, (3) testing the proposed framework on a system against its performance at different load and fault conditions, and finally, (4) validating the effectiveness and robustness of the proposed approach) it is compared with three other state-of-the-art techniques, i.e., Tabu Search (TS), Artificial Bee Colony (ABC), and ACO.

The subsequent sections of this work are organized as follows: Section 2 presents the proposed methodology and problem formulation for distinct network conditions. Section 3 examines the results achieved with the application of the proposed framework under different case scenarios. Finally, Section 4 gives the final conclusions and remarks on possible extensions of the presented work.

2. Research Methodology

The test system used to experimentally evaluate the performance of the proposed framework is shown in Figure 2. It consists of 132 kV grid stations supplying a 11 kV distribution feeder through a 26 MVA, 132/11 kV transformer. The feeder energizes different loads, i.e., industrial, commercial, and residential, rated at 1.1 MVA, 2.0 MVA, and 2.5 MVA with power factors of 0.94, 0.9, and 0.86, respectively. The residential load is connected through a 10 MVA, 11/0.4 kV step-down transformer. A synchronous generator (DG) of 3.33 MVA at a power factor of 0.953 is installed in the distribution network using a transformer of 8 MVA, 0.4/11 kV at distinct locations, determined by the search algorithm, as marked by the respective buses in Figure 2.

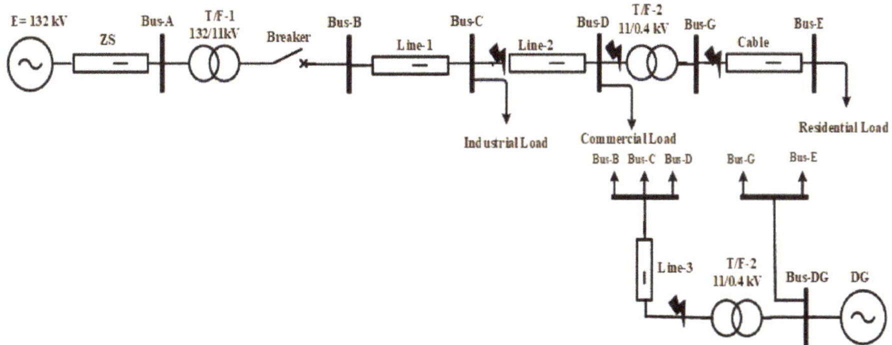

Figure 2. One line diagram of the test system.

However, in this research, a radial system is considered to analyze the performance of the proposed framework owing to the nature and characteristics of existing feeders in distribution networks. However, the proposed framework can also be applied to mesh networks, having the benefits of ensuring the reliability and continuity of supply under the contingency condition in contrast to radial systems with minor modifications. The network characteristics determined and the load flow analysis carried out in subsequent sections for the radial system in Figure 2 need to be changed according to [22].

2.1. Network Characteristics

Based on the same voltage regulation of the utility network and the new DG, the magnitude of the voltage sag at the point of connection in Figure 2 can be calculated using Equation (1) [23].

$$V_{sag} = V_N - (\frac{Z_N}{Z_{DG}}) \times I_F \tag{1}$$

where V_{sag} is the voltage sag due to system contingencies as reported in Figure 3, V_N and Z_N are the voltage and impedance of the utility network, respectively, Z_{DG} is the impedance of the installed DG system, and I_F is the magnitude of the fault current. The voltage sag eventually results in a phase angle jump and that is determined by Equation (2), with X_N, R_N, X_{DG}, and R_{DG} representing the reactances and resistances of the utility network and installed DG, respectively.

$$\Delta\theta = \tan^{-1}(\frac{X_N}{R_N}) - \tan^{-1}(\frac{X_N + X_{DG}}{R_N + R_{DG}}) \tag{2}$$

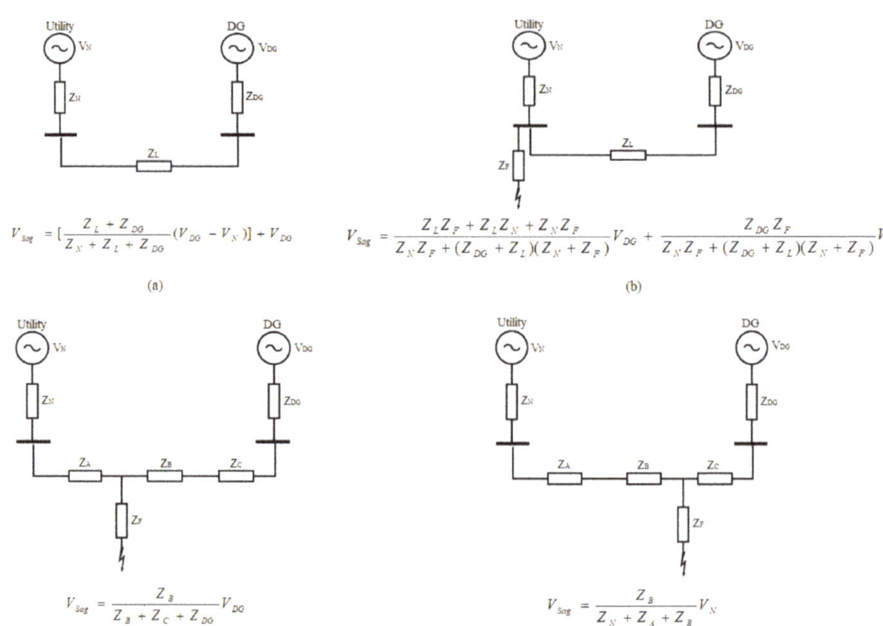

Figure 3. Different system contingencies and associated voltage sag estimations. (**a**) Normal condition, (**b**) fault condition at the utility bus, (**c**) fault condition for the transmission network near the utility, and (**d**) fault condition for the transmission network near the DG.

If $\frac{X_N}{R_N} = \frac{X_{DG}}{R_{DG}}$, it is a so-called zero phase angle jump condition. The jump in angle occurs when the X/R ratio of the utility network and DG system are mismatched during the system operation. Thus, it is desirable to have the least phase angle jump in the voltage to ensure the maximum contribution of the DG system in the network. With the maximum contribution of the DG system, the total system losses occurring are governed by Equation (3) [24], with P_L and Q_L representing active and reactive line losses, respectively, i the respective bus number, and I_N and I_{DG} the network and DG currents, respectively.

$$P_L + jQ_l = (I_N + I_{DG})^2 \times [(R_N + R_{DG}) + j(X_N + X_{DG})] \tag{3}$$

2.2. Distributed Load Flow Algorithm

A high R/X ratio, unbalanced loading, and the distributed nature of utility networks make some load flow techniques such as Gauss–Seidel, Newton–Raphson, etc., inefficient [25]. Hence, specialized algorithms such as the Backward/Forward Sweep (BFS) method are required for load flow analysis of radial distribution networks. Such techniques are based on Kirchhoff's Current Law (KCL), which defines the current injections into the network as Equation (4) [26].

$$I_i^j = \left(\frac{P_i^j + jQ_i^j}{V_i^j} \right) \tag{4}$$

where I_i^j, V_i^j, P_i^j, and Q_i^j represent the current, voltage, and active and reactive power, respectively, at the ith bus during the jth iteration. Applying KCL to the test system in Figure 2 results in the Bus Injection to Branch Current (BIBC) matrix in Equation (5) [26]:

$$\begin{bmatrix} B_A \\ B_B \\ B_C \\ B_D \\ -B_E \end{bmatrix} = \begin{bmatrix} 1 & 1 & 1 & -1 & 1 \\ 0 & 1 & 1 & -1 & 1 \\ 0 & 0 & 1 & -1 & 1 \\ 0 & 0 & 0 & -1 & 1 \\ 0 & 0 & 0 & 0 & 1 \end{bmatrix} \begin{bmatrix} I_A \\ I_B \\ I_C \\ I_D \\ I_E \end{bmatrix} \tag{5}$$

The current variations at all the buses due to either load changes or faults causing corresponding voltage changes are defined by the Branch Current to Bus Voltage (BCBV) matrix in Equation (6) [26].

$$\begin{bmatrix} V_A \\ V_A \\ V_A \\ V_A \\ V_A \end{bmatrix} - \begin{bmatrix} V_B \\ V_C \\ V_D \\ V_E \\ -V_F \end{bmatrix} = \begin{bmatrix} Z_{AB} & 0 & 0 & 0 & 0 \\ Z_{AB} & Z_{BC} & 0 & 0 & 0 \\ Z_{AB} & Z_{BC} & Z_{CD} & 0 & 0 \\ Z_{AB} & Z_{BC} & Z_{CD} & Z_{DE} & 0 \\ Z_{AB} & Z_{BC} & Z_{CD} & 0 & Z_{DF} \end{bmatrix} \begin{bmatrix} B_A \\ B_B \\ B_C \\ B_D \\ B_E \end{bmatrix} \tag{6}$$

2.3. Problem Formulation

The first objective of this research is to find the optimal location and size of a new DG in the utility network, which reduces the overall system losses and improves the voltage profile. Therefore, the objective function of this problem can be formulated as Equation (7),

$$F(X_{DG}, P_{DG}, Q_{DG}) = min(\gamma_1 V_{sag} + \gamma_2 S_L + \gamma_3 \Delta\theta) \tag{7}$$

where $\sum_i^3 \gamma_i = 1$ and $\gamma \in [0,1]$ is the weighted coefficient. The optimization problem defined in Equation (7), with its members defined in Equations (1)–(3), has equality, inequality, and bound constraints. The equality constraints are explained using Equation (8), which states that the sum of the utility power (S_N) and DG power (S_{DG}) should be equal to the sum of the total system load (S_{Load}) and losses (S_L) in order to maintain the balance of power and ensure the system stability. However, the load profile considered in Equation (8), which in this case is a combination of industrial, commercial, and residential entities with a cumulative capacity of 5.6 MVA at different power factors, is assumed to be fluctuating at a steady rate from 2.5 MW to 5.0 MW within the scheduled time frame.

$$S_N + S_{DG} = S_{Load} + S_L \tag{8}$$

With the successful integration of the DG with the utility network, the system losses need to be lower than the allowed thermal limits of the lines. Therefore, the X/R ratio of the new DG ($\frac{X_{DG}}{R_{DG}}$) should be less than the cumulative X/R ratio of the load and utility ($\frac{X_L}{R_L} + \frac{X_N}{R_N}$), satisfying the phase angle jump condition, as described by the inequality constraint in Equation (9).

$$\frac{X_{DG}}{R_{DG}} < \frac{X_L}{R_L} + \frac{X_N}{R_N} \tag{9}$$

Further, it is assumed that the location of the new DG (X_{DG}) is near the load center, so the iteration of the GA always starts from Bus-B (X_{Bus-B}) and ends at Bus-E (X_{Bus-E}), as defined by Equation (10).

$$X_{Bus-B} \leq X_{DG} \leq X_{Bus-E} \tag{10}$$

Moreover, the size of the new DG (S_{DG}) may be less than or equal to the total connected load (S_{Load}), but should be always greater than the system losses (S_L), as explained by Equation (11).

$$S_L < S_{DG} \leq S_{Load} \tag{11}$$

The voltage at all the buses (V_{Bus}) should be within the defined lower (V_{Low}) and upper (V_{High}) tolerance limits of $\pm 10\%$, which define the bound constraint using Equation (12) together with Equations (10) and (11) on the objective function in Equation (7).

$$V_{Low} \leq V_{Bus} \leq V_{High} \tag{12}$$

As a part of the optimization problem, the financial constraints also need to defined for the optimal placement of new DGs and the possible return in terms of loss reduction. The capital cost (C_c) in $/h upon installing a new DG is defined by Equation (13):

$$C_c = \alpha_1 S_{DG}^2 + \alpha_2 S_{DG} + \alpha_3 \tag{13}$$

where α_1, α_2, and α_3 are the cost coefficients, and their values not only depend on the size and nature of the incoming DG, but also the characteristics of the utility network under investigation. In this research, their values are chosen as 0.1, 0.23, and 0.34, respectively. Based on these values, the cost of installing a given synchronous DG of size 3.33 MVA is about 2.125 $/h. Hence, the daily return (R) achieved in $ on the loss reduction after an optimal integration of DG is determined by Equation (14):

$$R = \Delta P_L \times E_R \times T_D \tag{14}$$

where ΔP_L is the change in the active power loss of the network, E_R is the tariff rate of supplied energy, and T_D is the time duration of DG operation. The values of E_R and T_D selected for this research are 0.015 $/kWh and 24 h, respectively.

2.4. Genetic Algorithm

The Genetic Algorithm (GA) is a biologically-inspired method based on Darwin's principle, which evaluates the best possible set of solutions for the fitness of human beings [27]. The main reason for using GA for our application is its parallel search for points from the population. Therefore, it cannot be trapped into a local optimum like conventional techniques, which search for a single point in each iteration. The GA is a probabilistic approach rather than a deterministic and does not involve any derivatives or auxiliary data, but uses fitness parameters to search for the optimal solution.

The GA outputs different chromosomes as shown in Figure 4 for the optimal location and size of the DG based on the constraints of the objective function in Equation (7) for maximizing the voltage profile of the system and minimizing the power losses. Evaluating the objective function using two operators, mutation and crossover, the new set of chromosomes is generated. The crossover operator is used to find the best possible parameter space, and the mutation operator guards the resultant information such that it is encrypted. The stopping criteria for GA are based on two conditions, i.e., either the value of the objective function calculated from the new set of chromosomes is less than the pre-defined error or the maximum number of iterations for the system has been reached. In this research, the crossover and mutation operators (C and M) are assigned probabilities of 0.7 and 0.2, respectively. The algorithm is initialized with a set of 15 chromosomes for 36 iterations, and the final set of chromosomes in Figure 4 defines the best possible location and size of the new DG, obtained with the optimal solution of the objective function in Equation (7).

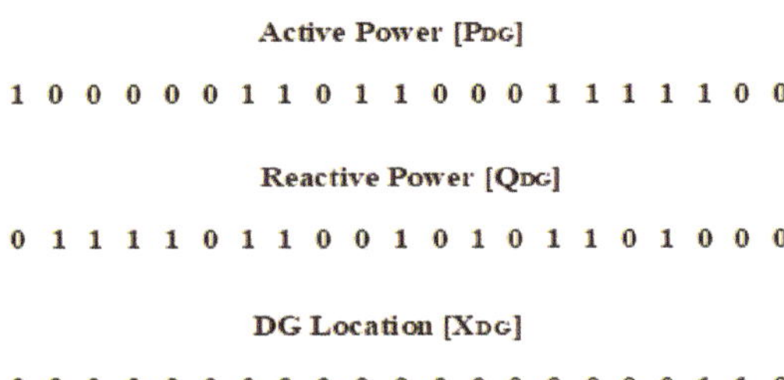

Figure 4. Final set of chromosomes for the optimal location and size of the new DG determined with the application of the GA.

2.5. Fuzzy Inference System

A Fuzzy Inference System (FIS) is based on if-then conditions to make approximations of complex nonlinear functions, representing the trends of network variables. The FIS mainly consists of five major functionalities for exploiting qualitative features of human reasoning and decisions in terms of data sets without considering their quantitative aspects in particular, as shown in Figure 5, where the fuzzification module transforms the numerical values of the system variables, i.e., current, voltage, and power, into fuzzy values by leveraging a knowledge base having pre-defined rules with the respective correspondence of the variables defined using membership functions. Such fuzzy values are given as the input to the decision making unit, which acts as an inference engine to generate switching sequences according to the network conditions. The resultant switching sequences are de-fuzzified using information from the knowledge base, to make them compatible with the tripping operation of appropriate breaker.

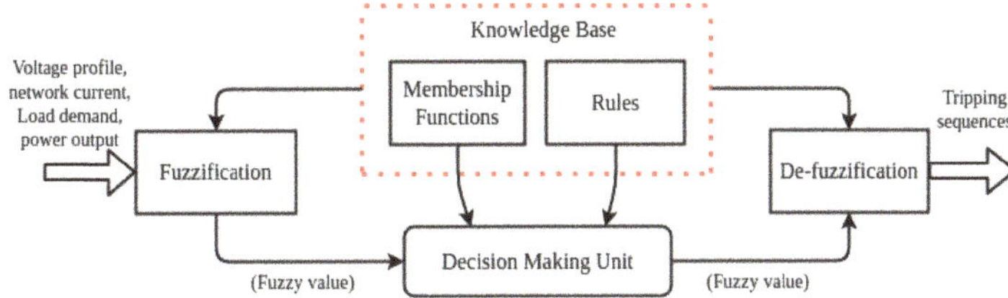

Figure 5. Proposed fuzzy inference system for effective energy management strategies in utility networks.

The input variables to the FIS in this research use the Gaussian membership function, which is defined by Equation (15):

$$f_{i,j}(x_i) = exp\{-\frac{1}{2}(\frac{x_i - \mu_j}{\sigma_j})^2\} \tag{15}$$

where x_i represents the system variables and μ_j and σ_j are the mean and variance of the Gaussian function, respectively. Further, the set of rules is defined considering the first order Sugeno formulation, and the corresponding output is thus given by Equation (16) [28].

$$y = \frac{\sum_{m=1}^{N} \tau_m (rV_p + sI_p + tP_{load} + uP_o + v)}{\sum_{m=1}^{N} \tau_m} \tag{16}$$

where N is the total number of rules developed, which in this case is 12. V_p, I_p, P_{load}, and P_o are the voltage profile, system current, load demand, and cumulative power output of the sources, respectively. r, s, t, u, and v are the parameters of the inference engine, and τ_m is the triggering instant of the respective rule.

The proposed framework for the optimal placement of new DGs and effective energy management under normal and contingency conditions is summarized in Algorithm 1.

Algorithm 1: DG placement and energy management.

$\quad$**input** $: V_p, S_L \longrightarrow$ voltage profile and system losses
$\quad$**output**$: y \longrightarrow$ optimal control strategy
$\quad$**while** $(V_{sag}|S_L|\Delta\theta \leq limits) \longrightarrow F$ **do**
$\quad\quad function(X_{DG}, P_{DG}, Q_{DG}) \longrightarrow (Location, Size)$
$\quad\quad X_{DG} := Location$
$\quad\quad p_{DG}, Q_{DG} := Size$
$\quad\quad$**if** $S_{Losses} < S_{DG} \leq S_{demand}|X_{source} \leq X_{DG} \leq X_{load}$ **then**
$\quad\quad\quad O(location, size);$
$\quad\quad$**else**
$\quad\quad\quad \phi \longrightarrow$discard chromosome;
$\quad\quad$**end**
$\quad$**end**
$\quad$**foreach** $(X_{DG}, P_{DG}, Q_{DG}) \in E \longrightarrow cumulative\ energy\ production$ **do**
$\quad\quad function(\tau_m, V_p, I_p, P_{load}, P_o) = y;$
$\quad\quad$**if** $I_N = I_F|S_N + S_{DG} + S_{battery} = S_{load} + S_{losses} \longrightarrow contingency\ detection$ **then**
$\quad\quad\quad return(t, \psi) \longrightarrow$ switching patterns;
$\quad\quad$**else**
$\quad\quad\quad \phi \longrightarrow$no abnormality;
$\quad\quad$**end**
$\quad\quad y = transformation(t, \psi)$
$\quad$**end**

In Algorithm 1, the GA and FIS are leveraged together in a unified framework to not only increase the benefits of integrating new DGs, but also to regulate the power flow under different network conditions to optimally meet the load demands. The input to the framework is the voltage profile, which includes variations in its magnitude (V_{sag}) and phase angle ($\Delta\theta$) and the system losses (S_L), and the output is the desired switching profile. For all the input variables of the test network in Figure 2 determined using distributed load flow algorithms using Equations (4)–(6), if their values are less than or equal to the predefined norms, the GA computes the optimal location and size of the incoming DG using Equation (7) provided all the constraints in Equations (8)–(12) are satisfied; otherwise, the GA output is discarded, and the loop repeats until it achieves the desired outcome. Once the optimal location and size of the new DG is determined, then different contingencies are simulated similar to real-time conditions, i.e., load deviations, islanding, faults, etc. Under such scenarios, the cumulative supply from the DGs and grid sources is checked against the desired load profile and losses to generate controlled switching sequences using the FIS. Abnormality is detected using the current profile of the system, and if it is found, the FIS either trips the associated breakers or regulates the energy flow to minimize the adverse effects of the given contingency. Conversely, if the system operates in the normal condition given that the sum of the supply from the DGs and grid is equal to the sum of the load demand and system losses, the FIS remains inactive.

3. Results and Discussion

In the proposed framework, the environment is initially created in Simulink, as a first step to test the idea, and it is then validated on a Power Hardware-In-the-Loop (PHIL) setup in real time as a proof of concept for practical results, as shown in Figure 6. In Figure 6, the power system simulator replicates the characteristics of the utility network defined in Figure 2. The PXI platform simulates the response of the DG technology, while the grid control server implements the optimization techniques (FIS, GA, etc.) to regulate the system operation and performance, and the communication system simulator is used to develop a communication path between different interfaces. Further, it uses Phasor Measurement Units (PMUs) that measure the magnitudes of the voltage, current, and phase angle jump. The results presented in the following subsections are the final outcomes from the PHIL setup in Figure 6.

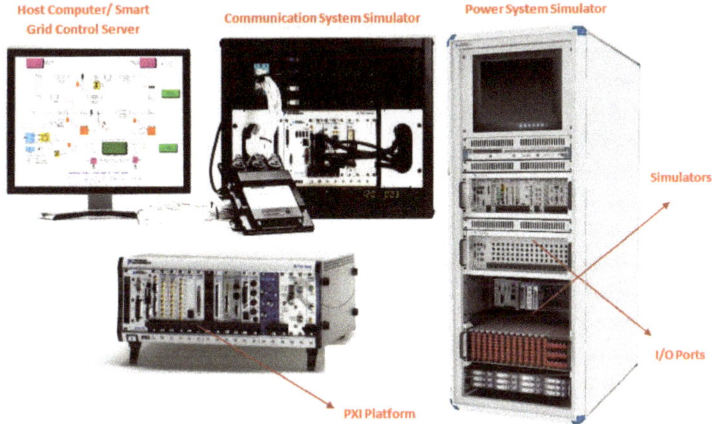

Figure 6. Power Hardware-In-the-Loop (PHIL) setup for DG placement and energy management in the utility network under normal and contingency conditions.

3.1. Optimal Location and Size of the DG

The installation of a new DG however has positive impacts on the network performance, thereby reducing the power losses and improving the voltage profile. However, based on the optimal location and size of the DG, the characteristics of the utility network change accordingly. In order to evaluate the performance of the test network in Figure 2 against the power losses, voltage sag, and Phase Angle Jump (PAJ), the following three distinct scenarios are considered.

3.1.1. Normal Condition

Under the normal condition, the integration of the DG improves the system performance as shown in Figure 7. The power loss at Bus-C beforehand is about 154.654 kW, as shown in Figure 7a, and reduces to 103.789 kW when the DG is connected at Bus-E near the load center, which gives an instant return of approximately \$0.793 per hour. Due to the radial nature of the test network in Figure 2, the voltage drop however should increase, as we move away from the sources. However, due to the presence of the DG, it is still under the permissible limits of $\pm 10\%$, as illustrated in Figure 7b. Moreover, the value of the PAJ varies with the deviations in the combined X/R ratio of the feeder and DG with respect to the source, and its minimum value occurs at Bus-E, as indicated in Figure 7c. Based on all such network characteristics, the best location of the new DG during the normal condition is found to be at Bus-E near the load center with the optimal size of 3.1735 MW and 1.136 MVAr, determined using the GA, as shown in Figure 7d. It is evident from Figure 7d that the network characteristics intersect at Bus-E after 30 iterations, specifying the optimal location and size of the new DG.

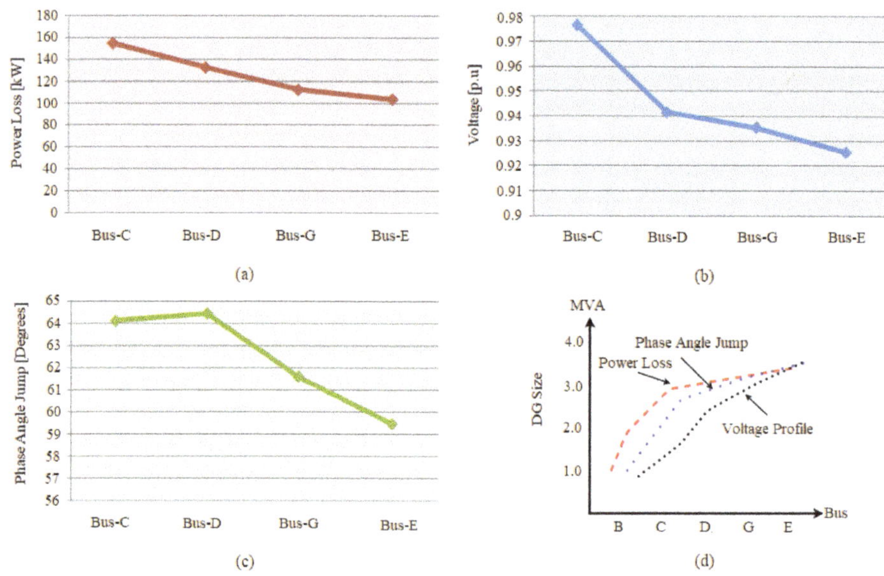

Figure 7. Characteristics of the system under the normal condition for proper placement and sizing of the new DG: (**a**) power losses at different buses, (**b**) voltage profile of the system, (**c**) Phase Angle Jump (PAJ) variations, and (**d**) GA outcome over 30 iterations.

3.1.2. Fault Condition

With the inception of symmetrical fault in the network, a huge amount of current starts to flow. Under such conditions, the DG should not be operated, otherwise it will contribute to the fault current and may even not only harm the network, but also itself. The isolation of the DG during fault conditions depends on its dedicated protection scheme, and if it fails to operate, this results in the DG continuing to supply power, which is hazardous for the network personnel. For fault conditions, if the fault occurs near the source, the amount of power loss, that is 278.93 kW, is higher than that occurring near the load center at Bus-G, i.e., 195.75 kW, as shown in Figure 8a, which gives a savings of approximately $ 1.248 per hour. It can be seen from Figure 8a that there is not a big difference between the losses at Bus-G and Bus-E. Further, the voltage limits are also violated at both, and among all, the best voltage profile is maintained at Bus-E, i.e., 0.7812 p.u., as illustrated in Figure 8b. Moreover, the PAJ excursions, shown in Figure 8c, are highest at Bus-C and lowest at Bus-E, i.e., 73.7509 radians. In view of such system characteristics, the optimal location of the DG is determined to be at Bus-E with a size of 3.1735 MW and 1.009 MVAr. However, during the fault condition, the desired solution using GA is obtained after 32 iterations, when the network characteristics intersect each other, which in this case again is at Bus-E, as illustrated in Figure 8d.

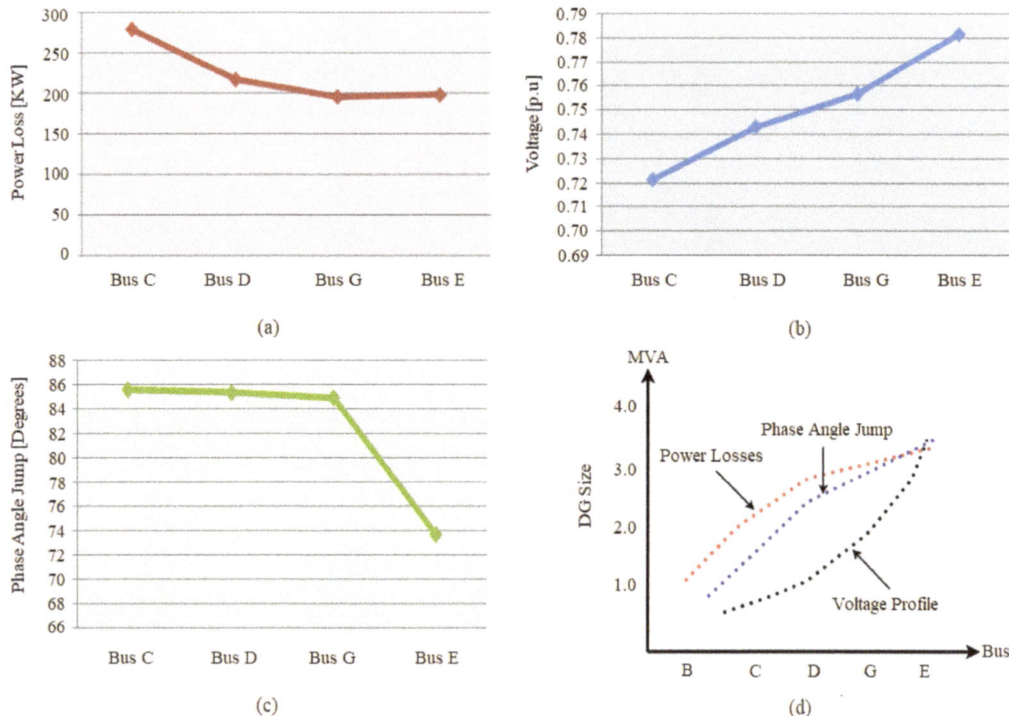

Figure 8. System performance under fault condition for the optimal location and size of the new DG: (**a**) power losses at different system buses, (**b**) system voltage profile, (**c**) the trend of PAJ at various buses, and (**d**) GA output over 32 iterations.

3.1.3. Islanding Condition

During the islanding condition, a power imbalance occurs owing to the fact that the DG alone is unable to supply all three types of connected loads, i.e., industrial, commercial, and residential, as shown Figure 2. Under such a scenario, the power loss in the system is low (with a reduction of about 26.707 kW, which gives a return of $0.4 per hour) because a lesser amount of current flows, as shown in Figure 9a, but unfortunately, the voltage profile is worse, violating the allowable tolerance limits due to the extra loading on the DG, as illustrated in Figure 9b. Further, the PAJ variations are least at Bus-C under such a condition, as presented in Figure 9c. Due to the network characteristics, a compromise decision is made to decide the best possible location and size of the new DG. However, the amount of power losses is least and the voltage profile is better at Bus-E; on the other hand, the PAJ deviations are small at Bus-C. Therefore, the best location during the islanding found using GA is at Bus-D with a size of 3.1735 MW and 1.009 MVAR, as shown in Figure 9d. It is evident from Figure 9d that the desired results are obtained after 35 successful iterations of the GA.

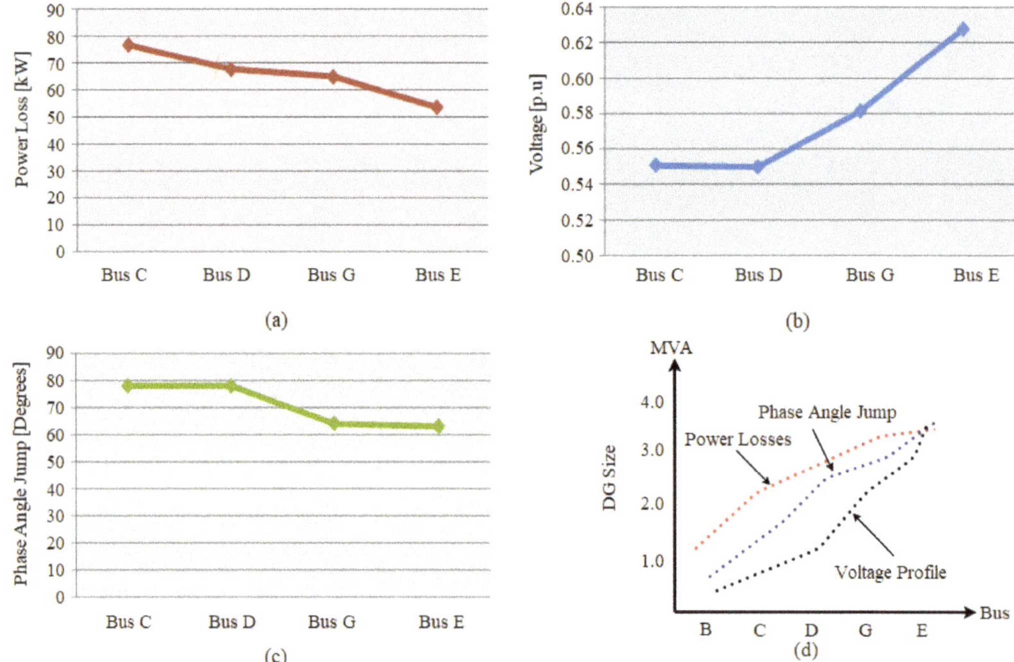

Figure 9. System profile during the islanding condition for the optimal location and size of the new DG: (**a**) power losses at different buses, (**b**) voltage profile of the system, (**c**) PAJ variations at different buses, and (**d**) GA output over 35 iterations.

3.2. Energy Management System

The performance of the utility network during the fault and islanding conditions is extremely compromised if corrective measures are not taken in a timely and efficient manner. However, even with the proper placement and size of the new DG, such adverse effects diminish, but still prevail in the system for a long time if proper strategies are not designed. Further, the reliability and robustness of the system against the fault and islanding conditions are lost and thus require designing an adaptive and efficient EMS to deal with them effectively, with the aim of improving the system performance. The energy management strategies proposed, designed, and validated under normal and contingency states are presented in Figure 10. The FIS is used for generating the different switching schemes, as shown in Figure 11 for the proposed strategies in Figure 10.

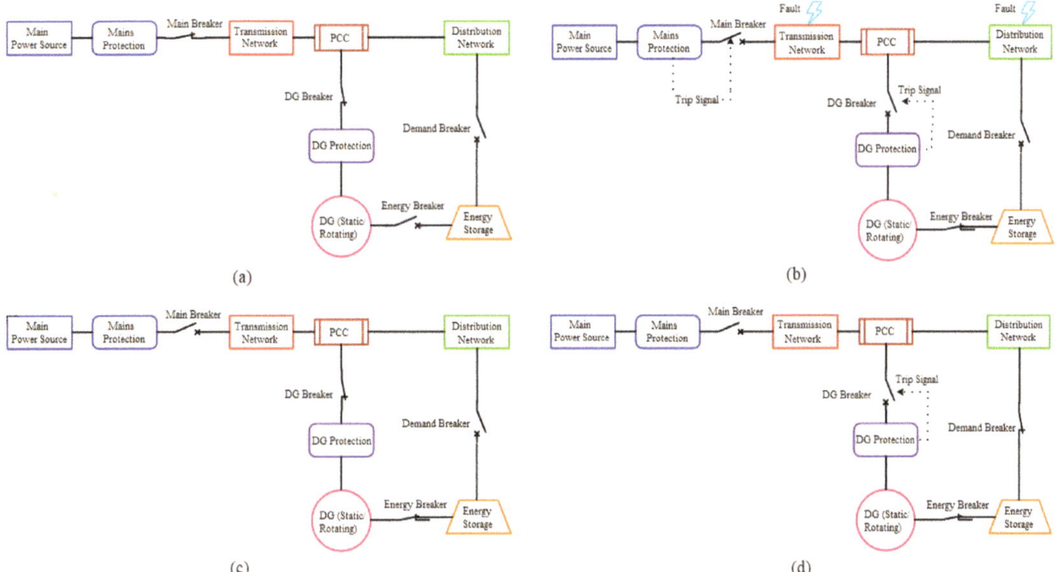

Figure 10. Energy management strategies for different operating conditions: (**a**) normal state, (**b**) fault condition, (**c**) balanced islanding condition, and (**d**) imbalanced islanding condition.

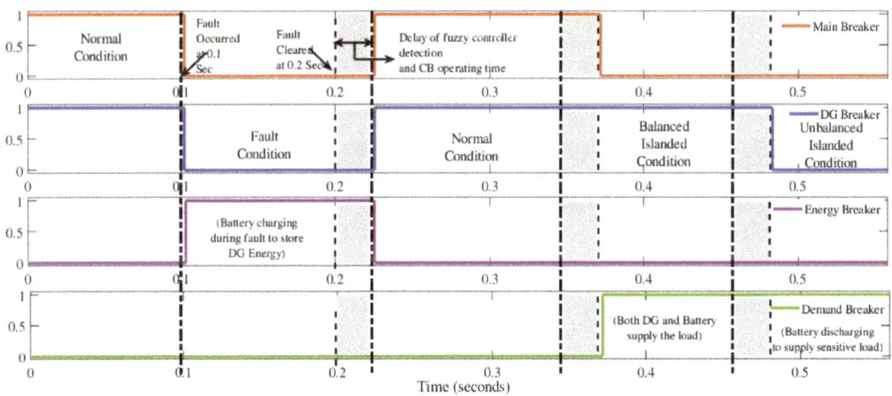

Figure 11. Switching patterns generated using the fuzzy inference system for circuit breakers under different operating conditions.

3.2.1. Normal Condition

Under the normal condition, when all the network constraints are within the pre-defined limits and the system is operating efficiently, the proper placement of the DG reduces the system losses and improves the voltage profile. The energy management strategy designed for such a network condition is presented in Figure 10a. In this scenario, the protection schemes of the DG and the main supply are inactive, and the energy breaker is responsible for energy management under abnormal conditions, while the demand breaker is used to meet the increased load demands during the critical situations. Both of these breakers are disconnected during the normal condition. In the normal state, the main supply and DG collectively meet the power demands of end-users, as shown in Figure 11, with both the energy and demand breakers being switched off.

3.2.2. Fault Condition

When the faults occur in the system either at the transmission or at the distribution side, the protection schemes of the mains and DG need to function in timely manner and thus isolate the sources. Under this condition, the energy generated by the DG may be stored for later use when the faults are removed, and this necessitates using an Energy Storage System (ESS). The energy management strategy designed to deal with different fault conditions is proposed in Figure 10b. In this condition, the energy breaker remains closed, and the demand breaker is open until a fault is present in the system. With the fault occurring near the load center at 0.1 s in Figure 12, the fuzzy logic controller instantly detects it in Figure 11 due to the abrupt changes in the network current and trips the main and DG circuit breakers, while the energy breaker remains closed to store the DG's energy. When the fault is cleared at 0.2 s, the main and DG breakers get closed after a 1.5 cycle delay, caused by the fuzzy logic controller and circuit breaker operating time sequences, as shown in Figure 13. In Figure 13, at 0.3 s, the network is restored to its normal state after the fault elimination. Further, at 0.35 s, the grid is disconnected, and the network enters into an islanding mode; thus, the power flow from the grid drops to approximately zero.

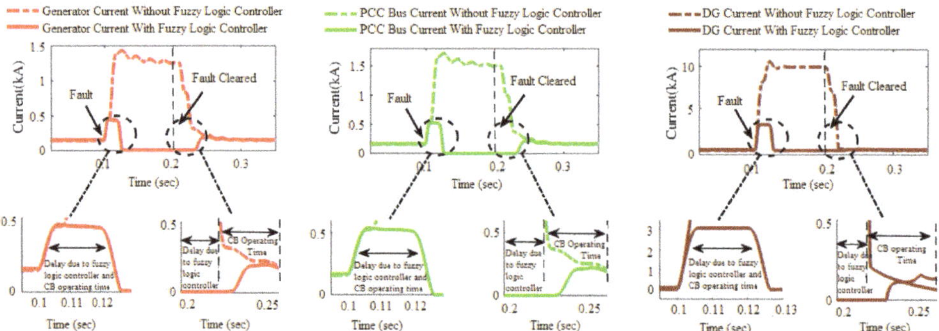

Figure 12. Current flows from different power system actors when fault occurs near the load center with the proposed energy management system using the fuzzy inference system. The dotted plots illustrate the short circuit current with no energy management strategy being active.

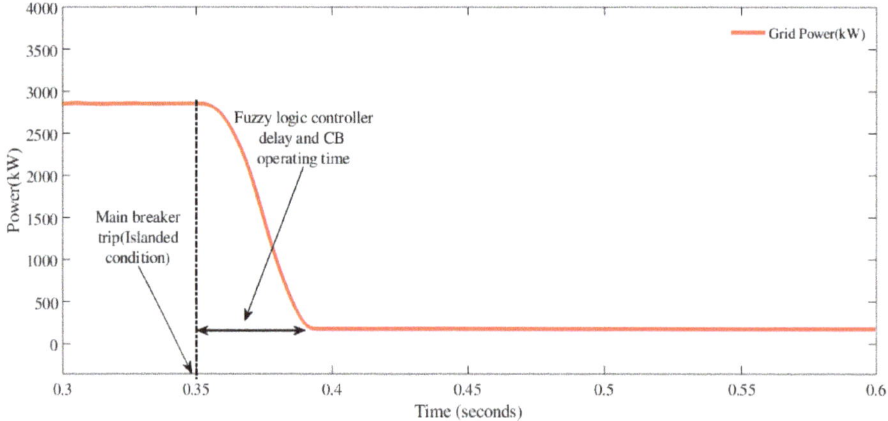

Figure 13. Profile of the power flow from the mains. At 0.3 s, the system returns to its normal state. At 0.35 s, the grid is disconnected, and the system enters the islanding condition, while the power flow from the grid drops approximately to zero.

3.2.3. Islanding Condition

When the main supply to the network is lost, this results in islanding conditions, and the DG may be allowed to supply the load alone, if its output matches the energy demand. However, if a power mismatch occurs, the DG needs to be taken out of the system, and the reliability of the network to supply the load is thus lost. To address this problem, the energy management strategy is designed and discussed in Figure 10c. This scenario is called balanced islanding, which represents the condition that the cumulative energy of the DG and ESS are able to cope with the given load requirements, as shown in Figure 11. In Figure 11, the main breaker trips after 0.35 s; the grid supply is disconnected, and the network operates in the balanced islanding condition. Initially, the system loading is larger than the DG output, and the ESS switches on to meet the additional energy demand, as shown in Figure 14. In Figure 14, after 0.35 s, the system operates in islanding mode, with both the DG and ESS supplying the load, as illustrated in Figure 15. The close-up view of the voltage profile in Figure 14 confirms that the cumulative supply from the DG and ESS is more stable than the grid and DG together. This is due to the obvious reason that the main source is at distance from the load center and causes more power losses, while the ESS and DG are installed near the load premises.

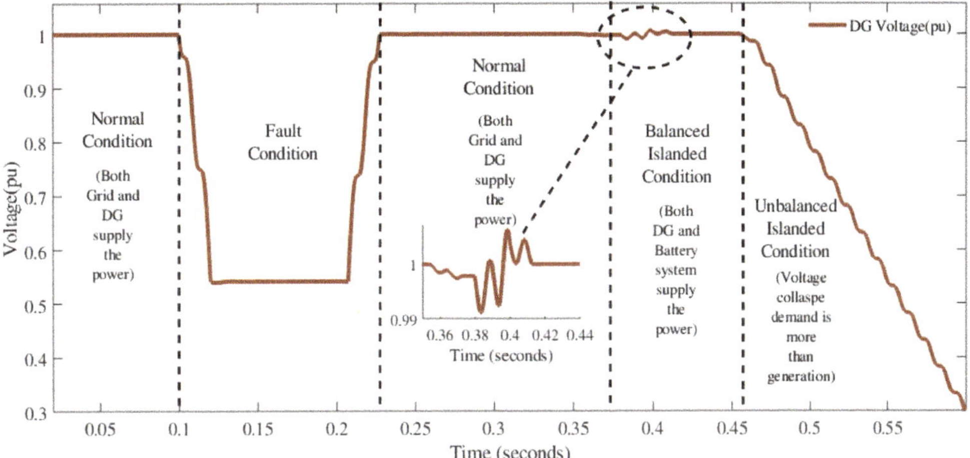

Figure 14. Voltage profile at the DG bus under different system conditions. The voltage under the normal condition is 1 p.u., and with the fault inception it dips down to 0.52 p.u. After 0.35 s, the system is in islanding mode with both the DG and battery system supplying the load. After 0.45 s, the load is increased and results in a voltage collapse.

During the imbalance islanding in Figure 10d, when the cumulative power of the ESS and DG is unable to meet the load demand, the DG needs to be disconnected from the system using its dedicated protection scheme. However, the ESS, having the fewer stability issues as compared to the DG system, continues to supply energy to emergency and critical loads, i.e., street lights, hospitals, etc., via its direct link. It is evident from Figure 11 that when the load on the distribution side is increased after 45 s in Figure 15, it causes unbalanced islanding and huge voltage collapse in Figure 14. Therefore, the DG breaker trips at 49 s while the ESS continues to supply the sensitive loads.

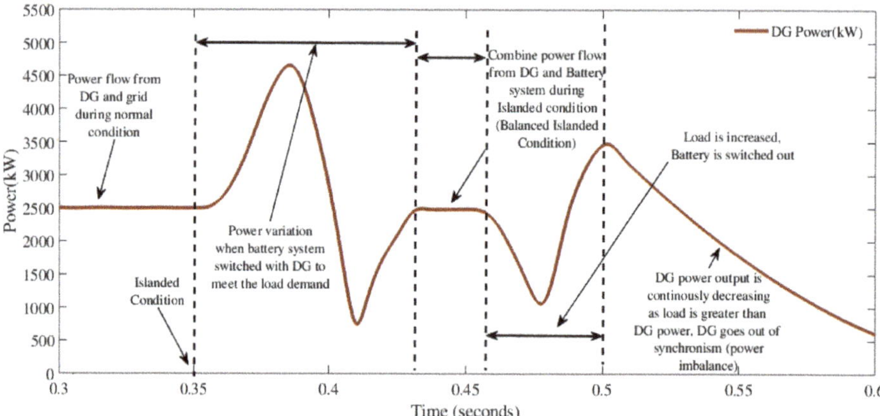

Figure 15. DG output under different system conditions. Until 0.35 s, a load of 5000 kW is shared equally among the mains and DG. At 0.35, the mains are tripped, and the system enters the islanding mode with the load greater than the DG, and thus, the ESS is switched in, to tackle the energy difference. At 0.45 s, the load is further increased, which causes the power imbalance with the DG's output constantly decreasing, which represents its being out of synchronism.

3.3. Comparative Analysis

To validate the performance of the proposed framework, it is compared with other existing approaches in Table 1. In Table 1, the Tabu Search (TS) algorithm, which is commonly used for evaluating the combinatorial optimization task, is exploited for the optimal placement of the DG in [29], based on improving the system voltage and reducing the line losses under a uniformly distributed load profile. This search algorithm has a higher convergence rate and hence takes less processing time to reach the desired results within fewer iterations. On the other hand, the Artificial Bee Colony (ABC) algorithm, used for the optimal sizing and location of the DG in [30], takes a longer time to converge and is also computationally expensive because it is based on the intuitive foraging behavior of honey bees. However, the Ant Colony Optimization (ACO) algorithm, inspired by the behavior of ants of reaching the target while following the shortest path, was used in [31] for the proper placement of new DGs. This algorithm has a compromising performance in terms of reducing the voltage deviations; otherwise, it is better than ABC. Overall, the suggested technique is not only robust against system abnormalities using the FIS, but it also tries to maximize the system performance, thereby reducing line losses and voltage variations (sag and phase angle jump) with the optimal placement of the DG using the GA. Further, the proposed framework is also computationally efficient and converges in a reasonable time to the desired outcomes.

Table 1. Comparative analysis for the optimal placement of the DG and energy management control. TS, Tabu Search; ABC, Artificial Bee Colony.

Technique	Loss Reduction (%)	Voltage Deviation (p.u.)	Number of Iterations	Energy Imbalance (%)	Detection Period (Cycles)	Computational Time (s)
TS [29]	51.47	0.02142	26	NA	NA	9.87
ABC [30]	53.63	0.03468	73	NA	NA	12.43
ACO [31]	35.55	0.04213	62	NA	NA	11.56
Proposed Framework	45.32	0.01176	34	13.28	2.50	11.65

NA—Not Applicable; loss reduction (%)—decrease in the percentage of existing line losses with the integration of each new DG; voltage deviation (p.u.)—the ability to maintain the voltage profile within the pre-defined IEEE regulation limits, i.e., ±5%; number of iterations—steps taken by the search algorithm to reach the final outcome; energy imbalance (%)—exploiting all the available energy sources to meet the load demand effectively; detection period (cycles)—operating duration of switching-sensing elements to detect the abnormality in the system; computational time (s)—processing resources consumed by the respective technique to reach the desired end goal.

4. Conclusions

This research proposed a framework that in the first steps exploited the genetic algorithm in order to determine the optimal placement and sizing of the distributed generators in the utility network under distinct system conditions. The GA together with load flow analysis made decisions on the basis of three critical parameters of the network: power losses, voltage profile, and phase angle jump. Under all investigated system conditions (normal operation, fault, and islanding), the best possible location was always decided to be near the load center, at Bus-E, with the optimal size of 3.33 MVA, in accordance with the system constraints. Even though the proper placement and sizing of the newly installed DG were not guaranteed, the excellent performance of the investigated power system—especially under the fault and islanding conditions—thus make it necessary to design a proper energy management system.

All the energy management strategies, under all the investigated system conditions, were designed using a fuzzy inference system. Amongst all proposed strategies, it was observed that the FIS with the proper membership functions and rules implemented was able to correctly detect, monitor, and regulate the abnormalities in a way that allowed the performance of the network to be optimized and the energy to be stored for emergency use. The switching patterns, generated by fuzzy controllers, were used to operate all four different circuit breakers (the main one, DG, energy, and demand breakers) sequentially, according to the network characteristics—determined through a load flow study. Eventually, in order to validate the effectiveness and robustness of the proposed approach against abnormalities, it was compared with other existing state-of-the-art methods, and we found that it integrated new DGs optimally and also regulated the energy flow according to the network conditions.

A possible extension, to the above presented research, is to use the Markov decision process for the optimum location and sizing of mixed (inverter- and non-inverter-based) DGs in a complex distribution network involving the combination of radial and mesh topologies. Furthermore, for a autonomous and adaptive energy management system, deep neural networks together with the forest of trees approach can be applied to deal with the environmental and network uncertainties and the dynamic nature of PVs or wind-based DG systems at run-time under severe network contingencies.

Author Contributions: The major contributions of all the authors are summarized as: conceptualization—S.K.; methodology—S.K. and M.F.S.; software—S.K. and M.F.S.; validation—S.K., F.S., S.S., A.M., and Z.H.K.; formal analysis—S.K., A.M., and V.B.; investigation—S.K. and L.S.; resources—Z.H.K. and L.Z.; data curation—Z.H.K. and M.A.S.; writing, original draft preparation—S.K., L.S., and S.A.S.; writing, review and editing—L.Z. and M.A.S.; visualization—Z.H.K.; supervision—L.S. and L.Z.; project administration—L.Z.; funding acquisition—L.Z. All authors have read and agreed to the published version of the manuscript.

Funding: This research received no external funding.

Acknowledgments: The authors are thankful to Sukkur IBA University, Pakistan, for providing state-of-the-art facilities to conduct this research. Additionally, the first author is also grateful to Jan Izykowski from Wroclaw University of Science and Technology, Poland, for his guidance and support during the Master's studies.

Conflicts of Interest: The authors declare no conflict of interest.

References

1. Katyara, S.; Staszewski, L.; Leonowicz, Z. Protection coordination of properly sized and placed distributed generations–methods, applications and future scope. *Energies* **2018**, *11*, 2672. [CrossRef]
2. Katyara, S.; Shah, M.A.; Iżykowski, J. Power loss reduction with optimal size and location of capacitor banks installed at 132 kV grid station qasimabad Hyderabad. *Present Probl. Power Syst. Control.* **2016**, 53–64.
3. Paliwal, P.; Patidar, N.P.; Nema, R.K. Planning of grid integrated distributed generators: A review of technology, objectives and techniques. *Renew. Sustain. Energy Rev.* **2014**, *40*, 557–570. [CrossRef]

4. Katyara, S.; Staszewski, L.; Chachar, F.A. Determining the Norton's Equivalent Model of Distribution System with Distributed Generation (DG) for Stability Analysis. *Recent Adv. Electr. Electron. Eng. Former. Recent Patents Electr. Electron. Eng.* **2019**, *12*, 190–198. [CrossRef]
5. Sultana, U.; Khairuddin, A.B.; Aman, M.M.; Mokhtar, A.S.; Zareen, N. A review of optimum DG placement based on minimization of power losses and voltage stability enhancement of distribution system. *Renew. Sustain. Energy Rev.* **2016**, *63*, 363–278. [CrossRef]
6. Jang, S.I.; Kim, K.H. An islanding detection method for distributed generations using voltage unbalance and total harmonic distortion of current. *IEEE Trans. Power Deliv.* **2004**, *19*, 745–752. [CrossRef]
7. Katyara, S.; Hashmani, A.; Chowdhary, B.S.; Musavi, H.A.; Aleem, A.; Chachar, F.A.; Shah, M.A. Wireless Networks for Voltage Stability Analysis and Anti-islanding Protection of Smart Grid System. *Wirel. Pers. Commun.* **2020**. [CrossRef]
8. Kanchev, H.; Lu, D.; Colas, F.; Lazarov, V.; Francois, B. Energy management and operational planning of a microgrid with a PV-based active generator for smart grid applications. *IEEE Trans. Ind. Electron.* **2011**, *58*, 4583–4592. [CrossRef]
9. Palma-Behnke, R.; Benavides, C.; Lanas, F.; Severino, B.; Reyes, L.; Llanos, J.; Sáez, D. A microgrid energy management system based on the rolling horizon strategy. *IEEE Trans. Smart Grid* **2013**, *4*, 996–1006. [CrossRef]
10. Öner, A.; Abur, A. Voltage stability based placement of distributed generation against extreme events. *Electr. Power Syst. Res.* **2020**, *189*, 106713. [CrossRef]
11. Katyara, S.; Shah, M.A.; Chowdhary, B.S.; Akhtar, F.; Lashari, G.A. Monitoring, control and energy management of smart grid system via WSN technology through SCADA applications. *Wirel. Pers. Commun.* **2019**, *106*, 1951–1968. [CrossRef]
12. Ogunjuyigbe, A.S.O.; Ayodele, T.R.; Akinola, O.A. Optimal allocation and sizing of PV/Wind/Split-diesel/Battery hybrid energy system for minimizing life cycle cost, carbon emission and dump energy of remote residential building. *Appl. Energy* **2016**, *171*, 153–171. [CrossRef]
13. Prommee, W.; Ongsakul, W. Optimal multiple distributed generation placement in microgrid system by improved reinitialized social structures particle swarm optimization. *Int. Trans. Electr. Energy Syst.* **2011**, *21*, 489–504. [CrossRef]
14. Thirugnanam, K.; Kerk, S.K.; Yuen, C.; Liu, N.; Zhang, M. Energy Management for Renewable Microgrid in Reducing Diesel Generators Usage with Multiple Types of Battery. *IEEE Trans. Ind. Electron.* **2018**, *65*, 6772–6786. [CrossRef]
15. Merabet, A.; Ahmed, K.T.; Ibrahim, H.; Beguenane, R.; Ghias, A.M.Y.M. Energy Management and Control System for Laboratory Scale Microgrid Based Wind-PV-Battery. *IEEE Trans. Sustain. Energy* **2017**, *8*, 145–154. [CrossRef]
16. Gözel, T.; Hocaoglu, M.H. An analytical method for the sizing and siting of distributed generators in radial systems. *Electr. Power Syst. Res.* **2009**, *79*, 912–918. [CrossRef]
17. Prabha, D.R.; Jayabarathi, T. Optimal placement and sizing of multiple distributed generating units in distribution networks by invasive weed optimization algorithm. *Ain Shams Eng. J.* **2016**, *7*, 683–694. [CrossRef]
18. Khalesi, N.; Rezaei, N.; Haghifam, M.-R. DG allocation with application of dynamic programming for loss reduction and reliability improvement. *Int. J. Electr. Power Energy Syst.* **2011**, *33*, 288–295. [CrossRef]
19. Valencia, F.; Collado, J.; Sáez, D.; Marín, L.G. Robust energy management system for a microgrid based on a fuzzy prediction interval model. *IEEE Trans. Smart Grid* **2015**, *7*, 1486–1494. [CrossRef]
20. Adhikari, S.; Sinha, N.; Dorendrajit, T. Fuzzy logic based on-line fault detection and classification in transmission line. *SpringerPlus* **2016**, *5*, 1002. [CrossRef]
21. Katyara, S.; Akhtar, F.; Solanki, S.; Leonowicz, Z.; Staszewski, L. Adaptive fault classification approach using digitized fuzzy logic (dfl) based on sequence components. In Proceedings of the 2018 IEEE International Conference on Environment and Electrical Engineering and 2018 IEEE Industrial and Commercial Power Systems Europe (EEEIC/I&CPS Europe), Palermo, Italy, 12–15 June 2018; pp. 1–7.
22. Mohsin Ajeel, K. Optimal location of a single distributed generation unit in power systems. *MS&E* **2018**, *433*, 012087.
23. Gomez, J.C.; Morcos, M.M. Voltage sag and recovery time in repetitive events. *IEEE Trans. Power Deliv.* **2002**, *17*, 1037–1043. [CrossRef]
24. Shukla, T.N.; Singh, S.P.; Srinivasarao, V.; Naik, K.B. Optimal sizing of distributed generation placed on radial distribution systems. *Electr. Power Components Syst.* **2010**, *38*, 260–274. [CrossRef]
25. Muruganantham, B.; Gnanadass, R.; Padhy, N.P. Performance analysis and comparison of load flow methods in a practical distribution system. In Proceedings of the 2016 National Power Systems Conference (NPSC), Bhubaneswar, India, 19–21 December 2016.
26. Thakur, T.; Dhiman, J. A new approach to load flow solutions for radial distribution system. In Proceedings of the 2006 IEEE/PES Transmission and Distribution Conference and Exposition: Latin America, Caracas, Venezuela, 15–18 August 2006 .
27. Augugliaro, A.; Dusonchet, L.; Favuzza, S.; Ippolito, M.G.; Mangione, S.; Sanseverino, E.R. A modified genetic algorithm for optimal allocation of capacitor banks in MV distribution networks. *Intell. Ind. Syst.* **2015**, *1*, 201–212. [CrossRef]
28. Heydari, G.; Gharaveisi, A.; Vali, M. New formulation for representing higher order tsk fuzzy systems. *IEEE Trans. Fuzzy Syst.* **2015**, *24*, 854–864. [CrossRef]
29. Nara, K.; Hayashi, Y.; Ikeda, K.; Ashizawa, T. Application of tabu search to optimal placement of distributed generators. In Proceedings of the 2001 IEEE Power Engineering Society Winter Meeting. Conference Proceedings (Cat. No. 01CH37194), Columbus, OH, USA, 28 January–1 February 2001; Volume 2, pp. 918–923.
30. Abu-Mouti, F.S.; El-Hawary, M.E. Heuristic curve-fitted technique for distributed generation optimisation in radial distribution feeder systems. *IET Gener. Transm. Distrib.* **2011**, *5*, 172–180. [CrossRef]
31. Wang, L.; Singh, C. Reliability-constrained optimum placement of reclosers and distributed generators in distribution networks using an ant colony system algorithm. *IEEE Trans. Syst. Man Cybern. Part C Appl. Rev.* **2008**, *38*, 757–764. [CrossRef]

Article

A Hybrid Approach Based on SOCP and the Discrete Version of the SCA for Optimal Placement and Sizing DGs in AC Distribution Networks

Oscar Danilo Montoya [1,2], Alexander Molina-Cabrera [3], Harold R. Chamorro [4,*], Lazaro Alvarado-Barrios [5] and Edwin Rivas-Trujillo [1]

1   Facultad de Ingeniería, Universidad Distrital Francisco José de Caldas, Carrera 7 No. 40B-53, Bogotá D.C. 11021, Colombia; odmontoyag@udistrital.edu.co (O.D.M.); erivas@udistrital.edu.co (E.R.-T.)
2   Laboratorio Inteligente de Energía, Universidad Tecnológica de Bolívar, km 1 vía Turbaco, Cartagena 131001, Colombia
3   Facultad de Ingeniería, Universidad Tecnológica de Pereira, Pereira 660003, Colombia; almo@utp.edu.co
4   Department of Electrical Engineering at KTH, Royal Institute of Technology, SE-100 44 Stockholm, Sweden
5   Department of Engineering, Universidad Loyola Andalucía, 41704 Sevilla, Spain; lalvarado@uloyola.es
*   Correspondence: haroldrene.chamorrovera@kuleuven.be

Abstract: This paper deals with the problem of the optimal placement and sizing of distributed generators (DGs) in alternating current (AC) distribution networks by proposing a hybrid master–slave optimization procedure. In the master stage, the discrete version of the sine–cosine algorithm (SCA) determines the optimal location of the DGs, i.e., the nodes where these must be located, by using an integer codification. In the slave stage, the problem of the optimal sizing of the DGs is solved through the implementation of the second-order cone programming (SOCP) equivalent model to obtain solutions for the resulting optimal power flow problem. As the main advantage, the proposed approach allows converting the original mixed-integer nonlinear programming formulation into a mixed-integer SOCP equivalent. That is, each combination of nodes provided by the master level SCA algorithm to locate distributed generators brings an optimal solution in terms of its sizing; since SOCP is a convex optimization model that ensures the global optimum finding. Numerical validations of the proposed hybrid SCA-SOCP to optimal placement and sizing of DGs in AC distribution networks show its capacity to find global optimal solutions. Some classical distribution networks (33 and 69 nodes) were tested, and some comparisons were made using reported results from literature. In addition, simulation cases with unity and variable power factor are made, including the possibility of locating photovoltaic sources considering daily load and generation curves. All the simulations were carried out in the MATLAB software using the CVX optimization tool.

Keywords: distributed generation; mixed-integer nonlinear programming; optimal power flow; second-cone programming; discrete-sine cosine algorithm; metaheuristic optimization

Citation: Montoya, O.D.; Molina-Cabrera, A.; Chamorro, H.R.; Alvarado-Barrios, L.; Rivas-Trujillo, E. A Hybrid Approach Based on SOCP and the Discrete Version of the SCA for Optimal Placement and Sizing DGs in AC Distribution Networks. *Electronics* **2021**, *10*, 26. https://dx.doi.org/10.3390/electronics10010026

Received: 16 November 2020
Accepted: 22 December 2020
Published: 27 December 2020

1. Introduction

Electrical distribution networks are entrusted with providing electricity services to the end users in medium- and low-voltage level in rural or urban areas [1]. These grids are typically operated with a radial configuration to reduce investment, maintenance and operative costs [2]. However, the radial configuration produces higher power losses in contrast to meshed configurations; also, the nodal voltage rapidly worsens, as the nodes are far from the substation [3]. To mitigate these higher power losses, the literature proposes multiple approaches to know: (i) optimal placement of shunt capacitors [4], (ii) optimal reconfiguration of the distribution grid [5], (iii) optimal selection/substitution of the calibers of the conductors [6,7], (iv) optimal placement and sizing distributed generators [8–10], among others. Each one of these approaches allow dealing with power losses minimization; nevertheless, the most effective approach for dealing with this power loss corresponds

39

to the optimal placement and sizing of DGs since reductions higher than 50% have been reported for this methodology [11].

The optimal placement and sizing of DGs in electric distribution networks is a complex and large-scale mixed-integer nonlinear programming (MINLP) problem. This MINLP structure of the optimization problem complicates the possibility of finding the global optimal solution due to the non-convexity shape of the solution space [12]. For this reason, in this research, we propose a combination of a metaheuristic approach with a second-order cone programming (SOCP) formulation to address this problem with excellent numerical performance as will be presented in the results section.

Due to the importance of having mathematical optimization in distribution systems analysis, here, we propose a new hybrid optimization approach based on the discrete version of the sine–cosine algorithm, i.e., (DSCA) added to the SOCP formulation to solve the exact mixed-integer nonlinear programming (MINLP) formulation of the problem of the optimal location and sizing of DGs in AC distribution networks [13]. This hybrid optimization approach called DSCA-SOCP is motivated by the following facts: (i) the exact MINLP structure makes it impossible to find the global optimal solution for this problem with the current optimization approaches even using metaheuristic; this situation occurs since the studied problem contains binary variables regarding the placement of the DGs and the continuous part associated with their sizing, which is formulated as an optimal power flow problem being non-convex due to the presence of trigonometric functions in its formulation where it is not possible to ensure global solution with non-convex methods [14,15]. The union of both problems (integer and nonlinear continuous) increases the possibility of branch and bound methods or metaheuristics to be stuck in local optimal solutions [16]; and (ii) the conventional metaheuristic approaches to solve the MINLP problem deals with the optimal power flow problems using controlled random procedures [8], which are inadequate approaches (they do not guarantee the global optimal solution); in opposition, the convex optimization allows to find it with duality zero gap [17].

Based on the aforementioned problems with conventional metaheuristic approaches, we propose a hybrid DSCA-SOCP programming to solve the studied problem using a master–slave optimization strategy, where the master stage is entrusted with determining the subset of nodes where DGs will be located, and the slave stage solves the resulting optimal power flow problem to determine their optimal sizes. The main advantage of the proposed approach is that the SOCP programming ensures the global optimal solution for each nodal combination provided by the DSCA [18], which implies that if the best subset of nodes is identified by the master stage, the global optimal solution for the problem of the optimal placement and sizing of DGs in AC distribution networks will be guaranteed (this will be confirmed in the results section) [19].

The problem of the optimal placement and sizing of distributed generation in AC distribution networks to minimize active power losses in all the branches of the grid has been largely studied in the last two decades [20]. Most of the proposed approaches in literature work with master–slave algorithms based on metaheuristic optimization techniques [8]. Some of the recent approaches in this field of study are listed in Table 1.

The common denominator of these approaches is that these references work with hybrid master–slave optimization approaches to solve the exact MINLP model in two stages, i.e., a discrete part of the algorithm is entrusted with determining the location of the DGs and the continuous part deals with the dimensioning problem via optimal power flow analysis [21]. However, no evidence about the combination of the convex optimization approach for the continuous part and the discrete sine–cosine algorithm for the integer part has been found after the revision of the state-of-the-art, and this gap has been exploited in this paper as an opportunity of research.

Table 1. Recent optimization methods for optimal placement and sizing distributed generators (DGs) in alternating current (AC) distribution networks.

Acronym	Optimization Method	Reference	Year
GA-PSO	Genetic algorithm and particle swarm optimization	[22]	2012
LSFSA	Loss sensitivity factor simulated annealing	[23]	2013
MINLP	Mixed-integer nonlinear programming formulation	[9]	2014
TBLO	Teaching learning based optimization	[11]	2014
QOTBLO	Quasi-oppositional teaching learning based optimization	[11]	2014
HSA-PABC	Harmony search algorithm and particle artificial bee colony algorithm	[4]	2014
RBFNN-PSO	Radial basis function neural network and particle swarm optimization	[24]	2015
GA-IWD	Genetic algorithm and intelligent water drops	[25]	2016
AHA	Algorithmic heuristic approach	[26]	2016
KHA	Krill-herd algorithm	[27]	2016
PBIL-PSO	Population-Based Incremental Learning and particle swarm optimizer	[8]	2018
ABCA	Artificial bee colony algorithm	[28]	2018
HTLBOGWO	Hybrid teaching–learning based optimization-grey wolf optimizer	[29]	2019
MSSA	Mutated salp swarm algorithm	[30]	2019
CHVSA	Constructive heuristic vortex search algorithm	[31]	2019
GAMS	General algobraic modeling system	[12]	2020
CBGA-VSA	Chu and Beasley genetic algorithm and vortex search algorithm	[21]	2020

Remark 1. *In the revision of the state-of-the-art, only the methodologies called MINLP proposed in [9] and GAMS presented in [12] work with the exact model of the problem by implementing branch and bound in conjunction with interior point methods to solve the problem. However, due to the non-convexities of the solution space, these are stuck in local optimums.*

To avoid being stuck in local optimum solutions, our approach combines the efficiency of conic programming with easily implementable metaheuristic to find the global optimal solution of the problem using a master–slave optimization approach. The main advantage of the SOCP is that if the combination of the nodes where DGs will be located is fixed, the optimal sizing provided by the SOCP approach remains equal (repeatability property), which is not ensured with conventional metaheuristics used for optimal power flow analysis.

Based on the review of the state-of-the-art presented in the previous section, the main contributions of our proposal can be summarized as follows:

✓　The reformulation of the exact mixed-integer nonlinear programming model into a mixed-integer one by transforming its continuous, i.e., optimal power flow, into a convex formulation via second-order cone programming.

✓　The presentation of the discrete version of the sine–cosine algorithm to address the integer part of the MISOCP approach by using an integer codification that contains the nodal numbers as decision variables.

✓　The hybridization of the SCA and the SOCP programming has the capability of finding the global optimal solution with low computational effort in both test feeders studied. Numerical results show improvements regarding classical mataheuristic methods available in literature, including exact MINLP approaches.

It is worth mentioning that the proposed optimization approach deals with the optimal placement and sizing of DGS in AC distribution networks considering the load peak conditions by assuming that the distributed generators are fully dispatchable as recommended in [9]. In addition, no considerations are made regarding the total distributed generation since we are interested in finding the best possible reduction in the active power losses in the distribution network without penetration limitations. Finally, we consider the possibility of installing three distributed generators since this is the most common assumption in literature [32]. In addition, three simulations cases are analyzed: (i) the optimal location and sizing of the DGs considering unity power factor, (ii) variable power factor, and (iii) daily load and photovoltaic solar curves.

The remainder of this document is organized as follows: Section 2 presents the exact mixed-integer nonlinear problem formulation of the optimal location and sizing of DGs in AC distribution networks with radial structure. Section 3 presents the proposed hybrid optimization methodology with master–slave structure, where the master slave is entrusted with solving the location problem by implementing the discrete version of the sine–cosine algorithm, and the slave stage is entrusted with determining the optimal sizes of the DGs by using a SOCP formulation. Section 4 presents the main features of the test feeders which are composed of 33 and 69 nodes, with radial structure and operated with 12.66 kV at the substation node. Section 5 presents the numerical achievements of the proposed optimization approach regarding the optimal location and sizing of DGs with their corresponding analysis and discussion. Section 6 shows the main concluding remarks as well as some possible future works derived from this research.

2. MINLP Formulation

The problem of the optimal location and sizing of distributed generation in AC distribution networks can be formulated as a mixed-integer nonlinear programming (MINLP) problem. The objective function of this problem corresponds to the minimization of the active power losses in the distribution network, which is subjected to a set of nonlinear constraints regarding active and reactive power balance equations, device capabilities and voltage regulation bounds, among others. Here, we present the MINLP formulation in the complex domain in order to simplify the proposed optimization approach that will be presented in Section 3. The complete MINLP model is presented below.

Objective function: The objective function that represents the problem of the optimal placement and sizing of DGs in AC distribution networks corresponds to the total power losses caused by the current flow in all the branches of the network. This objective function is formulated as presented in Equation (1).

$$\min p_{\text{loss}} = \text{real}\left\{ \sum_{i\in\mathcal{N}} \sum_{j\in\mathcal{N}} \mathbb{V}_i^\star \mathbb{Y}_{ij} \mathbb{V}_j \right\}, \tag{1}$$

where p_{loss} is the objective function value, $\mathbb{V}_i$ and $\mathbb{V}_j$ are the voltage values (magnitude and angle) in the nodes i and j, respectively; $\mathbb{Y}_{ij}$ is the complex admittance value of the nodal admittance matrix that relates nodes i and j. Note that $\mathcal{N}$ represents the set that contains all the nodes of the network, and $(\cdot)^\star$ represents the complex conjugate operator applied to the argument.

Set of constraints: The set of constraints that intervene in the problem of the optimal placement and sizing of DGs in AC distribution networks are described as follows:

$$\mathbb{S}_i^{s,\star} + \mathbb{S}_i^{dg,\star} - \mathbb{S}_i^{d,\star} = \mathbb{V}_i^\star \sum_{j\in\mathcal{N}} \mathbb{Y}_{ij} \mathbb{V}_j, \quad \{i\in\mathcal{N}\}, \tag{2}$$

where $\mathbb{S}_i^{s,\star}$ is the apparent power generation in the slack node connected at bus i, $\mathbb{S}_i^{dg,\star}$ corresponds to the apparent power generation provided by the DG connected at node i, and $\mathbb{S}_i^{d,\star}$ represents the apparent power consumption at node i.

Expression (3) is associated to the voltage regulation bounds in all the nodes of the network.

$$\|\mathbb{V}_i - 1\| \leq \gamma, \ \forall i \in \mathcal{N}, \tag{3}$$

where γ is the maximum deviation given by the regulatory policies, which is usually between 0.05 pu and 0.10 pu. Note that in the case of the substation, $\mathbb{V}_i = 1 + j0$ pu.

The capacity of the existing and newly distributed generators is upper and lower bounded as follows:

$$\underline{S_i^s} \leq \mathbb{S}_i^s \leq \overline{S_i^s}, \ \forall i \in \mathcal{N}, \tag{4}$$

$$x_i \underline{S^{gd,\text{new}}} \leq \mathbb{S}_i^{dg} \leq x_i \overline{S^{gd,\text{new}}}, \ \forall i \in \mathcal{N}, \tag{5}$$

where

$$x_i \in \{0,1\}, \ \forall i \in \mathcal{N}, \tag{6}$$

which denotes the binary variable of the problem, which has a value of 1 if a DG is installed at node i or 0. There is a limit to the number of DGs that can be installed in the system, which is given by (7),

$$\sum_{k \in \mathcal{N}} x_i \leq N_{\text{DGs}}, \tag{7}$$

where N_{DGs} is the total number of distributed generators available for installation in the AC distribution network.

Remark 2. *The structure of the optimization model (1) to (7) exhibits a nonlinear non-convex structure with the presence of binary variables regarding the location of the DGs in a particular node of the grid. However, the nonlinear structure of the power balance equations in (2) is the most challenging constraint since it does not guarantee the global optimum finding even if all the binary variable combinations are explored.*

Figure 1 summarizes the main characteristics of the MINLP model that represents the problem of the optimal placement and sizing of DGs in AC radial distribution networks.

Convex Equations
(1), (3)–(5) and (7)

MINLP

$x_i \in \{0,1\}$
Binary

Non-convex
Equation (2)

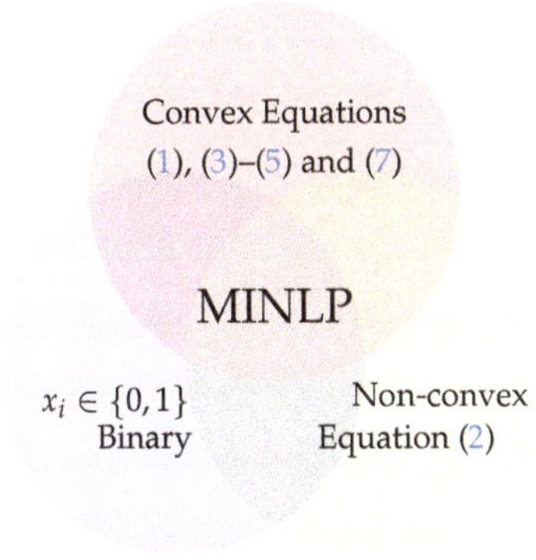

Figure 1. Characterization of the optimization model.

To address the nonlinear part of the optimization model described in Figure 1, we propose the reformulation of the nonlinear part of the model (i.e., power balance equations) into a second-order cone equivalent, while the binary part of the model is addressed through a metaheuristic approach as is presented in the following section.

3. Proposed Hybrid Optimization Approach

To solve the problem of the optimal placement and sizing of DGs in AC distribution networks, we propose a hybrid master–slave optimization algorithm. The master stage employs the metaheuristic sine–cosine algorithm (SCA) to solve the binary problem, i.e., the location of the distributed generators on the grid. In the slave stage the optimal power flow problem is reformulated as a second-order cone programming (SOCP) in order to guarantee the global optimum finding for each nodal combination providing for the SCA.

3.1. Slave Stage: SOCP Approach

The SOCP approach corresponds to a branch of the convex optimization where conic constraints allow for the reformulation of products between variables in order to transform nonlinear optimization problems into convex ones [18]. In the case of the optimal power flow analysis, the SOCP formulation permits to find the global optimal solution with zero gap when this is compared to the exact nonlinear programming power flow formulation [17]. Here, the SOCP formulation is presented to address the problem of the optimal sizing of DGs supposing that their locations have been previously informed by the master stage. To obtain the SOCP model, let us define a new auxiliary variable as follows

$$\mathbb{V}_{ij} = \mathbb{V}_i^\star \mathbb{V}_j, \tag{8}$$

where if we multiply in both sides for $\mathbb{V}_{ij}^\star$, we have

$$\|\mathbb{V}_{ij}\|^2 = \|\mathbb{V}_i\|^2 \|\mathbb{V}_j\|^2, \tag{9}$$

Now, if we define a new vector of U with entries $v_i = \|\mathbb{V}_i\|^2$, then we reach the following result

$$\|\mathbb{V}_{ij}\|^2 = v_i v_j, \tag{10}$$

which can be rewritten as follows

$$\begin{aligned}
\|\mathbb{V}_{ij}\|^2 &= u_i u_j, \\
\|\mathbb{V}_{ij}\|^2 &= \tfrac{1}{4}\left(u_i + u_j\right)^2 - \tfrac{1}{4}\left(v_i - v_j\right)^2, \\
\|\mathbb{V}_{ij}\|^2 + \tfrac{1}{4}\left(v_i - v_j\right)^2 &= \tfrac{1}{4}\left(v_i + v_j\right)^2, \\
\left\|\begin{matrix} 2\mathbb{V}_{ij} \\ v_i - v_j \end{matrix}\right\| &= v_i + v_j.
\end{aligned} \tag{11}$$

Note that Equation (11) is still a non-convex equality constraint, however, as recommended in [18], this can be relaxed as a second-order constraint by replacing the equality symbol by an inequality one as presented below:

$$\left\|\begin{matrix} 2\mathbb{V}_{ij} \\ v_i - v_j \end{matrix}\right\| \leq v_i + v_j. \tag{12}$$

Now, to rewrite the continuous part of the studied problem, let us substitute (8) into (1) and (2), which produces the following linear objective function and constraint, respectively.

$$\min p_{\text{loss}} = \text{real}\left\{ \sum_{i \in \mathcal{N}} \sum_{j \in \mathcal{N}} \mathbb{Y}_{ij} \mathbb{V}_{ij} \right\}, \tag{13}$$

$$\mathbb{S}_i^{s,\star} + \mathbb{S}_i^{dg,\star} - \mathbb{S}_i^{d,\star} = \sum_{j \in \mathcal{N}} \mathbb{Y}_{ij} \mathbb{V}_{ij}, \quad \{i \in \mathcal{N}\}, \tag{14}$$

Remark 3. *The SOCP reformulation allows reaching the global optimal solution of the optimal power flow problem associated with the optimal sizing of the DGs, since the resulting optimization model is essentially linear with an only conic constraint.*

Note that the characteristics of the studied optimization model depicted in Figure 1 can be redefine by eliminating the non-convex constraint based on the proposed SOCP formulation as presented in Figure 2.

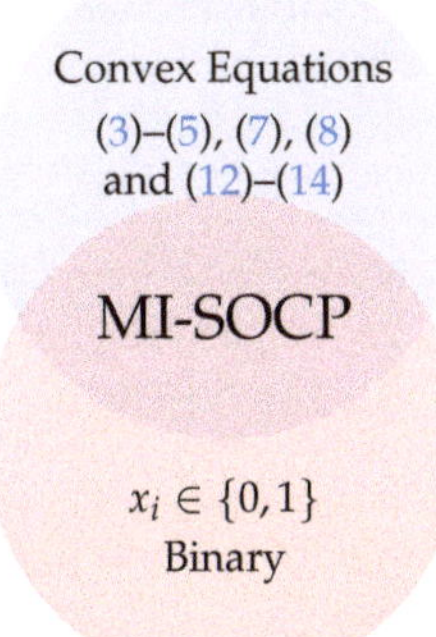

Figure 2. Mixed-integer second-order cone programming (SOCP) equivalent model for the problem of the optimal location and sizing of distributed generators in distribution networks.

Note that the SOCP approximation is given as a function of $\mathbb{V}_{ij}$ and v_i instead of the voltages V_i. Notwithstanding, it is possible to recover the original voltages by the following two-step procedure: First, the voltage magnitude is computed as $V_i = \sqrt{v_i}$. This value exists, and it is real since $u_i \geq 0$. Second, the angle of the voltages is calculated from $\theta_{ij} = \text{ang}(\mathbb{V}_{ij})$ in a forward iteration, starting from $\theta_1 = 0$. Therefore, a power flow calculation is not required after the optimization problem is solved.

3.2. Master Stage: Discrete SCA

The master stage is entrusted with solving the integer part of the optimization problem, i.e., to define the location of all the DGs. Here, we adopt the discrete version of the sine–cosine algorithm, which works with a reduced population by using an integer codification to represent the optimization problem [21].

The SCA is an optimization technique that works with a population which evolves by using trigonometric functions and variable radius in order to explore and exploit the solution space [33]. This optimization algorithm has been employed to solve different continuous domain problems, such as optimal power flow in power and distribution systems [34,35], parameter estimation in photovoltaic modules [36], optimal design of bend photonic crystal waveguides [37], and general solution of nonlinear non-convex optimization problems [38] among others. The main aspects of the implementation of the discrete SCA are described in the following subsections.

Initial Population

The SCA is a metaheuristic optimization technique that works with an initial population that is evolving through the iterative procedure by sine and cosine rule. The structure of the initial population for the proposed SCA is defined as follows

$$N^t = \begin{bmatrix} n_{11} & n_{12} & \cdots & n_{1N_{\text{DGs}}} \\ n_{21} & n_{22} & \cdots & n_{2N_{\text{DGs}}} \\ \vdots & \vdots & \ddots & \vdots \\ n_{M1} & n_{M2} & \cdots & n_{MN_{\text{DGs}}} \end{bmatrix} \tag{15}$$

where t is the iterative counter, which is fixed as zero for the initial population, and M is the number of individuals in the population. Remember that N_{DGs} represents the number of DGs available for installation.

Note that each element inside of the initial population is created as follows:

$$n_{ij} = \text{round}(2 + \text{rand}(1)(n-2)) \tag{16}$$

where n is the total number of nodes in the AC distribution network. Observe that the function $\text{round}(\cdot)$ takes the near integer part of the number and rand is a random number between 0 and 1 generated with a normal distribution. It is worth mentioning that node 1 is not considered in the population since it corresponds to the slack node. In addition, this codification guarantees the feasibility in the integer part of the solution space.

Remark 4. *To maintain the feasibility of the solution space during the generation of the initial population we ensure that each one of the components of the individual N_i^t is different to the remainder components, i.e., $n_{ij} \neq n_{ik}$, $\forall k = 1, 2, ..., N_{DGs}$, and $k \neq j$.*

3.3. Fitness Function Evaluation

The SCA evolves through the solution space typically using a modification of the objective function named fitness function [39]. This helps deal with possible infeasibilties of the decision variables [40]. However, due to the continuous part for the problem is formulated as a SOCP model; most of the constraints are directly fulfilled during the solution procedure via interior point methods. In this sense, the structure of the fitness function selected in this research takes the same form of the objective function. Note that this function is evaluated for each individual in the population, i.e., $p_{\text{loss}}(N_i^t)$, in order to identify the best individual in the current population. This individual is called N_{best}^t. Observe that in this research the best individual corresponds is the one who has the lower objective function value.

3.4. Evolution of the Population

The evolution of the of the population in the SCA algorithm is governed by trigonometric functions with a simple evolution rule as presented in Algorithm 1. Note that this evolution strategy takes the probability of 50% to evolve with sine or cosine trigonometric function (see r_1 parameter). In addition, r_2 controls the effect of the iteration counter in the modification of the population by presenting a linear decreasing rule; r_3 allows the evaluation of the sine or cosine function in all the points of the unitary circle, and r_4 introduces the importance of the best current individual in the evolution of the individual N_i^t to generate the next population.

3.5. Stopping Criterion

To finalize the searching procedure of the discrete version of the SCA, one of the following two conditions must be satisfied.

- ✓ If the total iterations t_{max} are reached, the SCA ends its iterative search and reports the best solution in the current population, i.e., $N_{\text{best}}^{t_{\text{max}}}$.
- ✓ If during k_{max} consecutive iterations the objective function does not improve, the iterative search of the SCA ends, and the best solution in the current population is reported, i.e., N_{best}^t.

3.6. Proposed Master–Slave Optimization Algorithm

The proposed master–slave optimization strategy to solve the problem of the optimal location and sizing of DGs in AC distribution networks based on the hybridization of the discrete version of the sine–cosine algorithm and the SOCP reformulation is summarized in Algorithm 2.

Algorithm 1: Evolution steps in the sine–cosine algorithm (SCA).

Result: Evolution of the Individuals in the Population
$i = 1$;
while $i \leq M$ **do**
> $r_1 = \text{rand}, r_2 = 1 - \frac{1}{t_{max}}$;
> $r_3 = 2\pi\text{rand}, r_4 = \text{rand}$;
> **if** $r_1 \leq \frac{1}{2}$ **then**
>> $Y_i = N_i^t + r_2 \sin(r_3)|r_4 N_{best}^t - N_i^t|$;
>
> **else**
>> $Y_i = N_i^t + r_2 \cos(r_3)|r_4 N_{best}^t - N_i^t|$;
>
> **end**
> **for** $j = 1 : N_{DGs}$ **do**
>> **if** $\left(Y_{ij} < 2 \ \ || \ \ Y_{ij} > n\right)$ **then**
>>> $Y_{ij} = \text{round}(2 + \text{rand}(1)(n - 2))$;
>>
>> **end**
>> **if** $\left(Y_{ij} < 2 \ \ || \ \ Y_{ij} > n\right)$ **then**
>>> $Y_{ij} = \text{round}(2 + \text{rand}(1)(n - 2))$;
>>
>> **end**
>> Evaluate the objective function value for the potential individual, i.e., $p_{loss}(Y_i)$;
>
> **end**
> **if** $p_{loss}(Y_i) < p_{loss}(N_i^t)$ **then**
>> $N_i^{t+1} = Y_i$;
>
> **end**

end

Algorithm 2: Proposed master–slave optimization approach.

Result: Optimal location and sizing of DGs
Define the AC grid parameters;
Define t_{max}, k_{max}, M and make $t = 0$ and $k = 0$;
while $t \leq t_{max}$ **do**
> **if** $t = 1$ **then**
>> Create the initial population, i.e., N^t;
>> Evaluate the fitness function of each individual, i.e., $p_{loss}(N_i^t)$;
>> Select the best current solution individual, i.e., N_{best}^t;
>
> **end**
> **while** $i \leq M$ **do**
>> Apply the evolution strategy defined in Algorithm 1 to update the current population, i.e., to obtain N_i^{t+1};
>
> **end**
> **if** $N_{best}^t = N_{best}^{t+1}$ **then**
>> k = k+1;
>
> **else**
>> k = 0;
>
> **end**
> **if** $t \geq t_{max} \ \ || \ \ k \geq k_{max}$ **then**
>> Report the best solution of the current population, i.e., N_{best}^t and solves the SOCP for it to determine the optimal sizes of the DGs.
>
> **end**

end

Remark 5. *Since the proposed hybrid SCA-SCOP depends on a metahueristic search in the master stage, statistical evaluation is required to determine its efficiency regarding the optimal solution finding capabilities. Here, we adopt 100 consecutive evaluations to the determine the general* *distribution of the solution findings by using maximum, minimum, mean and standard deviation indicators [21].*

4. Test Feeders

The computational validation of the proposed master–slave hybrid optimization algorithm to the optimal location and sizing of DGs in AC distribution is made in two classical distribution networks tests: 33 and 69 nodes. These grids works 12.66 kV at substation. The electrical connection between nodes in these test feeders are presented in Figures 3 and 4, respectively, while its parametric information can be consulted in [12]. It is worth mentioning that these test feeders are considered urban distribution networks that fed industrial users modeled as constant power consumption [8].

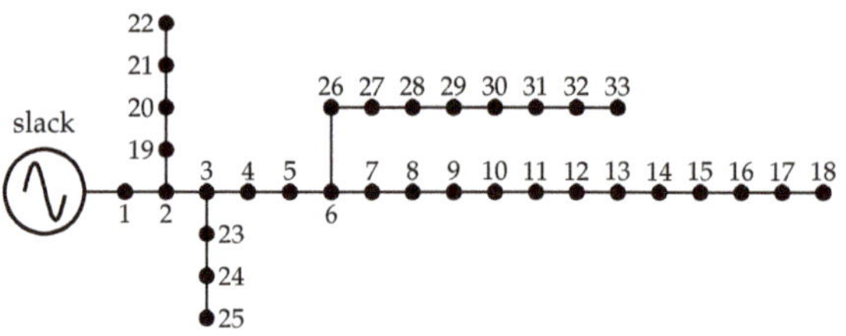

Figure 3. Electrical connection of nodes in the 33 node test feeder.

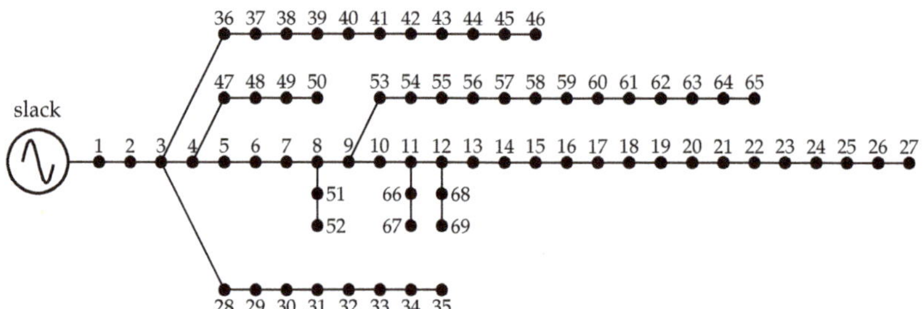

Figure 4. Electrical connection of nodes in the 69 node test feeder.

For both test feeders we consider as recommended in [21] the possibility of locating three distributed generators which will be sized at the peak load condition, we considered the voltage and power base values of 12.66 kV and 1000 kW, respectively. In addition, for the 33-node test feeder each DG was limited from 300 kW to 1200 kW, while for the 69-node test feeder these bounds were relaxed from 0 kW to 2000 kW, respectively.

5. Computational Validation

This section presents the computational validation of the proposed hybrid optimization approach based on the discrete version of the sine–cosine algorithm and the second-order cone programming model to deal with the problem of the optimal placement and sizing of distributed generators in AC distribution networks. We implement the proposed solution methodology on a personal computer AMD Ryzen 7 3700U, 2.3 GHz, 16 GB RAM with 64-bits Windows 10 Home Single Language using the MATLAB programming environment.

To compare the proposed hybrid optimization algorithm regarding objective function performance, we selected multiple metaheuristic optimization techniques reported in literature. These methodologies have been listed in Table 1. In the implementation of the proposed DSCA-SOCP approach, we have considered 50 iterations and a population of four individuals; in addition, 100 consecutive evaluations are made to validate the efficiency of the algorithm to reach the optimal solution and calculate the average processing time. Note that these parameters were found after multiple simulations that have allowed to identify an adequate trade-off between simulation times and the quality of the final solution.

5.1. Numerical Validation Considering Unity Power Factor

5.1.1. Results in the 33-Node Test Feeder

Table 2 reports the optimal placement and sizing of the distributed generators located in the 33-node test feeder after applying the proposed hybrid DSCA-SOCP (see last row) as well as the comparison with the literature reports.

Table 2. Optimal location and sizing of DGs in the 33-node test feeder for the proposed and comparative approaches.

Method	p_{loss} (kW)	Location Node	Size (MW)
GA-PSO	103.3600	{11,16,32}	{0.9250,0.8630,1.2000}
LSFSA	82.0525	{6,18,30}	{1.1124,0.4874,0.8679}
MINLP	72.7862	{13,24,30}	{0.8000,1.0900,1.0500}
TLBO	75.5400	{10,24,31}	{0.8246,1.0311,0.8862}
QOTLBO	74.1008	{12,24,29}	{0.8808,1.0592,1.0714}
HSA-PABC	72.8129	{14,24,30}	{0.7550,1.0730,1.0680}
GA-IWD	110.5100	{11,16,32}	{1.2214,0.6833,1.2135}
AHA	72.8340	{13,24,30}	{0.7920,1.0680,1.0270}
KHA	75.4116	{13,25,30}	{0.8107,0.8368,0.8410}
MSSA	72.7854	{13,24,30}	{0.8010,1.0910,1.0530}
CHVSA	78.4534	{6,14,31}	{1.1846,0.6468,0.6881}
CBGA-VSA	72.7853	{13,24,30}	{0.8018,1.0913,1.0536}
GAMS	72.8129	{14,24,30}	{0.7550,1.0730,1.0680}
DSCA-SOCP	72.7853	{13,24,30}	{0.8018,1.0913,1.0536}

The results in Table 2 illustrate that:

- There are three methods that finds the best optimal solution for the 33-node test feeder which are the MSSA, the CBGA-VSA and the DSCA-SOCP, which find a final power losses of about 72.7853 kW by installing DGs in nodes 13, 24 and 30 with power generations of 801.8 kW, 1091.3 kW and 1053.6 kW, respectively.
- The maximum reduction of the power loss is achieved by the aforementioned three methods (including our proposal) with a total reduction of about 65.50% regarding the base case, i.e., 210.9876 kW, while the worst approach corresponds to the GA-IWD with a reduction of about 47.62%. These results imply that all the remainder literature methods (see Table 1) have found solutions contained between these extreme solutions.
- The best solutions show that the best nodes to locate DGs are 13, 24 and 30; however, the second-best solution reported in this table is found when node 13 is changed by node 14 as reported by the GAMS and HSA-PABC algorithms with a small variation regarding final power losses of about 27.60 W.

It is worth mentioning that the results in Table 2 show that some methods identify the best optimal nodes for optimal locating DGs (see the AHA and the MINLP methods), however, due to the non-convexities in the dimensioning stage, i.e., optimal power flow, these methods present sub-optimal solutions since the nonlinear search approach (in some cases continuous metaheuristics) is stuck in local optimums. This situation does not occur at least with our proposal since each potential location for generators is optimally solved via SOCP which guarantees the optimal finding based on its convex structure. This implies that if we evaluate the same combination of nodes multiple times the optimal sizes of the DGs will be equal for each one of the evaluations (optimal solution), which confirms the efficiency of the convex optimization, i.e., SOCP, in power systems analysis.

5.1.2. Results in the 69-Node Test Feeder

The numerical behavior of the proposed DSCA-SOCP method for the 69-node test feeder is reported in Table 3 (see last row), where it is compared with multiple literature reports.

Table 3. Optimal location and sizing of DGs in the 69-node test feeder for the proposed and comparative approaches.

Method	p_{loss} (kW)	Location Node	Size (MW)
GA-PSO	84.5909	{21,61,63}	{0.9105,1.1926,0.8849}
LSFSA	72.1120	{18,60,65}	{0.4204,1.3311,0.4298}
MINLP	69.4090	{11,17,61}	{0.5300,0.3800,1.7200}
TLBO	72.4157	{15,61,63}	{0.5919,0.8188,0.9003}
QOTLBO	71.6345	{18,61,63}	{0.5334,1.1986,0.5672}
HTLBOGWO	71.7281	{18,61,62}	{0.5330,1.0000,0.7730}
GA-IWD	80.9100	{20,61,64}	{0.9115,1.3926,0.8059}
AHA	69.6669	{12,21,61}	{0.4710,0.3120,1.6890}
KHA	69.5730	{12,22,61}	{0.4962,0.3113,1.7354}
MSSA	69.4077	{11,18,61}	{0.5260,0.3800,1.7180}
CHVSA	69.4088	{11,17,61}	{0.5284,0.3794,1.7186}
GAMS	72.7900	{12,61,64}	{0.8131,1.4447,0.2896}
CBGA-VSA	69.4077	{11,18,61}	{0.5268,0.3801,1.7190}
DSCA-SOCP	69.4077	{11,18,61}	{0.5268,0.3801,1.7190}

The numerical values in Table 3 help conclude that:

✓ The same three methods found in the 33-node test feeder has global optimization capabilities, i.e., the MSSA, the CBGA-VSA and the proposed DSCA-SOCP, since these reach the best solution for the 69-node test feeder with a final power loss of 69.4077 kW by installing the DGs in nodes 11, 18 and 61 with power injections of 526.80 kW, 380.10 kW, and 1719.00 kW, respectively.

✓ Some near optimal solutions are found with other approaches such as MINLP, AHA, KHA and CHVSA since all of them provide solutions lower than 70 kW in the final power losses. However, these methods are stuck in local optimums since, in the case of the MINLP and the CHVSA approach, the node 18 (in the global optimal solution) was changed for the node 17. In addition, the AHA and KHA methods identify nodes 12, 21(22) and 61 as the best possible generators location, which also implies that in the solution of the discrete problem (i.e., master optimization stage), these are also stuck in local optimums.

✓ Regarding the total improvement of the power losses, we can observe that the proposed method allows reaching a total power loss reduction of about 69.15% (the same result for the MSSA and the CBGA-VSA methods), while the worst behavior regarding power losses minimization occurs for the GA-PSO with 62.40%. These extremes imply that all the other solutions are contained on this interval with a bandwidth of about 6.75%.

5.1.3. Additional Comments

For both test feeders it is important to mention that: (i) the proposed optimization method reaches the solution of the optimal problem of placement and sizing of DGs in AC distribution networks in the 33-node test feeder after 350 s of simulation, and in the case of the 69-node test feeder, this processing time was about 580 s; (ii) after 100 consecutive evaluations in both test feeders, the proposed DSCA-SOCP approach finds with the 30% of effectiveness in the 33-node test feeder and 20% in the case of the 69-node test system; and (iii) the differences between the best and the worst solution in both test feeders are about 2 kW, which implies that most of these solutions are indeed better than the current literature solutions presented in Tables 2 and 3.

Regarding voltage profiles, it is important to highlight that the minimum voltage regulation in the 33-node test feeder is 9.62% and in the 69-node test feeder is about 9.08%

previous to the optimal location of the DGs; however, after solving the MISOCP model with the proposed DSCA-SOCP approach, these regulations have improved until 3.13% and 2.10 % (note that the best possible regulation in a distribution is 0%, which implies that percentages close to zero are high-quality solutions). These results confirm the effectiveness of including DGs in AC distribution networks for improving voltage profiles since these are close to 1.00 pu in contrast to the base case.

It is worth motioning that, numerically speaking, the proposed DSCA-SOCP is equivalent to the CBGA-VSA approach reported in [21]; however, note that the main difference between both methods is associated with the continuous part of the MINLP model, i.e., the sizing of the DGs, since our approach solves these using an exact optimization method based on convex optimization, which implies that the sizes of the DGs are optimal; nevertheless, in the case of the VSA approach, this optimal property cannot be ensured due to the heuristic nature of this algorithm.

5.2. Numerical Validation Considering Variable Power Factor

To verify the effectiveness of the proposed hybrid DSCA-SOCP approach to determine the optimal location and sizing of DGs in radial distribution networks, here we consider the possibility of installing from 1 to 3 DGs, leaving free the total amount of reactive power injection as recommended in [9]. Tables 4 and 5 present the optimal solutions reported in literature for the improved analytical (IA) method, the particle swarm optimization (PSO) and the exact MINLP approach, all of which have been reported in [9] for the 33- and 69-node test feeders.

Table 4. Optimal location and sizing of the DGs considering variable power factor capabilities in the 33-node test feeder.

Method	Nodes	Active Power (MW)	Reactive Power (MVAr)	p_{loss} (kW)
IA	6	{2.6370}	{1.6340}	68.1570
MINLP	6	{2.5580}	{1.7610}	67.8540
PSO	6	{2.5570}	{1.7460}	67.8570
DSCA-SOCP	6	{2.5585}	{1.7614}	67.8560
IA	{6, 30}	{1.8000, 0.9000}	{1.1150, 0.5570}	44.8400
MINLP	{13, 30}	{0.8190, 1.5500}	{0.4340, 1.2400}	29.3100
PSO	{12, 29}	{0.8180, 1.6990}	{0.5660, 1.1910}	39.1000
DSCA-SOCP	{13, 30}	{0.8457, 1.1377}	{0.3988, 1.0643}	28.5040
IA	{6, 14, 30}	{0.9000, 0.6300, 0.9000}	{0.5570, 0.3900, 0.5570}	23.0500
MINLP	{13, 24, 30}	{0.7660, 1.0440, 1.1460}	{0.4110, 0.5520, 0.8590}	12.7400
PSO	{13, 24, 30}	{0.7640, 1.0680, 1.0160}	{0.5350, 0.6130, 0.6910}	15.0000
DSCA-SOCP	{13, 24, 30}	{0.7940, 1.0700, 1.0297}	{0.3734, 0.5172, 1.0115}	11.7400

Table 5. Optimal location and sizing of the DGs considering variable power factor capabilities in the 33-node test feeder.

Method	Nodes	Active Power (MW)	Reactive Power (MVAr)	p_{loss} (kW)
IA	61	{1.8390}	{1.2840}	23.2480
MINLP	61	{1.8280}	{1.3000}	23.3150
PSO	61	{1.8180}	{1.2500}	23.2480
DSCA-SOCP	61	{1.8285}	{1.3006}	23.1460
IA	{17, 61}	{0.5400, 1.7990}	{0.3770, 1.2563}	7.4564
MINLP	{17, 61}	{0.5220, 1.7350}	{0.3590, 1.2380}	7.2086
PSO	{17, 61}	{0.5240, 1.7430}	{0.3710, 1.1840}	7.4564
DSCA-SOCP	{17, 61}	{0.5221, 1.7347}	{0.3532, 1.2385}	7.2013
IA	{17, 50, 61}	{0.6300, 0.9000, 0.9000}	{0.3900, 0.5570, 0.5570}	5.0911
MINLP	{11, 17, 61}	{0.4940, 0.3790, 1.6740}	{0.43540, 0.2570, 1.1950}	4.2801
PSO	{18, 50, 61}	{0.5078, 0.6996, 1.7351}	{0.3440, 0.4740, 1.1580}	5.01911
DSCA-SOCP	{11, 17, 61}	{0.4944, 0.3790, 1.6744}	{0.3534, 0.2515, 1.1955}	4.2682

From results in Tables 4 and 5 it is possible to observe that:

✓ The exact MINLP approach and the proposed DSCA-SOCP approach identify the same subset of nodes to locate DGs for all the simulation cases in the 33- and 69-node test feeders.

✓ The inclusion of the reactive power capability of the DGs significantly improves the total power losses minimization regarding the unity power factor case. For example, for the case of three distributed generators, the total power loss passes from 72.7853 kW to 11.7400 kW; while for the 69-nodes, this change is from 69.4077 kW to 4.2682 kW.

✓ The solution with only one generator in the 33- and 69-node test feeders shows that the inclusion of reactive power improves the total active power losses in both systems, when three DGs are located considering unity power factor. This situation can be attributed to the positive effects of the moderated reactive power injections in the voltage profile behavior of the grid [10].

It is worth mentioning that the proposed DSCA-SOCP approach allows to reach the best optimal solutions compared to the comparative methods even if the location of the generators is the same as can be seen in Tables 4 and 5, since this hybrid approach ensures the optimal solution finding of the OPF problem associated with the sizing of the DGs by using a SOCP formulation, which is not the case with the PSO and IA algorithms. However, in the case of the MINLP approach, we can observe that the results presented by this method in the 33- and 69-node test feeders are comparable with the proposed DSCA-SOCP approach, and the difference in some decimals can be attributed to precision errors between both methodologies.

To verify that under the peak load condition all the voltage profiles in both test feeders fulfill their bounds, these are depicted in Figure 5. In this picture it is possible to observe that, in the case of the of the 33-node test feeder (see Figure 5a), when the distributed generation with reactive power capabilities is installed in the network, all the voltages increase and overpass 0.95 pu, which implies that the voltage regulation in this network is about 4.20% with one and two DGs and less than 1.00% in the case of the three DGs. In the case of the 69-bus test feeder (see Figure 5b), when distributed generators are located considered reactive power injections, we can observe that for one DG the regulation of the grid is about 2.00% caused by voltage drops in nodes 66 to 69, while for the two and three DGs the voltage regulation is lower than 1.00%.

It is worth mentioning that for both test feeders, when two or three DGs are used, all the voltage profile are very close to the substation voltage, which causes the line voltage drops to be very small, producing low power losses as can be observed in Tables 4 and 5.

5.3. Optimal Location of Renewable Energy Sources

To observe the effectiveness and robustness of the proposed approach to deal with renewable energy resources and variable load profiles, here, we study the problem of the optimal location of renewable energy resources in radial distribution networks. To do so, we consider that in the 69-node test feeder the possibility of installing three photovoltaic distributed generators considering a daily generation and load curves. These curves are depicted in Figure 6.

In this simulation scenario, the proposed DSCA-SOCP is compared with the large-scale nonlinear optimization package widely known as GAMS and the MINLP solves BONMIN, COUENNE and DICOPT. The results of this comparison is reported in Table 6.

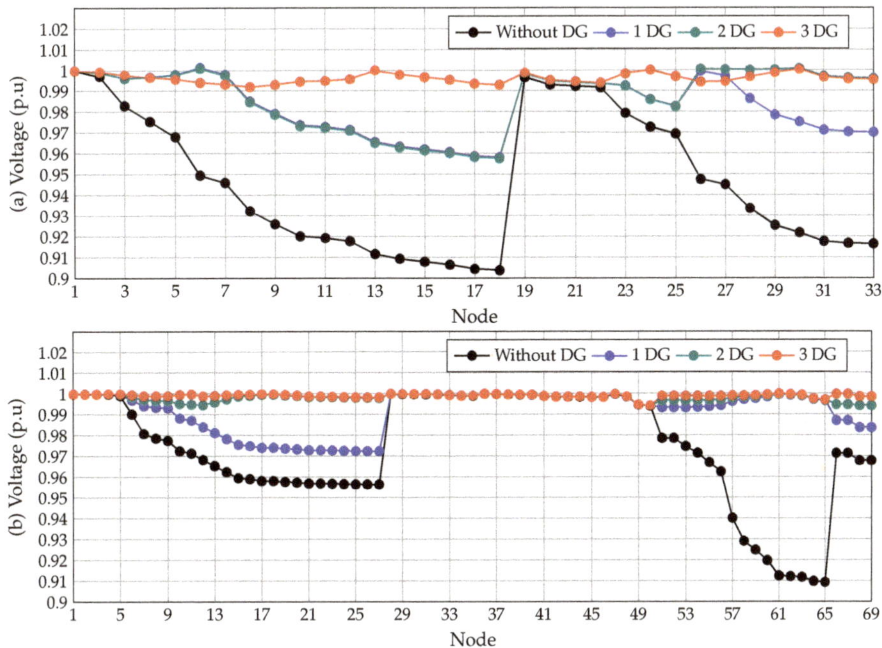

Figure 5. Voltage profile behavior in the 33- and 69-node test feeders when DGs with active and reactive power capabilities are installed: (**a**) 33-bus test feeder and (**b**) 69-bus test feeder.

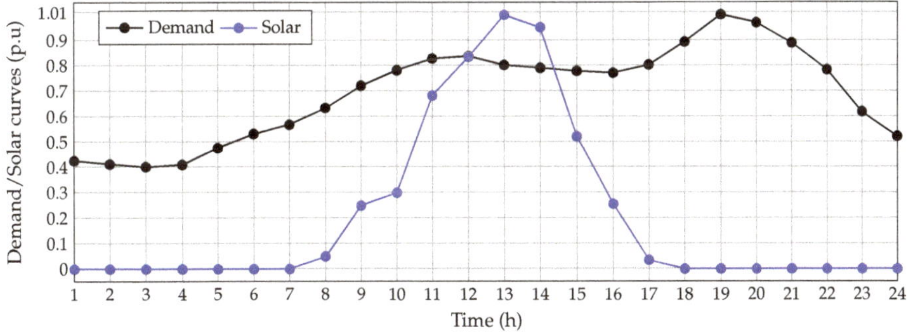

Figure 6. Daily behavior of the demand and solar photovoltaic generation.

Table 6. Optimal location and sizing of the DGs considering variable power factor capabilities in the 33-node test feeder.

Method	Nodes	Active Power (MW)	E_{loss} (kWh/day)
Base case	—	—	2666.2860
GAMS-BONMIN	{27, 61, 64}	{0.4366, 1.6744, 0.3253}	2046.0656
GAMS-COUENNE	{12, 18, 61}	{0.4990, 0.3808, 1.9254}	2030.5272
GAMS-DICOPT	{11, 17, 64}	{0.6900, 0.4194, 1.6626}	2074.0086
DSCA-SOCP	{11, 18, 61}	{0.5384, 0.4200, 1.8818}	1747.1748

The results in Table 6 demonstrate that: (i) with the location of three photovoltaic sources the proposed approach, i.e., the DSCA-SOCP approach, reduces the daily energy loss per day to about 919.1112 kWh/day, i.e., 34.47%; while the best GAMS approach using the COUENNE solver finds a reduction of 23.84%. These solutions demonstrate that the MINLP solvers in GAMS are stuck in local optimal solution in comparison with the optimal solution found by the DSCA-SOCP and (ii) in all the solutions reported in Table 6 nodes higher than 60 show the high power injection regarding photovoltaic penetration, and it can be observed that these nodes are more sensitive to active power injections when compared with the remainder of buses.

6. Conclusions and Future Works

The problem of the optimal location and sizing of DGs in AC distribution networks was explored in this research from the point of view of the hybrid optimization by proposing a master–slave optimization algorithm. The original MINLP model was rewritten as a MISOCP problem, where the master stage was entrusted with determining the optimal location of the DGS (i.e., discrete optimization problem), while the slave stage is entrusted with solving the sizing problem, i.e., the optimal power flow problem. The master stage was addressed with a new formulation of the sine–cosine algorithm in its discrete form, while the slave stage was formulated as a SOCP problem. The main advantage of using convex optimization for the optimal sizing of the DGs is that this approach guarantees global optimal solution for each nodal combination provided in the master stage.

Numerical simulations demonstrate that the proposed hybrid DSCA-SOCP approach allowed reaching the global optimal solution for both test feeders, which implies power loss reductions to about 65.50% and 69.15% for the 33- and 69-node test feeders, respectively. It was possible to establish that those solutions are indeed the global optimal ones for the test feeders considered since an exhaustive approach was made, i.e., the evaluation of the complete solution space: this has been demonstrated.

Evaluations considering active and reactive power in the distributed generation for both test feeders demonstrates that apparent power injections improve the grid performance by reducing grid power losses more than 90% for two or three distributed generators, with voltage regulation lower than 1.00% in the case of installing three distributed generators. In addition, the possibility of installing photovoltaic generation considering daily production and demand curves was tested in the 69-bus test feeder for the DSCA-SOCP approach and MINLP solvers available in GAMS, where it was observed that the proposed approach allows reducing daily energy losses by about 34.47%, while GAMS solvers are stuck in local optimal solutions with reductions lower than 25%, which demonstrates the efficiency of the proposed optimization for installing renewable energy resources in AC distribution networks.

Regarding processing times, both test feeders have been solved using less than 600 s. The time consumed for our approach illustrates the efficiency to solve the complex MINLP formulation by using an MISOCP equivalent with capabilities of optimal finding after 100 consecutive evaluations.

Lastly, the following researches can be derived from this proposal: (i) the application of the proposed MISOCP model to the problem of voltage stability improvement in distribution networks by including renewable distribution generation; (ii) the solution of the MISOCP model with branch and bound methods to guarantee the global optimum finding without requiring consecutive evaluations and statistical tests; and (iii) to propose a MISOCP formulation for the problem of the optimal location and selection of battery energy storage systems and distributed generators in AC distribution networks, including devices' costs during the planning horizon.

Author Contributions: Conceptualization, methodology, software, and writing—review and editing, O.D.M., A.M.-C., H.R.C., L.A.-B. and E.R.-T. All authors have read and agreed to the published version of the manuscript.

Funding: This work was supported in part by the Laboratorio de Simulación Hardware-in-the-loop para Sistemas Ciberfísicos under Grant TEC2016-80242-P (AEI/FEDER), in part by the Spanish Ministry of Economy and Competitiveness under Grant DPI2016-75294-C2-2-R.

Institutional Review Board Statement: Not applicable.

Informed Consent Statement: Not applicable.

Data Availability Statement: No new data were created or analyzed in this study. Data sharing is not applicable to this article.

Acknowledgments: This work was supported in part by the Centro de Investigación y Desarrollo Científico de la Universidad Distrital Francisco José de Caldas under grant 1643-12-2020 associated with the project: "Desarrollo de una metodología de optimización para la gestión óptima de recursos energéticos distribuidos en redes de distribución de energía eléctrica." and in part by the Dirección de Investigaciones de la Universidad Tecnológica de Bolívar under grant PS2020002 associated with the project: "Ubicación óptima de bancos de capacitores de paso fijo en redes eléctricas de distribución para reducción de costos y pérdidas de energía: Aplicación de métodos exactos y metaheurísticos."

Conflicts of Interest: The authors declare no conflict of interest.

References

1. Shaw, R.; Attree, M.; Jackson, T. Developing electricity distribution networks and their regulation to support sustainable energy. *Energy Policy* **2010**, *38*, 5927–5937. [CrossRef]
2. Lavorato, M.; Franco, J.F.; Rider, M.J.; Romero, R. Imposing Radiality Constraints in Distribution System Optimization Problems. *IEEE Trans. Power Syst.* **2012**, *27*, 172–180. [CrossRef]
3. Pegado, R.; Ñaupari, Z.; Molina, Y.; Castillo, C. Radial distribution network reconfiguration for power losses reduction based on improved selective BPSO. *Electr. Power Syst. Res.* **2019**, *169*, 206–213. [CrossRef]
4. Muthukumar, K.; Jayalalitha, S. Optimal placement and sizing of distributed generators and shunt capacitors for power loss minimization in radial distribution networks using hybrid heuristic search optimization technique. *Int. J. Electr. Power Energy Syst.* **2016**, *78*, 299–319. [CrossRef]
5. Verma, H.K.; Singh, P. Optimal Reconfiguration of Distribution Network Using Modified Culture Algorithm. *J. Inst. Eng. Ser. B* **2018**, *99*, 613–622. [CrossRef]
6. Rao, R.S.; Satish, K.; Narasimham, S.V.L. Optimal Conductor Size Selection in Distribution Systems Using the Harmony Search Algorithm with a Differential Operator. *Electr. Power Compon. Syst.* **2011**, *40*, 41–56. [CrossRef]
7. Zhao, Z.; Mutale, J. Optimal Conductor Size Selection in Distribution Networks with High Penetration of Distributed Generation Using Adaptive Genetic Algorithm. *Energies* **2019**, *12*, 2065. [CrossRef]
8. Grisales-Noreña, L.; Montoya, D.G.; Ramos-Paja, C. Optimal Sizing and Location of Distributed Generators Based on PBIL and PSO Techniques. *Energies* **2018**, *11*, 1018. [CrossRef]
9. Kaur, S.; Kumbhar, G.; Sharma, J. A MINLP technique for optimal placement of multiple DG units in distribution systems. *Int. J. Electr. Power Energy Syst.* **2014**, *63*, 609–617. [CrossRef]
10. Kacejko, P.; Adamek, S.; Wydra, M. Optimal voltage control in distribution networks with dispersed generation. In Proceedings of the 2010 IEEE PES Innovative Smart Grid Technologies Conference Europe (ISGT Europe), Gothenberg, Sweden, 11–13 October 2010. [CrossRef]
11. Sultana, S.; Roy, P.K. Multi-objective quasi-oppositional teaching learning based optimization for optimal location of distributed generator in radial distribution systems. *Int. J. Electr. Power Energy Syst.* **2014**, *63*, 534–545. [CrossRef]
12. Montoya, O.D.; Gil-González, W.; Grisales-Noreña, L. An exact MINLP model for optimal location and sizing of DGs in distribution networks: A general algebraic modeling system approach. *Ain Shams Eng. J.* **2020**, *11*, 409–418. [CrossRef]
13. Hassan, A.S.; Sun, Y.; Wang, Z. Optimization techniques applied for optimal planning and integration of renewable energy sources based on distributed generation: Recent trends. *Cogent Eng.* **2020**, *7*. [CrossRef]
14. Bukhsh, W.A.; Grothey, A.; McKinnon, K.I.M.; Trodden, P.A. Local Solutions of the Optimal Power Flow Problem. *IEEE Trans. Power Syst.* **2013**, *28*, 4780–4788. [CrossRef]
15. Molzahn, D.K. Identifying and Characterizing Non-Convexities in Feasible Spaces of Optimal Power Flow Problems. *IEEE Trans. Circuits Syst. II Express Briefs* **2018**, *65*, 672–676. [CrossRef]
16. Melo, W.; Fampa, M.; Raupp, F. An overview of MINLP algorithms and their implementation in Muriqui Optimizer. *Ann. Oper. Res.* **2018**, *286*, 217–241. [CrossRef]
17. Lavaei, J.; Low, S.H. Zero Duality Gap in Optimal Power Flow Problem. *IEEE Trans. Power Syst.* **2012**, *27*, 92–107. [CrossRef]
18. Farivar, M.; Low, S.H. Branch Flow Model: Relaxations and Convexification—Part I. *IEEE Trans. Power Syst.* **2013**, *28*, 2554–2564. [CrossRef]
19. Gil-González, W.; Molina-Cabrera, A.; Montoya, O.D.; Grisales-Noreña, L.F. An MI-SDP Model for Optimal Location and Sizing of Distributed Generators in DC Grids That Guarantees the Global Optimum. *Appl. Sci.* **2020**, *10*, 7681. [CrossRef]

20. Sultana, U.; Khairuddin, A.B.; Aman, M.; Mokhtar, A.; Zareen, N. A review of optimum DG placement based on minimization of power losses and voltage stability enhancement of distribution system. *Renew. Sustain. Energy Rev.* **2016**, *63*, 363–378. [CrossRef]

21. Montoya, O.D.; Gil-González, W.; Orozco-Henao, C. Vortex search and Chu-Beasley genetic algorithms for optimal location and sizing of distributed generators in distribution networks: A novel hybrid approach. *Eng. Sci. Technol. Int. J.* **2020**. [CrossRef]

22. Moradi, M.; Abedini, M. A combination of genetic algorithm and particle swarm optimization for optimal DG location and sizing in distribution systems. *Int. J. Electr. Power Energy Syst.* **2012**, *34*, 66–74. [CrossRef]

23. Injeti, S.K.; Kumar, N.P. A novel approach to identify optimal access point and capacity of multiple DGs in a small, medium and large scale radial distribution systems. *Int. J. Electr. Power Energy Syst.* **2013**, *45*, 142–151. [CrossRef]

24. Gupta, S.; Saxena, A.; Soni, B.P. Optimal Placement Strategy of Distributed Generators based on Radial Basis Function Neural Network in Distribution Networks. *Procedia Comput. Sci.* **2015**, *57*, 249–257. [CrossRef]

25. Moradi, M.; Abedini, M. A novel method for optimal DG units capacity and location in Microgrids. *Int. J. Electr. Power Energy Syst.* **2016**, *75*, 236–244. [CrossRef]

26. Bayat, A.; Bagheri, A. Optimal active and reactive power allocation in distribution networks using a novel heuristic approach. *Appl. Energy* **2019**, *233–234*, 71–85. [CrossRef]

27. Sultana, S.; Roy, P.K. Krill herd algorithm for optimal location of distributed generator in radial distribution system. *Appl. Soft Comput.* **2016**, *40*, 391–404. [CrossRef]

28. Deshmukh, R.; Kalage, A. Optimal Placement and Sizing of Distributed Generator in Distribution System Using Artificial Bee Colony Algorithm. In Proceedings of the 2018 IEEE Global Conference on Wireless Computing and Networking (GCWCN), Lonavala, India, 23–24 November 2018. [CrossRef]

29. Nowdeh, S.A.; Davoudkhani, I.F.; Moghaddam, M.H.; Najmi, E.S.; Abdelaziz, A.; Ahmadi, A.; Razavi, S.; Gandoman, F. Fuzzy multi-objective placement of renewable energy sources in distribution system with objective of loss reduction and reliability improvement using a novel hybrid method. *Appl. Soft Comput.* **2019**, *77*, 761–779. [CrossRef]

30. Gholami, K.; Parvaneh, M.H. A mutated salp swarm algorithm for optimum allocation of active and reactive power sources in radial distribution systems. *Appl. Soft Comput.* **2019**, *85*, 105833. [CrossRef]

31. Bocanegra, S.Y.; Montoya, O.D. Heuristic approach for optimal location and sizing of distributed generators in AC distribution networks. *Wseas Trans. Power Syst.* **2019**, *14*, 113–121.

32. ChithraDevi, S.; Lakshminarasimman, L.; Balamurugan, R. Stud Krill herd Algorithm for multiple DG placement and sizing in a radial distribution system. *Eng. Sci. Technol. Int. J.* **2017**, *20*, 748–759. [CrossRef]

33. Yang, Q.; Chu, S.C.; Pan, J.S.; Chen, C.M. Sine Cosine Algorithm with Multigroup and Multistrategy for Solving CVRP. *Math. Probl. Eng.* **2020**, *2020*, 1–10. [CrossRef]

34. Attia, A.F.; Sehiemy, R.A.E.; Hasanien, H.M. Optimal power flow solution in power systems using a novel Sine–Cosine algorithm. *Int. J. Electr. Power Energy Syst.* **2018**, *99*, 331–343. [CrossRef]

35. Manrique, M.L.; Montoya, O.D.; Garrido, V.M.; Grisales-Noreña, L.F.; Gil-González, W. Sine–Cosine Algorithm for OPF Analysis in Distribution Systems to Size Distributed Generators. In *Communications in Computer and Information Science*; Springer International Publishing: Berlin/Heidelberg, Germany, 2019; pp. 28–39. [CrossRef]

36. Montoya, O.D.; Gil-González, W.; Grisales-Noreña, L.F. Sine–cosine algorithm for parameters' estimation in solar cells using datasheet information. *J. Phys. Conf. Ser.* **2020**, *1671*, 012008. [CrossRef]

37. Mirjalili, S.M.; Mirjalili, S.Z.; Saremi, S.; Mirjalili, S. Sine Cosine Algorithm: Theory, Literature Review, and Application in Designing Bend Photonic Crystal Waveguides. In *Nature-Inspired Optimizers*; Springer International Publishing: Berlin/Heidelberg, Germany, 2019; pp. 201–217. [CrossRef]

38. Qu, C.; Zeng, Z.; Dai, J.; Yi, Z.; He, W. A Modified Sine–Cosine Algorithm Based on Neighborhood Search and Greedy Levy Mutation. *Comput. Intell. Neurosci.* **2018**, *2018*, 1–19. [CrossRef] [PubMed]

39. Sahin, O.; Akay, B. Comparisons of metaheuristic algorithms and fitness functions on software test data generation. *Appl. Soft Comput.* **2016**, *49*, 1202–1214. [CrossRef]

40. Dahal, K.; Remde, S.; Cowling, P.; Colledge, N. Improving Metaheuristic Performance by Evolving a Variable Fitness Function. In *Evolutionary Computation in Combinatorial Optimization*; Springer: Berlin/Heidelberg, Germany, 2008; pp. 170–181. [CrossRef]

Article

On the Efficiency in Electrical Networks with AC and DC Operation Technologies: A Comparative Study at the Distribution Stage

Oscar Danilo Montoya [1,2,]*, Federico Martin Serra [3] and Cristian Hernan De Angelo [4]

[1] Facultad de Ingeniería, Universidad Distrital Francisco José de Caldas, Carrera 7 No. 40B-53, Bogotá D.C. 11021, Colombia

[2] Laboratorio Inteligente de Energía, Universidad Tecnológica de Bolívar, km 1 vía Turbaco, Cartagena 131001, Colombia

[3] Laboratorio de Control Automático (LCA), Facultad de Ingeniería y Ciencias Agropecuarias, Universidad Nacional de San Luis, Villa Mercedes, San Luis 5730, Argentina; fmserra@unsl.edu.ar

[4] Grupo de Electrónica Aplicada (GEA), Instituto de Investigaciones en Tecnologías Energéticas y Materiales (IITEMA)—CONICET, Facultad de Ingeniería, Universidad Nacional de Rio Cuarto, Rio Cuarto, Córdoba 5800, Argentina; cdeangelo@ieee.org

* Correspondence: odmontoyag@udistrital.edu.co

Received: 23 July 2020; Accepted: 18 August 2020; Published: 20 August 2020

Abstract: This research deals with the efficiency comparison between AC and DC distribution networks that can provide electricity to rural and urban areas from the point of view of grid energy losses and greenhouse gas emissions impact. Configurations for medium- and low-voltage networks are analyzed via optimal power flow analysis by adding voltage regulation and devices capabilities sources in the mathematical formulation. Renewable energy resources such as wind and photovoltaic are considered using typical daily generation curves. Batteries are formulated with a linear representation taking into account operative bounds suggested by manufacturers. Numerical results in two electrical networks with 0.24 kV and 12.66 kV (with radial and meshed configurations) are performed with constant power loads at all the nodes. These simulations confirm that power distribution with DC technology is more efficient regarding energy losses, voltage profiles and greenhouse emissions than its AC counterpart. All the numerical results are tested in the General Algebraic Modeling System widely known as GAMS.

Keywords: alternating current networks; direct current networks; optimal power flow; non-linear optimization; control of power electronic converters

---

1. Introduction

1.1. General Context

Presently electrical distribution networks are essential systems in economic development around the word [1,2]; these grids are also responsible for distributing energy from large-scale power systems to all end users at medium and low voltage levels [3], which implies that in terms of size, the distribution networks are the lengthiest infrastructure inside of the power system [4,5]. This is important since higher losses can be presented at distribution networks in comparison to power systems (transmission and sub-transmission networks), e.g., in the Colombian context, energy losses at distribution networks can be between 6% and 18% while losses at transmission networks can be between 1% and 2% [6]. Recent advancements in power electronics, renewable energy, and energy storage technologies have made distribution networks be focused on the current modernization of power systems. In this sense, three main tendencies can be identified as follows: (i) expanding the existing distribution networks using conventional AC technologies considering AC–DC inverters to interface the distributed energy resources [7]; (ii) Use of DC feeders to expand distribution networks taking the advantages of renewables and batteries that can work directly under the DC paradigm [8]; (iii) design hybrid distribution networks using AC and DC feeders taking the advantages of these technologies regarding reliability and security in the network operation, particularly in the new microgrids environment [3]. To analyze these possible distribution network configurations, power flow and optimal power flow models (steady-state analysis) appear to be essential tools in the literature [9]. These methods (convex and heuristics) determine the state variable (voltage magnitudes and angles) for a particular load condition being applicable to AC and DC networks with minimal changes [3].

1.2. Motivation

The analysis of electrical distribution networks from the point of view of power flow and optimal power flow is a fundamental step to validate the efficiency of these grids regarding energy losses, voltage profiles and conductor chargeability. In this sense, this research is motivated in the analysis of electrical distribution networks using AC or DC technologies in order to identify their performance regarding efficiency in terms of energy losses and greenhouse gas emissions, when it is selected one of both technologies for distributed electricity at medium-voltage level [2].

1.3. Brief Literature Survey

Electrical distribution networks had been designed under the AC paradigm for decades [10,11]; however, presently multiple reports can be found where distribution networks are analyzed under the DC paradigm, some of them are recompiled below.

The authors of [3] have presented an optimal power flow model for multi-terminal DC networks in medium-voltage levels where the energy losses in power converters have been added with quadratic constraints. These constraints allow the obtaining of an equivalent convex optimization model easily solved with semidefinite programming. In Reference [12] an optimization model for optimal phase-balancing in DC low-voltage distribution circuits with a bipolar configuration, which is represented through a Mixed-Integer Non-Linear Programming (MINLP) multi-objective model, has been presented. Numerical results demonstrated that phase-balancing reduces energy losses significantly when compared to the benchmark case. The authors of [13] have proposed the optimal location of photovoltaic sources in DC networks to minimize the total greenhouse gas emissions of CO_2 in rural networks. The proposed optimization model has an MINLP structure and it was solved through the General Algebraic Modeling System (GAMS) optimization software. In Reference [8,14] three approaches for optimal operation of battery energy storage systems in DC networks using day-ahead economic formulations, have been presented. The main idea of those works is the minimization of the daily energy purchasing costs in slack nodes by using metaheuristic and convex optimization methods with excellent results when these are compared to the benchmark cases. In Reference [15] a convex optimization model added to a branch and bound approach to solve the problem of optimal reconfiguration of DC networks, has been proposed. The main advantage of this approach is that the global optimal solution is guaranteed via second-order cone optimization, applied to a study case using real-time simulations. The authors in [16] have proposed a MINLP model for optimal location and reallocation of battery energy storage systems in DC grids to reduce the daily energy losses and the total energy purchasing costs in the conventional sources. Numerical results demonstrated that the location of the batteries is dependent on the performance index used as the objective function, i.e., energy losses or energy purchasing costs; in this sense, authors have demonstrated a multi-objective compromise between both objectives. All the simulations were carried-out in the GAMS optimization package.

Regarding AC distribution networks multiple works related with power system planning and operation have been proposed in scientific literature. Some of these works are: optimal reconfiguration of distribution networks [17–19], optimal location of shunt capacitor banks and distributed generators [20–22], optimal selection of wire gauges in radial distribution networks [23–25], optimal location and coordination of protective devices [26–28], optimal location and operation of battery energy storage systems [7,14,29], and optimal planning of AC distribution networks including new substations [5,30].

It is important to mention that for all aforementioned approaches regardless the operation technology, i.e., AC or DC paradigms, the concepts of power flow and optimal power flow analyses are essential to determine their operative conditions [31]. This clearly implies that these concepts can be used to address both technologies and compare them regarding greenhouse gas emissions, energy purchasing costs and grid energy losses as will be addressed in this contribution.

1.4. Contribution and Scope

The main contributions of this research can be summarized as follows:

✓ The comparison from the point of view of power flow and optimal power flow analyses of AC and DC technologies energy distribution at medium-voltage levels, considering renewable energies and battery energy storage systems under an economic dispatch environment.
✓ The derivation of the multiperiod power flow model for DC distribution networks from the classical and well-known AC model by presenting the necessary simplifications, which are also presented in a tutorial style for solution purposes through the general algebraic modeling system (i.e., GAMS) optimization package.
✓ The use of power electronic converters to interface the DC power distribution system with AC loads to provide the reactive power demanded by the load. This is made by using control strategies that guarantees asymptotic stability during closed-loop operation taking advantage on the passivity-based control design and the Hamiltonian model often found in power converters.

Additionally, the main considerations taken into account in the development of this work are: (i) to make distribution AC and DC technologies it is supposed that all the loads in the AC grid operate with unity power factor (only applicable for AC loads), (ii) in the case of loads connected in the DC grid that require reactive power support (e.g., motors), these are interfaced via power inverters that can provide this support without affecting the operative condition of the DC grid, and (iii) a low-voltage grid operating with 240 V and 60 kW of load is considered to present the effect of the AC or DC distribution network in residential applications, while a medium-voltage grid (12.66 kV) allows to compare AC and DC technologies when considerable reactive power appear in loads.

Observe that this research is focused on the efficiency comparison between AC and DC grids from the point of view of the distribution stage, i.e., when AC or DC technologies are used to transfer power from conventional and renewable generators to loads and batteries; for this reason, we assume that power losses in all the conversion stage are similar when these devices are interfaced in AC and DC grids, which allows us to consider them as equals in both scenarios for comparison purposes in the distribution layer.

Please note that we include ahead in this paper a section dedicated to the analysis of voltage source converters since these devices are essential in the DC distribution paradigm [3]. We introduced these devices with a classical passivity-based controller that operates these devices in the inversion mode, i.e., these are used to provide AC power to three-phase loads assuming a constant voltage source in the DC side [32]. However, these devices can also be used as the main sources when AC conventional networks interface with DC distribution feeders [33].

1.5. Organization of the Document

The remainder of this document is rearranged as follows: Section 2 presents the complete mathematical formulation of the multiperiod optimal power flow problem for AC grids as well as the necessary simplifications to derive the equivalent DC model. Section 3 presents the main characteristics of the GAMS software to solve non-linear non-convex optimization problems with a small test feeder composed by six nodes and five lines that operates with 240 V to to meet a total load about 60 kW. Section 4 presents the integration of three-phase loads in DC networks using voltage source converter interface. In addition, a general control design to guarantee sinusoidal voltage profile in the AC load regardless the active and reactive power consumption via passivity-based control methods is presented. Section 5 presents all the numerical information regarding the medium-voltage distribution network analyzed, which is composed by 33 nodes, 32 lines and operates with a nominal voltage of 12.66 kV. This system has four renewable generators (two photovoltaic-based generators and two wind turbines), and three battery energy storage systems (note that these renewable sources and battery energy storage systems are indeed composed by DC sub-networks interfaced with power electronic converters to manage the power transferred (absorbed) to (from) the DC or AC distribution networks). Section 6 presents all the numerical simulations on the 33-nodes test feeder considering multiple simulation cases. Section 7 shows the main conclusions derived from this work as well as some possible future works.

2. Mathematical Formulation

To compare electrical AC and DC distribution to provide electrical service in rural areas we assume that all the power consumptions have unity power factor and the voltage profile for both technologies is the same. The main characteristic of the proposed formulation is that the distribution network lacks of a voltage controlled node, since the operation is governed by the best coordination between renewable, batteries, and fossil fuels that guarantees the power supply to all the loads during a daily operative scenario. Regarding possible objective function in rural isolated areas two operative scenarios are considered: (i) minimization of the total grid energy losses, and (ii) minimization of the greenhouse emissions by diesel generators. Both objective functions are formulated under an optimal power flow environment.

In the case of the mathematical model of the battery energy storage systems we assume a linear representation to facilitate the implementation in GAMS environment based on the simplified model proposed in [34] which considers that in the conversion stage all the energy losses are neglected, i.e., 100% of efficiency in all the power electronic interface [7,14]; nevertheless, if more accurate battery models are required, then, references [35,36] can be consulted.

2.1. Optimal Power Flow Model in AC Grids

The Optimal power flow (OPF) problem in AC networks is a classical non-linear non-convex optimization problem due to the presence of the active and reactive power balance equations. Here, we consider OPF formulation presented in [7] to analyze distribution networks without constant voltage suppliers. The complete formulation of the OPF problem for AC grids is presented below:

Objective Functions

$$\min z_{\text{ge}}^{\text{ac}} = \sum_{t \in \mathcal{T}} \sum_{i \in \mathcal{N}} T_i^{ge} p_{i,t}^{dg} \Delta t, \tag{1a}$$

$$\min z_{\text{loss}}^{\text{ac}} = \sum_{t \in \mathcal{T}} \sum_{i \in \mathcal{N}} \sum_{ij \in \mathcal{N}} Y_{ij} v_{i,t} v_{j,t} \cos\left(\delta_{j,t} - \delta_{i,t} + \theta_{ij}\right) \Delta t, \tag{1b}$$

where $z_{\text{ge}}^{\text{ac}}$ and $z_{\text{loss}}^{\text{ac}}$ are the objective function values related to the amount of greenhouse emissions and energy losses per day, respectively. T_i^{ge} represents the quantity of CO_2 emitted to the atmosphere in $\frac{kg}{kWh}$ by a diesel generator connected at node i, $p_{i,t}^{dg}$ is the active power delivered by the diesel generator connected at node i in the period of time t; Δt is the length of the period of time considered (typically 1 h). Y_{ij} is the value of the admittance that relates nodes i and j, which have voltages $v_{i,t}$ and $v_{j,t}$ at each period of time t. $\delta_{i,t}$ ($\delta_{j,t}$) is the angle of the voltage at node i (j) in the interval of time t, and θ_{ij} is the angle of the admittance between nodes i and j. Please note that $\mathcal{N}$ and $\mathcal{T}$ are the sets that contains all the nodes in the grid and the total of periods of time of the operation horizon.

Set of Constraints

The power balance equations in the AC power flow are related to the amount of active and reactive power injected at each node i in each period of time t. These take the following form:

$$p_{i,t}^{dg} + p_{i,t}^{rs} + p_{i,t}^{b} - p_{i,t}^{d} = v_{i,t} \sum_{j \in \mathcal{N}} Y_{ij} v_{j,t} \cos\left(\delta_{i,t} - \delta_{j,t} - \theta_{ij}\right), \tag{2a}$$

$$q_{i,t}^{dg} + q_{i,t}^{rs} + q_{i,t}^{b} - q_{i,t}^{d} = v_{i,t} \sum_{j \in \mathcal{N}} Y_{ij} v_{j,t} \sin\left(\delta_{i,t} - \delta_{j,t} - \theta_{ij}\right), \tag{2b}$$

where $p_{i,t}^{rs}$ and $q_{i,t}^{rs}$ are the active and reactive power generation by renewable sources connected at node i in the period of time t; $p_{i,t}^{b}$ and $q_{i,t}^{b}$ are the active and reactive power capabilities in batteries and $p_{i,t}^{d}$ and $q_{i,t}^{d}$ represent the active and reactive power demands, respectively. Please note that in the literature it is recommended that batteries can operate with unity power factor which implies that $q_{i,t}^{b} = 0$ [14,37].

Constraints related with batteries are listed below:

$$SoC_{i,t+1}^{b} = SoC_{i,t}^{b} - \varphi_i^{b} p_{i,t}^{b} \Delta t, \tag{3a}$$

$$SoC_i^{b,\min} \leq SoC_{i,t}^{b} \leq SoC_i^{b,\max}, \tag{3b}$$

$$p_i^{b,\min} \leq p_{i,t}^{b} \leq p_i^{b,\max}, \tag{3c}$$

$$SoC_{i,t_0}^{b} = SoC_i^{b,\text{initial}}, \tag{3d}$$

$$SoC_{i,t_f}^{b} = SoC_i^{b,\text{final}}, \tag{3e}$$

where $SoC_{i,t}^{b}$ is the state-of-charge of the battery b connected at node i in the period of time t, which is bounded by $SoC_i^{b,\min}$ and $SoC_i^{b,\max}$; note that the state-of-charge can be interpreted as the quantity of energy stored in the battery in percentage. φ_i^{b} is the charging/discharging coefficient of the battery b. $p_i^{b,\min}$ and $p_i^{b,\max}$ represent the minimum and maximum power allowable draws/injections at node i for secure operation of the battery at each period of time. $SoC_i^{b,\text{initial}}$ and $SoC_i^{b,\text{final}}$ represent the initial and final state-of-charges defined by the utility to operate the battery, i.e., the initial condition of the economic dispatch problem at t_0 and the final operative consign at the end of the operation period t_f.

The complete interpretation of the mathematical models (1)–(3) is as follows: Expression (1a) determines the value of the objective function regarding greenhouse gas emissions to the atmosphere by diesel generation. Equation (1b) defines the total energy losses in all the conductors of the network during the operation horizon (i.e., typically 24 h). Expressions (2a) and (2b) define the power balance constraints regarding active and reactive components of the power at each node. Equation (3a) calculates future the state-of-charge in the battery for each period of time as function of the current charge and the power injection/absorption to/from the grid. Expressions (3b) and (3c) determine the maximum and minimum values allowed to the state-of-charge (energy stored) in the battery as well as its maximum power injection (discharging state) or absorption (charging state), respectively. Finally, Equations (3d) and (3e) determine the initial state-of-charge of the battery and the final operative consign defined by the utility. These are defined as function of the operational requirements of the network. Nevertheless, in specialized literature it is recommended for Ion-Lithium batteries to start and end the day with 50% of state of charge [14].

Remark 1. *The mathematical model for the optimal operation of AC networks with renewables and batteries defined from (1) to (3) is non-linear and non-convex due to the power balance constraints which makes difficult to reach the global optimum [8]. For this reason, here we recurred to the GAMS optimization software to solve this problem to make our results comparable to those obtained from the mathematical model regarding DC grids reported in next subsection.*

Remark 2. *The studied optimization model (1)–(3) corresponds to a single-phase representation of AC distribution network that can be used if: (i) the AC network is indeed a single-phase network which is the most typical scenario in low-voltage distribution environments, or (ii) it is a three-phase balanced distribution network that can be represented through a single-phase equivalent model [7].*

2.2. Optimal Power Flow Model in DC Grids

The mathematical formulation of the optimal power flow problem for DC networks can be obtained directly from the AC formulation by simplifying the objective function regarding energy losses minimization. Also, power balance constraints can be simplified as follows:

- ✓ The angle of the voltage do not exist ($\delta_{i,t} = \delta_{j,t} = 0$) in DC networks since no frequency concept is present in these networks.
- ✓ The reactive power constraint disappears in the DC paradigm, since it is a concept regarding inductors and capacitors in AC grids, and in DC grids they behave as short- and open-circuits, respectively.
- ✓ The admittance matrix in DC grids is only defined by real numbers ($\theta_{ij} = 0$) regarding only with resistive effects in branches and loads, i.e., $Y_{ij} = G_{ij}$.

With these assumptions, the objective function of the DC optimal power flow and the power balance constraint takes the following forms:

Objective Function Regarding Energy Losses

$$\min z_{\text{loss}}^{\text{dc}} = \sum_{t \in \mathcal{T}} \sum_{i \in \mathcal{N}} \sum_{ij \in \mathcal{N}} G_{ij} v_{i,t} v_{j,t} \Delta t, \tag{4}$$

where $z_{\text{loss}}^{\text{dc}}$ is the amount of energy losses in all the branches of the DC network.

Power Balance Constraint

$$p_{i,t}^{dg} + p_{i,t}^{rs} + p_{i,t}^{b} - p_{i,t}^{d} = v_{i,t} \sum_{j \in \mathcal{N}} G_{ij} v_{j,t}, \tag{5}$$

Remark 3. *The complete mathematical model of the optimal power flow problem in DC grids is composed by the greenhouse gas emission objective function (1a), the objective function regarding energy losses minimization defined by (4), the power balance constraint (5) and the remainder of constraints defined in (3).*

It is worthy to mention that the model of the optimal power flow in DC grids is also non-linear and non-convex due to the product between voltage variables in the power balance constraint, which makes necessary to use specialized software (i.e., GAMS) to solve it efficiently [16].

3. Solution Methodology

To solve the optimal power flow problems in AC and DC grids in this paper it is selected the GAMS software to implement their mathematical structures with a non-linear programming (NLP) solver that typically works with interior point methods to reach the optimal solution [36,38].

GAMS software has been largely used in specialized literature to address non-linear non-convex optimization problems in many areas of engineering, some of these works are: optimal planning and operation of power systems with batteries in AC and DC networks [36,37,39,40]; optimal location of distributed generators [13,22,41,42]; optimal design of osmotic power plants [43]; optimal design of water distribution networks [44]; stability analysis in DC networks [45]; optimal design of thermoacoustic engines [46]; optimal location of protective devices [47], and optimal planning of distribution networks [30], among others.

In general terms, GAMS software is a powerful optimization package that solves complex optimization problems focused on the mathematical formulation of the problem rather than the solution methodology [38]. This represents an ideal situation to introduce students and researchers with mathematical optimization [36]. The main characteristics of the GAMS software can be summarized as follows:

✓ It works with a compact model, i.e., by using the symbolic representation of the problem such as reported in model (1)–(3).
✓ The parametric information of the mathematical model is introduced via constants, vectors and matrices, which are named in GAMS as scalars, parameters and tables.
✓ Multiple optimization models can be selected depending on the nature of the problem under study. These models can be linear programming, non-linear programming, or mixed-integer combinations.

A numerical example is presented to understand the use of GAMS software to solve optimization problems with non-linear and non-convex structure. To do so, below it is presented the solution of the power flow problem for a small test feeder that can be operated with AC or DC technology. In the case of the AC technology, it is important to mention that this grid corresponds to a low-voltage distribution network with a single-phase structure, which is the most typical operation case in low-voltage applications [48]. The configuration of this test feeder is presented in Figure 1. This test feeder is composed by six nodes and five distribution lines (radial topology). The information of the branches and loads is reported in Table 1. Please note that to make both configurations comparable, we assume unity power factor in all the loads. In addition, this system operates with 240 V typically found in Colombian AC grids.

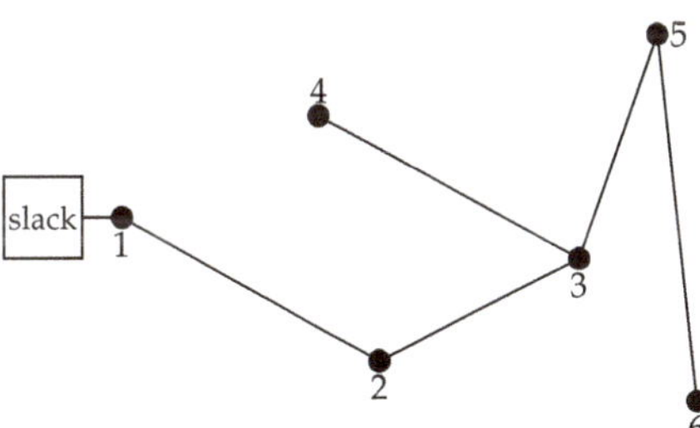

Figure 1. Electrical configuration for the 6-nodes test system used in the GAMS example.

Table 1. Branches and load information.

Node i	Node j	R_{ij} (Ω)	X_{ij} (Ω)	P_j (kW)
1	2	0.01233	0.01827	16
2	3	0.02467	0.03051	11
3	4	0.01469	0.02545	15
3	5	0.02984	0.03084	10
5	6	0.01325	0.01922	8

Please note that this example is a typical low-voltage distribution network where distribution transformers have nominal power of 75 kVA. The implementation of the optimal power flow problem for this test system considers:

✓ A unique hour of analysis.
✓ Operation with unity power factor in all the loads.
✓ No presence of distributed generators or batteries.

The simplified mathematical model for the optimal power flow in AC grids is presented below:

$$\min \; p_{\text{loss}} = \sum_{i \in \mathcal{N}} v_i \sum_{j \in \mathcal{N}} Y_{ij} v_j \cos \left(\delta_i - \delta_j - \theta_{ij} \right), \tag{6a}$$

$$p_i^{dg} - p_i^d = v_i \sum_{j \in \mathcal{N}} Y_{ij} v_j \cos \left(\delta_i - \delta_j - \theta_{ij} \right), \tag{6b}$$

$$q_i^{dg} - q_i^d = v_i \sum_{j \in \mathcal{N}} Y_{ij} v_j \sin \left(\delta_i - \delta_j - \theta_{ij} \right). \tag{6c}$$

The implementation of the simplified mathematical model (6) for the optimal power flow analysis in AC grids is presented in Listing 1.

In the case of the DC model the set of equations reported in (6) can be simplified as presented in Equation (7)

$$\min \; p_{\text{loss}} = \sum_{i \in \mathcal{N}} v_i \sum_{j \in \mathcal{N}} G_{ij} v_j, \tag{7a}$$

$$p_i^{dg} - p_i^d = v_i \sum_{j \in \mathcal{N}} G_{ij} v_j, \tag{7b}$$

The implementation of this simplified mathematical model (6) for the optimal power flow analysis in DC grids is presented in Listing 2.

Once both models (i.e., Equations (6) and (7)) are solved using the CONOPT solver in GAMS by using algorithms presented in Listings 1 and 2, we reach the solution of the voltage profiles reported in Table 2. Please note that for both cases the lowest voltage occurs at node 6 being 207.86 V for the AC grid and 208.24 V for the DC network. These results imply a difference of about 0.38 V between both networks.

Table 2. Voltage profile for the AC and DC grids.

Node i	AC Voltage (pu)	DC Voltage (pu)
1	1	1
2	0.983164996232027	0.984105190366298
3	0.958943129797591	0.960589773608504
4	0.954139851448628	0.955826686308744
5	0.947125787365783	0.948881700223591
6	0.944803809697248	0.946567991473344

Additionally, power losses in the AC grid are 2.40 kW, and 2.39 kW in the DC case (i.e., a difference about 10 W). These results demonstrate that even considering unity power factor at all the loads, the distribution using DC technology is more efficient than the AC counterpart. The voltage drop in lines is lower due to the irrelevance of inductive reactance of the lines.

An important fact when comparing AC and DC configurations is the amount of reactive power required by the AC grid to operate adequately. In this sense, in this small example the grid needs to generate about 3.21 kVAr, which corresponds to reactive power losses through all the lines. This obviously does not occurs in the DC paradigm since reactive effect is not presented as previously mentioned. An additional simulation case in this small test feeder is made in the case of the AC distribution network by considering that all the loads are operated with a lagging power factor of 0.95. In these conditions, the total grid losses exhibit by this network are about 2.77 kW and the minimum voltage profile is about 221.04 V at node 6. These results imply that in comparison to the unity power factor operation case, the reactive power consumption at all the loads makes that the power losses to be increased about 0.37 kW, requiring at the substation node 23.40 kVAr to supply the reactive power requirements at all the loads; and that the voltage at the node has worsened about 5.76 V. Please note that when reactive power is considered for loads in the AC grid, its behavior affects and worsens the results of the comparison with respect to the unity Please note that these results are particularly important when voltage increase to medium level, since these differences become in tens of volts and hundreds of watts. In Section 6 these impacts will be widely discussed under an economic dispatch environment in the 33-node test feeder taking into account higher penetration of renewables and battery energy storage systems.

Listing 1. Algorithm implemented in GAMS for OPF model (6).

```
1   SETS
2   i set of nodes /N1*N6/
3   g set of generators /G1/
4   map(g,i) Associates node with gen /G1.N1/;
5   alias(i,j);
6   SCALARS
7   vmax Maximum voltage bound /1.10/
8   vmin Minimum voltage bound /0.90/
9   v0   Slack voltage /1.00/
10  d0   Slack angle /0.00/;
11  PARAMETER Pd(i)
12  /N1 0,N2 0.16,N3 0.11,N4 0.15,N5 0.10,N6 0.08/;
13  TABLE Ybus(i,j,*)
14  Yij              Thij
15  N1.N1 21.9587119756815   -0.977131269610956
16  N2.N1 21.9587119756815    2.164461383978837
17  N1.N2 21.9587119756815    2.164461383978837
18  N2.N2 34.2649004085076   -0.946099794913962
19  N3.N2 12.3355889037129    2.250751567181446
20  N2.N3 12.3355889037129    2.250751567181446
21  N3.N3 39.8699085005043   -0.930090535155209
22  N4.N3 16.4707942748332    2.094290056213377
23  N5.N3 11.2786162705081    2.339716087328560
24  N3.N4 16.4707942748332    2.094290056213377
25  N4.N4 16.4707942748332   -1.047302597376417
26  N3.N5 11.2786162705081    2.339716087328560
27  N5.N5 31.9116612264763   -0.909021422256346
28  N6.N5 20.7328356619391    2.174363262762511
29  N5.N6 20.7328356619391    2.174363262762511
30  N6.N6 20.7328356619391   -0.967229390827282;
31  VARIABLES
32  ploss Power losses variable
33  v(i) Magnitude of the voltage at node i.
34  d(i) Angle of the voltage at node i.
35  p(g) Active power generation at node i.
36  q(g) Reactive power generation at node i.;
37  v.lo(i) = vmin; v.up(i) = vmax;
38  d.fx('N1') = d0;
39  v.fx('N1') = v0;
40  EQUATIONS
41  ObjFun Objective function
42  PowerA(i) Active power balance per node.
43  PowerR(i) Reactive power balance per node.;
44  * Mathematical structure
45  ObjFun.. ploss =E= SUM(i,v(i)*SUM(j,v(j)*Ybus(i,j,'Yij')*
46  cos(d(i)-d(j)-Ybus(i,j,'Thij'))));
47  PowerA(i).. sum(g$map(g,i),p(g)) - Pd(i)=E= v(i)*SUM(j,v(j)*
48  Ybus(i,j,'Yij')*cos(d(i)-d(j)-Ybus(i,j,'Thij')));
49  PowerR(i).. sum(g$map(g,i),q(g)) =E= v(i)*SUM(j,v(j)*
50  Ybus(i,j,'Yij')*sin(d(i)-d(j)-Ybus(i,j,'Thij')));
51  MODEL OPF /ALL/;
52  OPTIONS decimals   = 4;
53  SOLVE OPF us NLP min ploss;
54  DISPLAY ploss.l, v.l, p.l;
```

Listing 2. Algorithm implemented in GAMS for the OPF model (7).

```
1   SETS
2   i set of nodes /N1*N6/
3   g set of generators /G1/
4   map(g,i) Associates node with gen /G1.N1/;
5   alias(i,j);
6   SCALARS
7   vmax Maximum voltage bound /1.10/
8   vmin Minimum voltage bound /0.90/
9   v0   Slack voltage /1.00/;
10  PARAMETER Pd(i)
11  /N1 0, N2 0.16, N3 0.11, N4 0.15, N5 0.10, N6 0.08/;
12  TABLE Ybus(i,j,*)
13  Gij
14  N1.N1   39.2538523925385
15  N2.N1  -39.2538523925385
16  N1.N2  -39.2538523925385
17  N2.N2   58.8728228019427
18  N3.N2  -19.6189704094041
19  N2.N3  -19.6189704094041
20  N3.N3   68.7863929415566
21  N4.N3  -32.9475833900613
22  N5.N3  -16.2198391420912
23  N3.N4  -32.9475833900613
24  N4.N4   32.9475833900613
25  N3.N5  -16.2198391420912
26  N5.N5   52.7481410288836
27  N6.N5  -36.5283018867925
28  N5.N6  -36.5283018867925
29  N6.N6   36.5283018867925
30  VARIABLES
31  ploss Power losses variable
32  v(i) Magnitude of the voltage at node i.
33  p(g) Active power generation at node i.;
34  v.lo(i) = vmin; v.up(i) = vmax;
35  v.fx('N1') = v0;
36  EQUATIONS
37  ObjFun Objective function
38  PowerA(i) Active power balance per node.;
39  * Mathematical structure
40  ObjFun.. ploss =E= SUM(i,v(i)*SUM(j,v(j)*Ybus(i,j,'Gij')));
41  PowerA(i).. sum(g$map(g,i),p(g)) - Pd(i) =E= v(i)*SUM(j,v(j)*Ybus(i,j,'Gij'));
42  MODEL OPF /ALL/;
43  OPTIONS decimals  = 4;
44  SOLVE OPF us NLP min ploss;
45  DISPLAY ploss.l, v.l, p.l;
```

It is worth mentioning that as described in Section 5, the comparisons made in this research regarding the efficiency comparison between AC and DC paradigms are focused on the distribution stage, which implies that we will not consider power losses in the power electronic interfaces used for interfacing renewable sources and battery energy storage systems. Nevertheless, these will be able to be explored in future research regarding energy distribution technologies and power electronic interfaces for batteries and renewables.

Remark 4. *Regarding optimal implementation in GAMS environment of the OPF problem for AC and DC grids depicted in Listings 1 and 2, we can observe that the DC model is pretty simple with less variables (no angles), which makes this easiest to be solved since its nonlinearities are soft when compared to the AC model that contains trigonometric functions.*

4. Generation Reactive Power in DC Grids with Voltage Source Converter Interfaces

To provide apparent power to three-phase loads in medium- and low-voltage levels using DC distribution feeders, it is required a power electronic interface, i.e., voltage source converter (VSC) (A voltage source converter interface corresponds to a power electronic converter that is fed by a DC source that provides the required active power in the AC side, which is managed by controlling the switches (on or off) states via pulse-width modulation techniques to provide a sinusoidal voltage signal to AC loads [3,32]) that manages the active power interchange between the DC grid and the load at the same time that the reactive power is correctly provided to this load by the converter. In Figure 2 it is depicted the interconnection of a three-phase load to a DC distribution network with a voltage source converter and RLC filter.

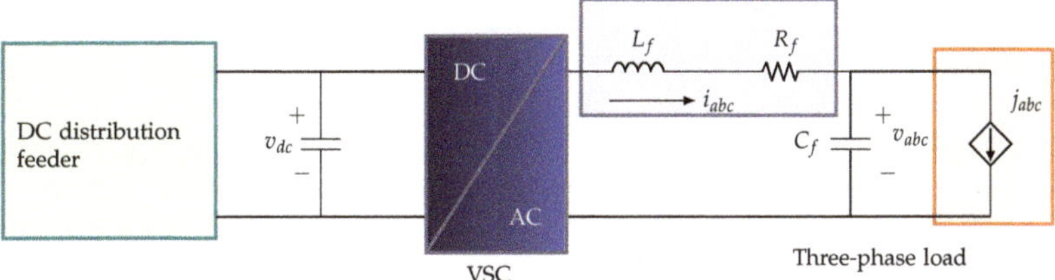

Figure 2. Interconnection of three-phase loads to DC distribution networks via VSCs.

Remark 5. *The VSC interface is a power electronic device that can provide active and reactive power to AC loads when at the DC side it is interconnected a constant voltage source (inversion mode of operation), as the case of the DC distribution paradigm [32]; however, the same device can also be employed to transform AC energy into DC energy when is operated in the conversion mode, as the case of high-voltage direct power transmission systems [33], i.e., the case of DC distribution is used to interface the conventional AC grid to the DC working as the main transformer [49,50].*

To demonstrate that it is possible to manage the active and reactive power consumption in the three-phase load we consider the following facts:

✓ The voltage provided by the DC distribution networks is constant, i.e., the dynamics of the capacitor in the DC side can be neglected, which implies that the dynamical model of the interface presented in Figure 2 can be considered linear.
✓ The current absorbed by the three-phase load (i.e., j_{abc}) is measurable and controllable to guaranteeing constant power absorption.
✓ All the state variables (i.e., i_{abc} and v_{abc}) in Figure 2 are measurable and all the parameters of the RLC filter are perfectly known.

Please note that the control objective in the power electronic interface presented in Figure 2 is to maintain the voltage across the capacitor C_f with purely sinusoidal form as presented below:

$$v_a^\star = \sqrt{2}V_{\mathrm{rms}}\sin(\omega t),$$
$$v_b^\star = \sqrt{2}V_{\mathrm{rms}}\sin\left(\omega t - 2\frac{\pi}{3}\right), \tag{8}$$
$$v_c^\star = \sqrt{2}V_{\mathrm{rms}}\sin\left(\omega t + 2\frac{\pi}{3}\right).$$

where V_{rms} is the root-mean-square value of the voltage in the point of load connection, ω is the angular frequency of the three-phase voltage which is defined as $2\pi f$, being f the electrical frequency en hertz.

Since the desired voltages are three sinusoidal signals defined with positive sequence, then, the control design for the VSC presented in Figure 2 can be designed using the Park's reference frame as demonstrated in [32]. The complete dynamical model of the three-phase VSC to support active and reactive power to a three-phase load as presented in Figure 2 takes the following form in the $dq-$reference frame:

$$L_f \frac{d}{dt} i_d = m_d v_{dc} - R_f i_d + \omega L_f i_q - v_d, \tag{9a}$$

$$L_f \frac{d}{dt} i_d = m_q v_{dc} - R_f i_q - \omega L_f i_d - v_q, \tag{9b}$$

$$C_f \frac{d}{dt} v_d = i_d - \omega C_f v_q - j_d, \tag{9c}$$

$$C_f \frac{d}{dt} v_q = i_q + \omega C_f v_d - j_q, \tag{9d}$$

where m_d and m_q are the modulation indexes in the $dq-$reference frame; i_{dq} and j_{dq} are the current that flow in the inductor of the filter and the current absorbed by the three-phase load, v_{dq} are the dq components of the voltage across the capacitor C_f.

The main characteristic of the dynamical model (9) is that it corresponds to a Hamiltonian system which can be easily controllable via passivity-based control theory as presented in [32]. The Hamiltonian model of (9) takes the following form:

$$\mathcal{D}\dot{x} = [\mathcal{J} - \mathcal{R}] x + g u + \zeta, \tag{10}$$

where $\mathcal{D}$ is known as the inertia matrix based on its similarities with mechanical systems, $\mathcal{J}$ corresponds to the interconnection matrix which is skew-symmetric, $\mathcal{R}$ is the damping matrix which is positive semidefinite, g is the control input matrix, and z is a vector that contains external inputs. Please note that x is the vector of state variables and u is a vector with control inputs, respectively. Each of the aforementioned parameters and variables can be easily defined by comparing (9) and (10).

Remark 6. *The dynamical system (10) can be asymptotically stabilized by using an incremental representation as presented in [32] with a proportional-integral strategy over the passive output $\tilde{y}$ as follows:*

$$\tilde{u} = -K_p \tilde{y} - K_i \int_0^{t_f} \tilde{y} dt, \tag{11}$$

where K_p and K_i are defined as diagonal positive definite matrices and $\tilde{y}$ is $g^T \tilde{x}$. In addition, the complete control is defined from the incremental model as $u = \tilde{u} + u^\star$, where $u^\star$ is obtained by evaluating the equilibrium point $x^\star$ in (10).

To show that the power electronic interface presented in Figure 2 is able to control active and reactive power in a three-phase load, let us consider the following parameters: $f = 50$ Hz, $L_f = 1.25$ mH, $R_f = 0.20\ \Omega$, $C_f = 45\ \mu$F, $V_{rms} = 100$ V and $v_{dc} = 311$ V. The three-phase load is modeled as a combination between a resistance of 2 Ω and an inductance of 7.958 mH connected in parallel. During the period of time between 0 s and 150 ms only the resistive load is connected, then, when time simulation is greater than 150 ms, the inductive load is also connected. It is important to mention that these parameters imply an equivalent active power consumption of about 5 kW added with 4 kVAr at each phase.

In Figure 3 the voltage and current profile provided to the a-phase by the VSC interface to the three-phase load are presented.

Please note that the behavior of the $a-$phase current depicted in Figure 3 demonstrates that if the voltage profile is supported at the load terminals via passivity-based control as defined in (11), then, the active and reactive power required by the load is guaranteed. This implies that the power electronic interface presented in Figure 2 can be used to interface three-phase loads to DC distribution networks with local reactive power support (i.e., unity power factor operation). The main advantage of this interface is that the reactive power is provided locally to the load and no power losses are observed by the DC grid caused by reactive currents, which clearly shows that DC grids are more efficient than AC grids when loads have power factors lower than unity, as it will be presented in section of Results.

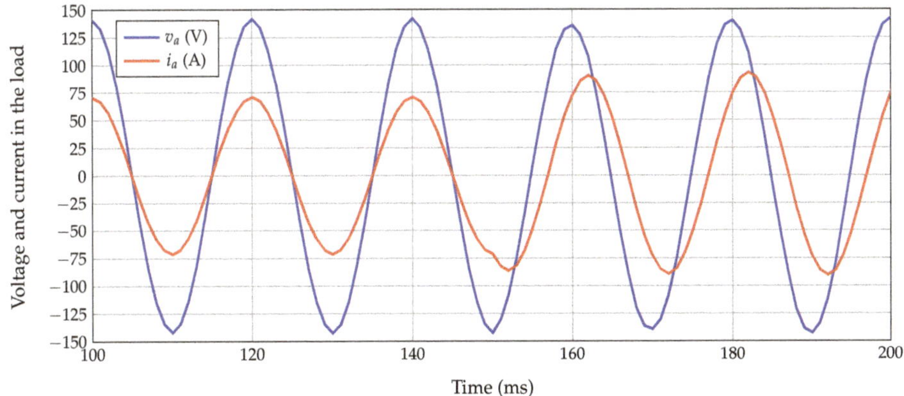

Figure 3. Behavior of the voltage and current in the *a*-phase of the three-phase load.

5. Test System

The comparison of the AC and DC technologies for energy distribution in medium-voltage levels is made by using the 33-nodes test feeder reported in [7]. This test feeder is designed to be operated at 12.66 kV with the connections depicted in Figure 4. The information regarding branches and loads for this test feeder are reported in Table 3. Please note that this test feeder corresponds to a three-phase distribution network typically used to study the problem of the optimal location of distributed generators in power distribution networks as reported in [51].

To evaluate the effect of the renewable generation in the daily operation of this test feeder, we consider four renewable generators previously located in this test system with the information reported in Table 4.

The connection of the generators for each test system is described as follows: at the node 13 it is connected the photovoltaic generator PV_1 and the wind turbine WT_1 with nominal rates of 450 kW and 825 kW, respectively. At the node 25, it is connected a second PV_2 with a nominal power rate of 1500 kW while at the node 30 it is connected the second wind generator WT_2 with the nominal capability of 1200 kW. The information regarding battery energy storage systems considered in this test feeder is reported in Table 5. We assume that the utility has located three batteries, which are distributed as follows: at node 6, a C-type battery is located; at node 14 an A-type battery is used, and at the node 31, a B-type battery is considered.

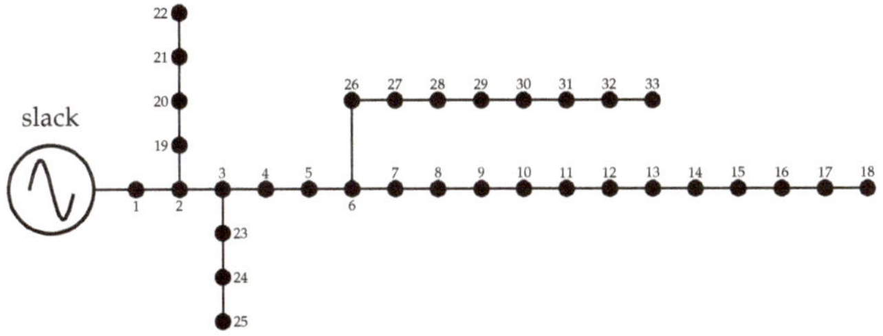

Figure 4. Electrical configuration for the 33-nodes test system.

Table 3. Parameters of the 33-nodes test feeder.

Node i	Node j	R_{ij} (Ω)	X_{ij} (Ω)	P_j (kW)	Q_j (kW)	Node i	Node j	R_{ij} (Ω)	X_{ij} (Ω)	P_j (kW)	Q_j (kW)
1	2	0.0922	0.0477	100	60	17	18	0.7320	0.5740	90	40
2	3	0.4930	0.2511	90	40	2	19	0.1640	0.1565	90	40
3	4	0.3660	0.1864	120	80	19	20	1.5042	1.3554	90	40
4	5	0.3811	0.1941	60	30	20	21	0.4095	0.4784	90	40
5	6	0.8190	0.7070	60	20	21	22	0.7089	0.9373	90	40
6	7	0.1872	0.6188	200	100	3	23	0.4512	0.3083	90	50
7	8	1.7114	1.2351	200	100	23	24	0.8980	0.7091	420	200
8	9	1.0300	0.7400	60	20	24	25	0.8960	0.7011	420	200
9	10	1.0400	0.7400	60	20	6	26	0.2030	0.1034	60	25
10	11	0.1966	0.0650	45	30	26	27	0.2842	0.1447	60	25
11	12	0.3744	0.1238	60	35	27	28	1.0590	0.9337	60	20
12	13	1.4680	1.1550	60	35	28	29	0.8042	0.7006	120	70
13	14	0.5416	0.7129	120	80	29	30	0.5075	0.2585	200	600
14	15	0.5910	0.5260	60	10	30	31	0.9744	0.9630	150	70
15	16	0.7463	0.5450	60	20	31	32	0.3105	0.3619	210	100
16	17	1.2890	1.7210	60	20	32	33	0.3410	0.5302	60	40

Table 4. Renewable energy behavior during a typical sunny day.

Time (s)	PV_1 (p.u)	PV_2 (p.u)	WT_1 (p.u)	WT_2 (p.u)	Time (s)	PV_1 (p.u)	PV_2 (p.u)	WT_1 (p.u)	WT_2 (p.u)
0.0	0	0	0.633118295	0.489955551	12.0	0.924486326	0.975683083	0.972218577	0.942224932
0.5	0	0	0.629764678	0.467954207	12.5	1	1	0.980049847	0.949956724
1.0	0	0	0.607259323	0.449443905	13.0	0.982041153	0.978264398	0.981135531	0.963773634
1.5	0	0	0.609254545	0.435019277	13.5	0.913674689	0.790055240	0.988644844	0.974977461
2.0	0	0	0.605557422	0.437220792	14.0	0.829407079	0.882557147	0.991393173	0.986750539
2.5	0	0	0.630055346	0.437621534	14.5	0.691912077	0.603658738	0.998815517	0.995058133
3.0	0	0	0.684246423	0.450949300	15.0	0.733063295	0.606324907	1	1
3.5	0	0	0.758357805	0.453259348	15.5	0.598435064	0.357393267	0.996070963	0.998107341
4.0	0	0	0.783719339	0.469610539	16.0	0.501133849	0.328035635	0.987258076	0.997690423
4.5	0	0	0.815243582	0.480546213	16.5	0.299821403	0.142423488	0.976519817	0.993076899
5.0	0	0	0.790557706	0.501783479	17.0	0.177117518	0.142023463	0.929542167	0.982629597
5.5	0	0	0.738679217	0.527600299	17.5	0.062736095	0.072956701	0.876413965	0.972084487
6.0	0	0	0.744958950	0.586555316	18.0	0	0.019081590	0.791155379	0.930225756
6.5	0	0	0.718989730	0.652552760	18.5	0	0.008339287	0.691292162	0.891253999
7.0	0.039123365	0.026135642	0.769603567	0.697699990	19.0	0.000333920	0	0.708839248	0.781950905
7.5	0.045414292	0.051715061	0.822376817	0.774442755	19.5	0	0	0.724074349	0.660094138
8.0	0.065587179	0.110148398	0.826492212	0.820205405	20.0	0	0	0.712881960	0.682715246
8.5	0.132615282	0.263094042	0.848620129	0.871057775	20.5	0	0	0.733954043	0.686617947
9.0	0.236870796	0.431175761	0.876523598	0.876973635	21.0	0	0	0.719897641	0.681865563
9.5	0.410356256	0.594273035	0.904128455	0.877065236	21.5	0	0	0.705502389	0.717315757
10.0	0.455017818	0.730402039	0.931213527	0.897955131	22.0	0	0	0.703007456	0.718080346
10.5	0.542364455	0.830347309	0.955557477	0.903245007	22.5	0	0	0.686551618	0.726890145
11.0	0.726440265	0.875407050	0.965504834	0.916903429	23.0	0	0	0.687238555	0.734452193
11.5	0.885104984	0.898815348	0.971037333	0.924757605	23.5	0	0	0.682569771	0.739699146

It is worth mentioning that each one of the renewable source or battery energy storage system corresponds to a DC sub-network interfaced with a power electronic converter that manages the power transferred(absorbed)/to(from) the distribution network regardless whether this is operated under AC or DC paradigm [3].

Table 5. Battery types.

Type	Nominal Energy (kWh)	Charge/Dis. Time (h)	Nominal Power (kW)
A	1000	4	250
B	1500	4	375
C	2000	5	400

To evaluate the daily variation of the active and reactive power consumption and emulating the hourly price behavior, we consider the load and cost curves reported in Figure 5. Furthermore, as the peak of the electricity, we assume the information reported by CODENSA utility from Colombia in May 2019, which is COP\$/kWh 479.3389.

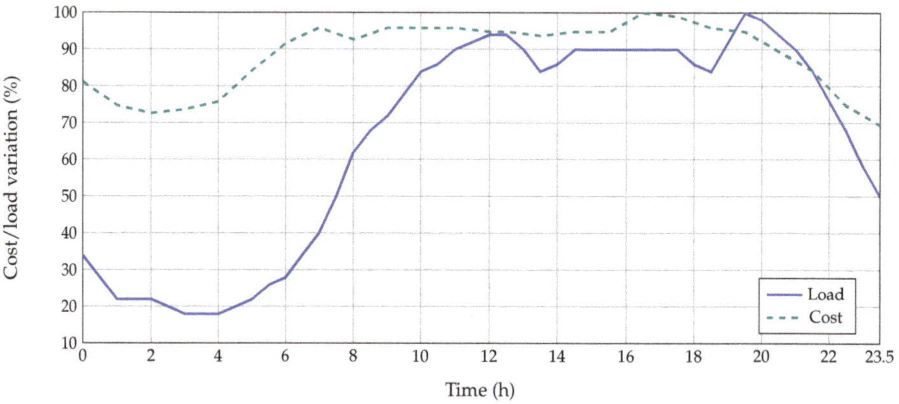

Figure 5. Typical behavior of load consumption and electricity spot market cost.

The numerical information about the demand and cost curves presented in Figure 5 can be consulted in [7].

6. Numerical Results and Discussion

In this section, all the numerical results reached by GAMS after implementing the OPF models for AC and DC networks are described. To make a fair comparison between both distribution technologies the following simulation scenarios are proposed:

- ✓ S1: It is considered that the 33-nodes test feeder operates with unity power factor, i.e., all the reactive power consumptions reported in Table 3 are considered zero in this scenario.
- ✓ S2: The complete information of the 33-nodes test feeder is considered to evaluate the effect of the reactive power demand in the operation of classical AC networks.

It is important to mention that as recommended in [14] all the batteries start and end the day with 50% of state-of-charge and during the day (for Ion-Lithium batteries) this variable can be moved from 10% to 90%.

6.1. Operation of the AC Network

In the operation of this grid, we consider three objective functions as follows: Case 1: minimization of the energy losses during the operation horizon, Case 2: the minimization of the energy purchase costs in the conventional generator (node 1) considering the cost curve reported in Figure 5, and Case 3: the minimization of the total CO_2 gas emissions in the slack source considering that the 33-nodes test feeder is a rural grid fed by a diesel generator. Here, we consider as reported in [13] that CO_2 emission coefficient is 1300 lb/MWh.

Table 6 presents all the numerical results regarding the three objective functions for the S1.

Table 6. Simulation results in the 33-nodes test feeder operated with unity power factor.

Obj. Fun.	Case 1 (kWh/day)	Case 2 (MCOP$)	Case 3 (lb/day)
Case 1	186.918	9.627	30,042.760
Case 2	449.949	4.152	13,731.976
Case 3	366.155	4.345	13,618.771

The results in the S1 reported in Table 6 allows to observe that:

✓ The minimization of the energy losses during the day (186.918 kWh/day) implies high costs regarding energy purchase at the slack node (MCOP$9.627). This is a logical result since in this case, the main idea is to define the optimal power injection in renewables (and batteries) and conventional sources to minimize the magnitude of the current flowing through the lines. The energy losses is a function of the square values of the current. Since the objective function is associated with lines the optimization model does not take into account the origin of the energy, which implies that the algorithm chooses whether to purchase energy in the spot market to optimize the objective function regarding energy losses without taking into account its acquisition costs.
✓ The second case shows that the minimization of the energy purchase costs in the slack node (MCOP$4.152) is directly related to the minimization of the total gas emissions of CO_2 (13731.976 lb/day) since both are involved with the amount of power injected at this node. Nevertheless, this situation produces the highest energy losses during the operation horizon (449.949 kWh/day). This is explained by the objective function selected since in this case, the goal is to minimize the energy production on the slack node regardless the final magnitude of the current flow through the lines, that accordingly increases the energy losses in the whole system.
✓ The third case shows that effectively the energy costs and the amount of CO_2 emissions are correlated objective functions, since minimum values of one of them produces minimum values in the other one with minimal variations; however, this objective function allows reaching a better performance regarding energy losses with 366.155 kWh/day, being an attractive solution since allows to reduce polluting emission gases with low energy purchase costs (MCOP$4.345) and acceptable daily energy losses.

In Table 7 the numerical results for the S2, i.e., the operation of the 33-nodes test feeder considering active and reactive power consumptions are presented. In general, numerical results presents the same behavior reported in the analysis of the S1. Nevertheless, we can notice that energy losses are drastically affected by the presence of the reactive power consumptions inside of the network. Note that in the first case, energy losses have been incremented at least 5 times in comparison to the unity power factor case. In addition, regarding the minimization of the energy purchase costs in the spot market and the greenhouse gas emissions, the increments are about 1.04 times for both cases, which allows concluding that reactive power practically does not produce effects in those objectives when compared to the operation with unitary power factor.

Table 7. Simulation results in the 33-nodes test feeder operated with active and reactive power loads.

Obj. Fun.	Case 1 (kWh/day)	Case 2 (MCOP$)	Case 3 (lb/day)
Case 1	948.979	9.774	30,501.450
Case 2	1219.420	4.347	14,329.381
Case 3	1134.773	4.535	14,215.115

6.2. Operation of the DC Network

To evaluate the performance of the 33-nodes test feeder, it was considered that the system can be represented by a DC equivalent as defined in the S1. In this sense, the reactive power loads and reactance of this model are removed. The implementation of optimal power flow for this DC medium-voltage distribution network is reported in Table 8.

Table 8. Simulation results in the 33-nodes test feeder operated with DC technology.

Obj. Fun.	Case 1 (kWh/day)	Case 2 (MCOP$)	Case 3 (lb/day)
Case 1	186.898	9.627	30,041.788
Case 2	449.365	4.152	13,731.910
Case 3	365.504	4.345	13,618.717

From the results reported in Table 8 we can note that:

✓ The behavior of the DC equivalent of the 33-node test feeder (see Table 8) is identical to the behavior of this system when unity power factor is assigned at all the loads as can be seen in Table 6. These results imply that when the AC grid is used to support only active power consumption (residential applications), its electrical efficiency is 100% comparable to the emerging DC distribution networks.

✓ The only difference between AC and DC distribution considering purely active power consumptions correspond to the need of generating small reactive power quantities in the slack source to support the reactive power losses throughout all the lines. In this context, when the 33-nodes test feeder is operated with unity power factor and considering the minimization of the total energy purchasing costs, the amount of reactive energy generated during the day is about 332.022 kVAr/day, which needs to be provided by the slack source; while in the DC distribution case this reactive power is inexistent; which can be considered and advantage of the DC technology when compared with the AC counterpart.

6.3. Efficiency Comparison for Different Power Factors

To demonstrate that the DC distribution network is an attractive alternative to provide electrical service to industrial users connected at medium-voltage levels, let us compare the efficiency of this technology with conventional AC grids for different percentages of reactive power consumptions.

Remark 7. *Recall that DC distribution networks are able to provide active and reactive power support to AC loads by using a voltage source converter (VSC) that interfaces the DC grid to the AC load as can be demonstrated in [32], where an isolated (i.e., rural) AC grid receives voltage and power support by interconnecting linear and non-linear loads to DC distribution grids via a VSC.*

In Figure 6 it is presented the amount of energy losses (objective function minimize) in the AC grid when the reactive power load changes from 0 to 120% of the peak value reported in Table 3; in addition, the total energy losses are also presented in the DC equivalent network, assuming that the energy losses at all the VSCs are about 10% of the total energy losses of the network. Please note that in the case of the DC grid the reactive power is provided directly at the load side, which implies that this power is provided by the power electronic interface as explained in Section 4. For this reason, the active power losses for DC grids remains constant for different percentage of reactive power demands at the load side, since no currents are associated with this power flow in DC distribution lines.

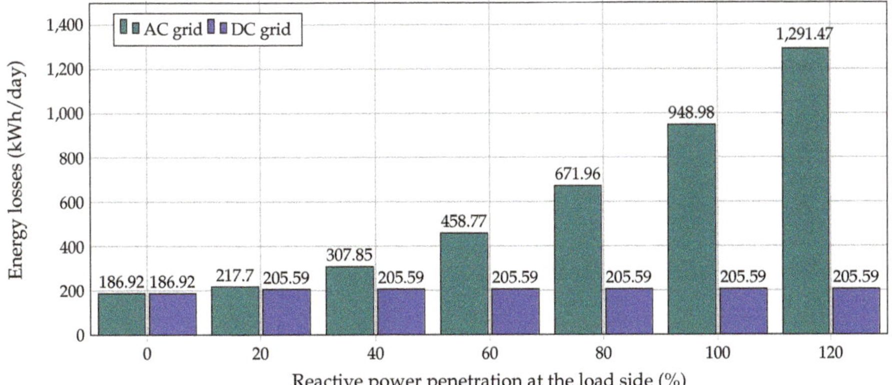

Figure 6. Amount of daily energy losses for different penetrations of reactive power consumptions at the load side.

The information regarding daily energy losses in the AC and DC equivalent networks make evident that the efficiency of the AC grid is deteriorated rapidly as a function of the amount of reactive power consumption at all the loads due to the exponential increment of the total energy losses. Nevertheless, in the case of the DC network the efficiency is always constant regardless the reactive power consumption. This is explained by the fact that the VSCs that interface the AC loads are able to locally generate reactive power, which implies that the DC distribution makes the node can sense their effects in its lines. Please note that when the loads in the 33-nodes test feeders are 100% reactive power consumption, the AC grid has 948.979 kWh/day of energy losses, while in the case of the DC equivalent these losses are about 205.588 kWh/day. These results imply that the AC grid has at least 4.6 times more energy losses than the DC equivalent, which confirms that the DC technology is a promissory alternative to provide the electricity service at medium- and low-voltage distribution levels with higher efficiency levels regarding energy losses, compared at the distribution stage, i.e., involving energy losses in conductors used in the electricity distribution. It is important to mention that this high relation (i.e., 4.6 times for the 33-nodes test feeder) is largely influenced by the relation between active and reactive power demand in the distribution network under study. This implies that this data can be considered to be an indicator, but more studies regarding energy efficiency at all the electronic interfaces (renewable generators, energy storage devices and controllable loads) are needed to determine the overall energy efficiency at distribution levels.

6.4. Effect of Renewable Energy Variations in the Economic Dispatch

In this subsection it is explored the possible operation scenario where renewable energy has important variations regarding weather conditions such as cloudy and rainy days, including very low-speed winds. In this case, we consider as objective function the total energy losses minimization during daily operation, i.e., the Case 1. To consider all the possible operation scenarios in a real network, we consider that the amount of renewable energy varies from 0% to 100% in steps of 20%. In addition, we consider that all the loads in the 33-nodes test feeder operate under normal conditions, i.e., 100% of active and reactive power consumption.

Table 9 presents the behavior of the AC and DC distribution networks when there are higher variations in renewable energy production.

Table 9. Daily energy losses as a function of the renewable energy variations.

Penetration (%)	AC Network (kWh/day)	DC Network (kWh/day)	Diff. (kWh/day)	Relation AC/DC
0	2355.704	1639.291	716.4130	1.4370
20	1705.694	969.892	735.8020	1.7586
40	1271.238	524.844	746.3940	2.4221
60	1039.554	291.767	747.7870	3.5630
80	963.319	218.631	744.6880	4.4061
100	948.979	205.588	743.3910	4.6159

From results reported in Table 9 we can observe that: (i) the difference regarding daily energy losses between AC and DC grids remains practically constant with an average value of 739.0792 kWh/day. This implies that at all the possible renewable energy penetration scenarios the DC grid has better behavior in terms of grid energy losses, which can be explained by the possibility to provide local reactive power with the VSC interface. This latter is not the case of the AC grid where the reactive energy flows from the substation to the loads; (ii) the division between the DC and AC energy losses presented in the last column of Table 9 shows that the efficiency of the DC grid in comparison to the AC case increases as a function of the renewable energy penetration in the grid; this behavior is mainly associated with the important reductions in the power flow through the lines caused by local injections of active power by renewable sources; and (iii) the total energy reduction in the AC grid when renewable energy penetration passes from 0 % to 100% is about 59.72%, while in the case of the DC network this reduction is about 84.46%, which entails that the same level of renewable energy penetration provides more positive impacts in a distribution network designed under the DC paradigm in contrast with the conventional AC grids.

7. Conclusions and Future Works

A comparative study regarding energy efficiency in AC and DC electrical networks for power distribution from the point of view of optimal power flow analysis was presented in this paper. This study allowed to confirm that AC and DC technologies have identical performances in residential applications, i.e., unity power factor, since the amount of energy losses, greenhouse gas emissions of CO_2 or energy purchase costs are practically the same for both technologies. Nevertheless, in the case of high penetration of reactive power consumptions in AC networks (mainly in industrial applications), it was demonstrated that the performance of the AC grid is rapidly deteriorated compared with the DC equivalent, due to the need to transport this reactive power from the substation towards the loads. This increases the magnitude of the current through the lines, being translated into higher energy losses during the operation horizon. This situation does not happen in the case of the DC grids where reactive power is directly provided by the VSCs that interfaces all the AC loads, which implies that the efficiency of the DC distribution system remains constant regardless the reactive power requirements of the load.

To solve the optimal power flow models regarding the daily operation of AC and DC grids, we have introduced the GAMS software to efficiently solves both models with low computation effort, i.e., processing times about 5 s in all the simulation cases and scenarios. This low-computational time is important since multiple simulation cases can be analyzed before taking the final decision in regards with the day-ahead economic dispatch environment, which makes the GAMS software an attractive alternative for tertiary control in distribution networks. In addition, the GAMS package is a proper tool to solve complex optimization problems by focusing the attention on correctly developing the optimization models rather than the solution technique. This represents an ideal framework to easily introduce engineers and researchers in mathematical optimization; for this reason, this paper has been addressed in a tutorial form.

As future work it will be possible to analyze the following problems: (i) propose convex reformulations for optimal power flow analyses in AC and DC networks that will ensure reaching the global optimum of the problem under well-defined operative conditions, which are very attractive for real economic dispatch applications, (ii) make a comparative study between AC and DC grids considering transient operation scenarios such as suddenly load disconnections or short-circuit cases, which can be used in protective devices coordination studies for these grids, and extend the economic dispatch optimization model to three-phase distribution networks and bipolar DC configurations operated under unbalanced loads scenarios to analyze their efficiency in terms of power losses and voltage profiles.

Author Contributions: Conceptualization, methodology, software and writing—review and editing, O.D.M., F.M.S. and C.H.D.A. All authors have read and agreed to the published version of the manuscript.

Funding: This work was partially supported by the Universidad Tecnológica de Bolívar under grant CP2019P011 associated with the project: "Operación eficiente de redes eléctricas con alta penetración de recursos energéticos distribuidos considerando variaciones en el recurso energético primario".

Acknowledgments: The authors want to express to thanks to Universidad Nacional de San Luis and Consejo Nacional de Investigaciones Científicas y Técnicas (CONICET) in Argentina, and Universidad Distrital Francisco José de Caldas and Universidad Tecnológica de Bolivar in Colombia.

References

1. Poudineh, R.; Peng, D.; Mirnezami, S. *Electricity Networks: Technology, Future Role and Economic Incentives for Innovation*; Technical Report; The Oxford Institute for Energy Studies: Oxford, UK, 2017. [CrossRef]
2. Liu, Y.; Yu, Z.; Li, H.; Zeng, R. The LCOE-Indicator-Based Comprehensive Economic Comparison between AC and DC Power Distribution Networks with High Penetration of Renewable Energy. *Energies* **2019**, *12*, 4621. [CrossRef]
3. Simiyu, P.; Xin, A.; Wang, K.; Adwek, G.; Salman, S. Multiterminal Medium Voltage DC Distribution Network Hierarchical Control. *Electronics* **2020**, *9*, 506. [CrossRef]
4. Li, H.; Cui, H.; Li, C. Distribution Network Power Loss Analysis Considering Uncertainties in Distributed Generations. *Sustainability* **2019**, *11*, 1311. [CrossRef]
5. Montoya, O.D.; Grajales, A.; Hincapié, R.A.; Granada, M. A new approach to solve the distribution system planning problem considering automatic reclosers. *Ingeniare Rev. Chil. Ing.* **2017**, *25*, 415–429. [CrossRef]
6. Correa, K.A.; Cortes-Alonso, R.A. *Technical Analysis of Power Losses in Medium Voltage Levels*; Resreport; Universidad Tecnológica de Pereira: Risaralda, Colombia, 2019. (In Spanish)
7. Montoya, O.D.; Gil-González, W. Dynamic active and reactive power compensation in distribution networks with batteries: A day-ahead economic dispatch approach. *Comput. Electr. Eng.* **2020**, *85*, 106710. [CrossRef]
8. Gil-González, W.; Montoya, O.D.; Grisales-Noreña, L.F.; Cruz-Peragón, F.; Alcalá, G. Economic Dispatch of Renewable Generators and BESS in DC Microgrids Using Second-Order Cone Optimization. *Energies* **2020**, *13*, 1703. [CrossRef]
9. Wang, X.-F.; Song, Y.; Irving, M. Load Flow Analysis. In *Modern Power Systems Analysis*; Springer US: New York, NY, USA, 2008; pp. 71–128. [CrossRef]
10. Li, R.; Wang, W.; Chen, Z.; Jiang, J.; Zhang, W. A Review of Optimal Planning Active Distribution System: Models, Methods, and Future Researches. *Energies* **2017**, *10*, 1715. [CrossRef]
11. Wang, S.; Sun, Y.; Li, Y.; Li, K.; An, P.; Yu, G. Optimal Planning of Distributed Generation and Loads in Active Distribution Network: A Review. In Proceedings of the 2020 4th International Conference on Green Energy and Applications (ICGEA), Singapore, 7–9 March 2020; pp. 176–181.
12. Chew, B.S.H.; Xu, Y.; Wu, Q. Voltage Balancing for Bipolar DC Distribution Grids: A Power Flow Based Binary Integer Multi-Objective Optimization Approach. *IEEE Trans. Power Syst.* **2019**, *34*, 28–39. [CrossRef]
13. Montoya, O.D.; Grisales-Noreña, L.F.; Gil-González, W.; Alcalá, G.; Hernandez-Escobedo, Q. Optimal Location and Sizing of PV Sources in DC Networks for Minimizing Greenhouse Emissions in Diesel Generators. *Symmetry* **2020**, *12*, 322. [CrossRef]
14. Grisales-Noreña, L.; Montoya, O.D.; Ramos-Paja, C.A. An energy management system for optimal operation of BSS in DC distributed generation environments based on a parallel PSO algorithm. *J. Energy Storage* **2020**, *29*, 101488. [CrossRef]

15. Altun, T.; Madani, R.; Yadav, A.P.; Nasir, A.; Davoudi, A. Optimal Reconfiguration of DC Networks. *IEEE Trans. Power Syst.* **2020**. [CrossRef]

16. Montoya, O.D.; Gil-González, W.; Rivas-Trujillo, E. Optimal Location-Reallocation of Battery Energy Storage Systems in DC Microgrids. *Energies* **2020**, *13*, 2289. [CrossRef]

17. Jakus, D.; Čađenović, R.; Vasilj, J.; Sarajčev, P. Optimal Reconfiguration of Distribution Networks Using Hybrid Heuristic-Genetic Algorithm. *Energies* **2020**, *13*, 1544. [CrossRef]

18. Diaaeldin, I.; Aleem, S.A.; El-Rafei, A.; Abdelaziz, A.; Zobaa, A.F. Optimal Network Reconfiguration in Active Distribution Networks with Soft Open Points and Distributed Generation. *Energies* **2019**, *12*, 4172. [CrossRef]

19. Čađenović, R.; Jakus, D.; Sarajčev, P.; Vasilj, J. Optimal Distribution Network Reconfiguration through Integration of Cycle-Break and Genetic Algorithms. *Energies* **2018**, *11*, 1278. [CrossRef]

20. Injeti, S.K.; Shareef, S.M.; Kumar, T.V. Optimal Allocation of DGs and Capacitor Banks in Radial Distribution Systems. *Distrib. Gener. Altern. Energy J.* **2018**, *33*, 6–34. [CrossRef]

21. Nagaraju, S.K.; Sivanagaraju, S.; Ramana, T.; Satyanarayana, S.; Prasad, P.V. A Novel Method for Optimal Distributed Generator Placement in Radial Distribution Systems. *Distrib. Gener. Altern. Energy J.* **2011**, *26*, 7–19. [CrossRef]

22. Gil-González, W.; Montoya, O.D.; Grisales-Noreña, L.F.; Perea-Moreno, A.J.; Hernandez-Escobedo, Q. Optimal Placement and Sizing of Wind Generators in AC Grids Considering Reactive Power Capability and Wind Speed Curves. *Sustainability* **2020**, *12*, 2983. [CrossRef]

23. Rao, R.S.; Satish, K.; Narasimham, S.V.L. Optimal Conductor Size Selection in Distribution Systems Using the Harmony Search Algorithm with a Differential Operator. *Electr. Power Compon. Syst.* **2011**, *40*, 41–56. [CrossRef]

24. TÜRKAY, B.; ARTAÇ, T. Optimal Distribution Network Design Using Genetic Algorithms. *Electr. Power Compon. Syst.* **2005**, *33*, 513–524. [CrossRef]

25. Falaghi, H.; Ramezani, M.; Haghifam, M.R.; Milani, K. Optimal selection of conductors in radial distribution systems with time varying load. In Proceedings of the 18th International Conference and Exhibition on Electricity Distribution (CIRED 2005), Turin, Italy, 6–9 June 2005. [CrossRef]

26. Javadian, S.; Haghifam, M.R. Optimal placement of protective devices in distribution networks based on risk analysis. In Proceedings of the 2010 IEEE/PES Transmission and Distribution Conference and Exposition: Latin America (T&D-LA), Sao Paulo, Brazil, 8–10 November 2010. [CrossRef]

27. Montoya, O.D.; Hincapie, R.A.; Granada, M. Optimal Location of Protective Devices Using Multi-objective Approach. In *Communications in Computer and Information Science*; Springer International Publishing: Berlin/Heidelberg, Germany, 2018; pp. 3–15. [CrossRef]

28. Katyara, S.; Staszewski, L.; Leonowicz, Z. Protection Coordination of Properly Sized and Placed Distributed Generations–Methods, Applications and Future Scope. *Energies* **2018**, *11*, 2672. [CrossRef]

29. Fathima, H.; Palanisamy, K. Optimized Sizing, Selection, and Economic Analysis of Battery Energy Storage for Grid-Connected Wind-PV Hybrid System. *Model. Simul. Eng.* **2015**, *2015*, 1–16. [CrossRef]

30. Lavorato, M.; Franco, J.F.; Rider, M.J.; Romero, R. Imposing Radiality Constraints in Distribution System Optimization Problems. *IEEE Trans. Power Syst.* **2012**, *27*, 172–180. [CrossRef]

31. Montoya, O.D.; Gil-González, W.; Garces, A. Numerical methods for power flow analysis in DC networks: State of the art, methods and challenges. *Int. J. Electr. Power Energy Syst.* **2020**, *123*, 106299. [CrossRef]

32. Serra, F.M.; Fernández, L.M.; Montoya, O.D.; Gil-González, W.; Hernández, J.C. Nonlinear Voltage Control for Three-Phase DC-AC Converters in Hybrid Systems: An Application of the PI-PBC Method. *Electronics* **2020**, *9*, 847. [CrossRef]

33. Rodriguez, P.; Rouzbehi, K. Multi-terminal DC grids: challenges and prospects. *J. Mod Power Syst. Clean Energy* **2017**, *5*, 515–523. [CrossRef]

34. Luna, A.C.; Diaz, N.L.; Andrade, F.; Graells, M.; Guerrero, J.M.; Vasquez, J.C. Economic power dispatch of distributed generators in a grid-connected microgrid. In Proceedings of the 2015 9th International Conference on Power Electronics and ECCE Asia (ICPE-ECCE Asia), Seoul, Korea, 1–5 June 2015; pp. 1161–1168.

35. Macedo, L.H.; Franco, J.F.; Rider, M.J.; Romero, R. Optimal Operation of Distribution Networks Considering Energy Storage Devices. *IEEE Trans. Smart Grid* **2015**, *6*, 2825–2836. [CrossRef]

36. Soroudi, A. *Power System Optimization Modeling in GAMS*; Springer International Publishing: Berlin/Heidelberg, Germany, 2017. [CrossRef]

37. Montoya, O.D.; Grajales, A.; Garces, A.; Castro, C.A. Distribution Systems Operation Considering Energy Storage Devices and Distributed Generation. *IEEE Lat. Am. Trans.* **2017**, *15*, 890–900. [CrossRef]

38. Giraldo, O.D.M. Solving a Classical Optimization Problem Using GAMS Optimizer Package: Economic Dispatch Problem Implementation. *Ing. Cienc.* **2017**, *13*, 39–63. [CrossRef]

39. Aguado, J.; de la Torre, S.; Triviño, A. Battery energy storage systems in transmission network expansion planning. *Electr. Power Syst. Res.* **2017**, *145*, 63–72. [CrossRef]

40. Montoya, O.D.; Gil-González, W.; Grisales-Noreña, L.; Orozco-Henao, C.; Serra, F. Economic Dispatch of BESS and Renewable Generators in DC Microgrids Using Voltage-Dependent Load Models. *Energies* **2019**, *12*, 4494. [CrossRef]

41. Montoya, O.D.; Gil-González, W.; Grisales-Noreña, L. An exact MINLP model for optimal location and sizing of DGs in distribution networks: A general algebraic modeling system approach. *Ain Shams Eng. J.* **2020**, *11*, 409–418. [CrossRef]

42. Kumar, M.; Kumar, A.; Sandhu, K. Optimal Location of WT based Distributed Generation in Pool based Electricity Market using Mixed Integer Non Linear Programming. *Mater. Today Proc.* **2018**, *5*, 445–457. [CrossRef]

43. Naghiloo, A.; Abbaspour, M.; Mohammadi-Ivatloo, B.; Bakhtari, K. GAMS based approach for optimal design and sizing of a pressure retarded osmosis power plant in Bahmanshir river of Iran. *Renew. Sustain. Energy Rev.* **2015**, *52*, 1559–1565. [CrossRef]

44. Skworcow, P.; Paluszczyszyn, D.; Ulanicki, B.; Rudek, R.; Belrain, T. Optimisation of Pump and Valve Schedules in Complex Large-scale Water Distribution Systems Using GAMS Modelling Language. *Procedia Eng.* **2014**, *70*, 1566–1574. [CrossRef]

45. Amin, W.T.; Montoya, O.D.; Grisales-Noreña, L.F. Determination of the Voltage Stability Index in DC Networks with CPLs: A GAMS Implementation. In *Communications in Computer and Information Science*; Springer International Publishing: Berlin/Heidelberg, Germany, 2019; pp. 552–564. [CrossRef]

46. Tartibu, L.; Sun, B.; Kaunda, M. Multi-objective optimization of the stack of a thermoacoustic engine using GAMS. *Appl. Soft Comput.* **2015**, *28*, 30–43. [CrossRef]

47. Gallego-Londoño, J.P.; Montoya-Giraldo, O.D.; Hincapié-Isaza, R.A.; Granada-Echeverri, M. Ubicación óptima de reconectadores y fusibles en sistemas de distribución. *ITECKNE* **2016**, *13*, 113. [CrossRef]

48. Shen, T.; Li, Y.; Xiang, J. A Graph-Based Power Flow Method for Balanced Distribution Systems. *Energies* **2018**, *11*, 511. [CrossRef]

49. Gil-González, W.; Montoya, O.D.; Garces, A. Direct power control for VSC-HVDC systems: An application of the global tracking passivity-based PI approach. *Int. J. Electr. Power Energy Syst.* **2019**, *110*, 588–597. [CrossRef]

50. Siraj, K.; Khan, H.A. DC distribution for residential power networks-A framework to analyze the impact of voltage levels on energy efficiency. *Energy Rep.* **2020**, *6*, 944–951. [CrossRef]

51. Grisales-Noreña, L.; Montoya, D.G.; Ramos-Paja, C. Optimal Sizing and Location of Distributed Generators Based on PBIL and PSO Techniques. *Energies* **2018**, *11*, 1018. [CrossRef]

 *electronics*

Article

LC Impedance Source Bi-Directional Converter with Reduced Capacitor Voltages

Dogga Raveendhra *, Rached Dhaouadi, Habibur Rehman and Shayok Mukhopadhyay

College of Engineering, American University of Sharjah, Sharjah 26666, UAE; rdhaouadi@aus.edu (R.D.); rhabib@aus.edu (H.R.); smukhopadhyay@aus.edu (S.M.)
* Correspondence: doggaravi12@gmail.com

Received: 3 June 2020; Accepted: 17 June 2020; Published: 28 June 2020

Abstract: This paper proposes an LC (Inductor and Capacitor) impedance source bi-directional DC–DC converter by redesigning after rearranging the reduced number of components of a switched boost bi-directional DC–DC converter. This new converter with a conventional modulation scheme offers several unique features, such as a) a lower number of components and b) reduced voltage stress on the capacitor compared to existing topologies. The reduction of capacitor voltage stress has the potential of improving the reliability and enhancing converter lifespan. An analysis of the proposed converter was completed with the help of a mathematical model and state-space averaging models. The converter performance under different test conditions is compared with the conventional bi-directional DC–DC converter, Z-source converter, discontinuous current quasi Z-source converter, continuous current quasi Z-source converter, improved Z-source converter, switched boost converter, current-fed switched boost converter, and quasi switched boost converter in the Matlab Simulink environment. MATLAB/Simulink results demonstrate that the proposed converter has lesser components count and reduced capacitors' voltage stresses when compared to the topologies mentioned above. A 24 V to 18 V LC-impedance source bi-directional converter and a conventional bidirectional converter are built to investigate the feasibility and benefits of the proposed topology. Experimental results reveal that capacitor voltage stresses, in the case of proposed topology are reduced by 75.00% and 35.80% in both boost and buck modes, respectively, compared to the conventional converter circuit.

Keywords: bi-directional converter; LC impedance source converter; DC–DC power converter; bi-directional power flow

---

1. Introduction

The study, development, and applications of bidirectional power converters are gaining a lot of attention due to their vital role in areas like renewable energy systems, DC microgrids, hybrid energy storage systems, smart mobility, etc. A bidirectional DC-DC converter (BDC) allows power flow in both directions. This functionality is not available in a traditional unidirectional DC-DC converter. Because of this flexibility, BDCs are widely used in several applications, such as battery-powered electric vehicles (BEVs) or hybrid electric vehicles (HEVs), power trains, uninterruptable power supplies (UPS), smart grids, charging stations for BEVs and plug-in hybrid electric vehicles (PHEV), aerospace, defense, aerospace, and non-conventional energy sources such as photovoltaic (PV) arrays, fuel cells (FCs), and wind turbines. Specifically, BDCs are widely adopted by the electric vehicle industry to achieve objectives, such as battery charging/discharging and energy recovery during regeneration modes of operation in electric vehicles. In case of the BEVs, electric energy needs to flow in both directions, i.e., from the motor to the battery and vice versa in regenerative mode. To avoid pollutant emissions, the electric vehicle must be powered only by batteries or other electrical sources (fuel cells, solar panels, etc.) [1–4]. In all the above-mentioned applications, a BDC is preferred for saving space by eliminating a separate boost and buck converter. A BDC can offer some benefits, like cost reduction, improved power density, and effective utilization of the converter [4]. Figure 1 shows the typical structure of the bidirectional DC–DC converters. The BDC, shown in Figure 2, helps to enhance the system efficiency and performance by interfacing with power and energy storage devices [5]. It also avoids a couple of individual unidirectional converters for achieving bidirectional power flow. The BDC's mode of operation (buck or boost) is mainly decided by power flow direction and voltage levels of sources/energy storage elements. Accordingly, the controller must be designed to regulate the voltage/current of the system. While designing DC–DC converters, the main functional objectives are high power density and high efficiency. The high density can be achieved by increasing the switching frequency [6] due to the reduction in reactive components size. However, the problem is that increasing the switching frequency increases the switching losses, which leads to efficiency reduction. This problem can be addressed by adopting wide-bandgap power devices along with suitable gate drivers instead of conventional Si devices.

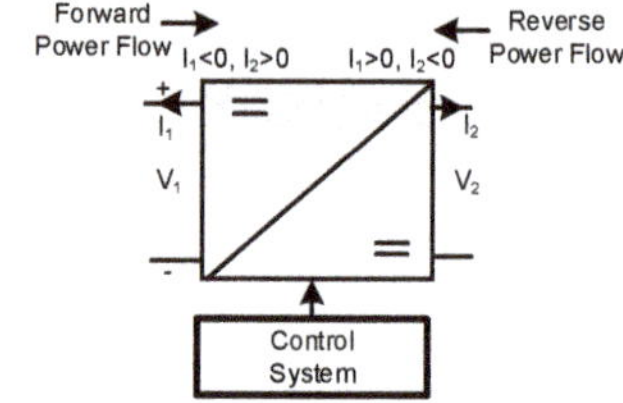

Figure 1. Structure of bi-directional DC–DC converters.

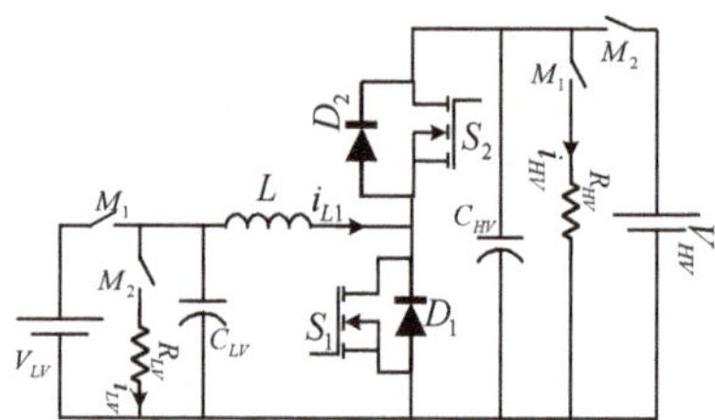

Figure 2. Conventional bidirectional converter (BDC).

In general, conventional step-up DC–DC converters are classified into isolated and non-isolated converters. Isolated converters like fly-back, push–pull, forward, half-bridge, and full-bridge converters have a high voltage gain by keeping a high enough transformer turns ratio. However, there is a problem with voltage spikes due to transformer leakage inductance, which leads to high power losses across the switch. On the other hand, in non-isolated converters, a high duty cycle is required to get a high voltage gain, which leads to decreasing efficiency due to reverse recovery problems [7]. In addition, non-isolated converters also have the problem of voltage stress nearly equal to the output voltage, causing a reduction of the device's reliability. Many DC–DC converter topologies are introduced to mitigate the problems mentioned above, such as interleaving topologies for the reduction of current ripple [8,9], soft-switching techniques to mitigate voltage spikes and efficiency improvement [10], and cascading boost converters [11] and incorporating a coupled inductor [12] in the conventional boost topology to get a high conversion gain. Input current ripples are reduced with the help of an interleaving concept, which leads to improving the source life. Additionally, it offers the flexibility of current sharing to enhance the power handling capacity [8,9].

On the other hand, several other converter topologies are suggested in the literature; most of these are designed to meet the various objectives, such as reliability, capacitor voltage reduction, and input current ripple reductions, by placing an impedance network between input DC source and switching network in various fashions. An X-shaped LC impedance network, as shown in Figure 3a, is placed to get the voltage boosting capability by operating a switching network in the shoot-through mode [13]. As an alternative to the Z-source converter, the same authors proposed a quasi Z-source (qZS) converter in two variants based on input current, namely continuous input current q-ZS (qZS-CC) and discontinuous input current q-ZS (qZS-DC) [14,15] with a reduced current and capacitor voltage stresses, respectively. The main variation between these two topologies is the input side inductor connection with the supply. In case of qZS-CC, the inductor is placed directly in series with the source, and it tries to always maintain constant input current, whereas the source current is of discontinuous nature in the case of qZS-DC, which increases the stress on the source [15]. Later, Yu Tang et al. proposed an improved Z-source (IZS) converter [16] with reduced capacitor stresses. In this paper, the authors claim that the utilization of a low voltage capacitor reduces the inrush current, the resonance between the Z-source inductor and capacitors, and the cost and volume of the system compared to a conventional Z-source converter [17]. The switched boost converter is proposed with a reduced passive components count, achieved by replacing one pair of LCs with power semi-conductor devices to have the same kind of buck-boost conversion, as shown in Figure 3b [18]. However, this topology uses more power semiconductor devices compared to the topologies mentioned above.

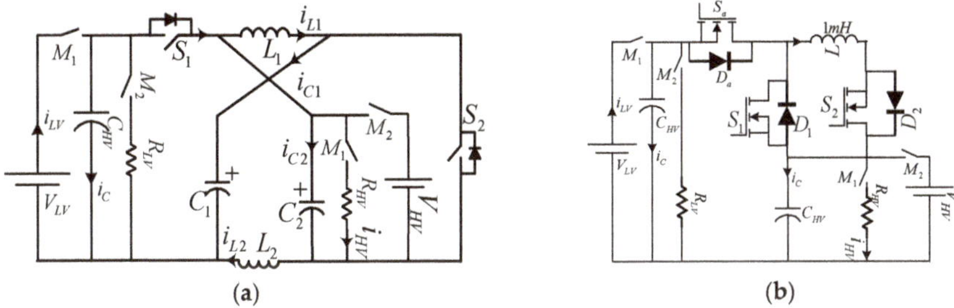

(a)   (b)

Figure 3. Impedance DC–DC converters. (**a**) Z-source DC–DC converter, (**b**) switched boost DC–DC converter.

A SL-ZS converter is proposed with an enhanced gain by placing switched inductors instead of inductors in the impedance network [19]. However, this topology suffers from a large component count (six power diodes and two inductors higher than the ZS converter) in the switching network. Alternatively, the SL-qZS converter proposed in [20], consists of switched inductors in place of standard inductors in the qZS converter to reduce the capacitor voltage and startup inrush current compared to the SL-ZS converter. However, the downside of this topology is a higher component count. Hossein Fathi et al. [21] proposed an enhanced boost ZS converter (EB-ZSC), achieved by replacing the impedance network with switched impedance to enhance the conversion gain further. Although this topology increases gain, it suffers from a higher component count (four inductors, four capacitors, and five power diodes). Additionally, this topology suffers from the usage of sophisticated control platforms to achieve smoother voltage control in the case of adjustable speed-controlled drive applications. Moreover, with a similar concept of variations in the impedance network either in ZS or qZS as discussed above, there are several other impedance source topologies, such as a diode-assisted qZS (DA-qZS) converter [22], a capacitor-assisted qZS (CS-qZS) converter [22], and an enhanced boost quasi ZS (EB-qZS) converter [23], which are proposed in the literature. Though these topologies are mainly proposed for DC–AC power conversion applications due to high reliability (operation during shoot-through mode), they are equally applicable for bi-directional applications and are widely used in micro/nano-grid applications [18,23].

For most of these topologies, it has been suggested to incorporate switched-inductor, switched-capacitor, and hybrid switched-capacitor/switched-inductor structures resulting in high boosting factors. However, the effect of nonlinearity can be increased by increasing the energy storage elements in the circuit, which leads to a higher output current and voltage distortion [24]. Additionally, introducing more energy storage elements in the circuit affects the control complexity, total cost, size, volume, losses, and weight of the converter [25,26]. Moreover, these topologies are suffering from the usage of more capacitors and higher capacitor voltage stresses. Additionally, the voltage across most of the capacitors is generally more than the supply voltage in the case of impedance source topologies in order to perform the voltage boost functionality. Hence, high-voltage Z-capacitors should be used, which may increase the volume and system cost. Capacitors are prone to failure in the field operation of power electronic converters [27]. Hence, due to the stricter reliability constraints brought by aerospace, automotive, defense, space, and energy industries, the stresses and usage of capacitors should be reduced to enhance the converter's reliability [28]. Therefore, to enhance the life and converter reliability, either reduction in capacitors usage or voltage stresses on the capacitor is highly recommended [29,30].

In this paper, the LC impedance bi-directional DC–DC converter (LC-BDC) is proposed by placing one inductor between source and half-bridge, and one capacitor between source and the load, as shown in Figure 4 [31]. These small passive components are arranged in such a way that the converter offers several features, such as lower capacitor voltages, which in turn reduces the cost, size, and volume of the converter and also increases the reliability while achieving the desired functionality. This topology reduces the voltage stresses on the device due to the usage of small passive components compared to existing converters in the case of SiC converters, which are less immune to parasitic components. The paper is organized as follows: the working principle, modes of operation, mathematical modeling, and state-space average models of the proposed topology are discussed in Section 2. The concept validation using simulation and experimentation, along with the respective results, are presented in Section 3. Additionally, to demonstrate the effectiveness of the proposed topology, a detailed comparative analysis of the proposed converter and conventional converter is carried out along with the results of the proposed converter. Moreover, a separate simulation-based comparative analysis of the proposed LC converter with eight similar boost/buck-boost converter topologies is presented in Section 4. Finally, conclusions are presented in Section 5 of this paper.

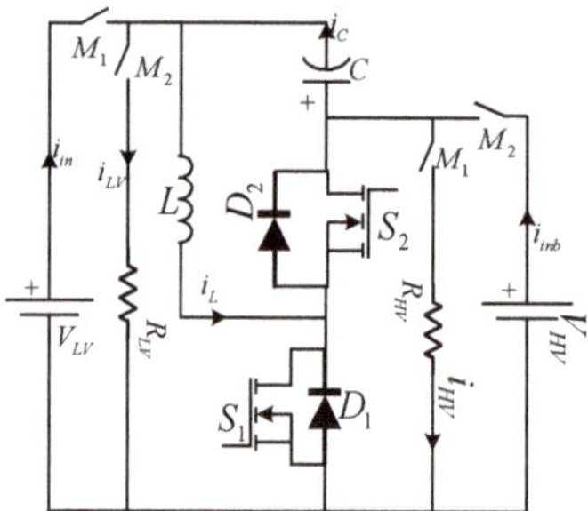

Figure 4. LC impedance bi-directional DC–DC converter (LC-BDC).

2. Proposed System

The LC bidirectional converter shown in Figure 4 is an advanced version of a conventional bidirectional converter and switched boost bidirectional converter, designed to reduce the voltage stresses on the capacitor. The primary function of the inductor is to store energy during the converter "on" period and release the stored energy during the "off" period of the primary device. The inductor is also used to eliminate the current ripple. Another energy storage element, the capacitor, is used to eliminate the ripple in the output voltage in both cases, namely the conventional BDC and the proposed BDC. Switches M_1 and M_2 are unidirectional switches used to realize the bidirectional power flow in the test setup which operate in a complementary fashion. For the forward direction of power flow, M_1 must be in the "on" state, and S_1 acts as the main switch operating at switching frequency, while D_2

acts as a freewheeling diode. Similarly, M_2 must be in the "on" position for the reverse direction of power flow, and S_2 acts as the main switch, which operates at switching frequency, while D_1 acts as a freewheeling diode. V_{LV} and R_{HV} are source and load in boost mode, and V_{HV} and R_{LV} are source and load in buck mode. The gating signals for boost switch (G_1) and buck switch (G_2) complement each other. The duty cycle of boost switch (S_1) and buck switch (S_2) is denoted as δ_1 and δ_2, respectively. T_s represents the switching period of switches S_1 and S_2.

2.1. Boost Operation

The equivalent circuit and idealized waveforms in boost mode of the LC-BDC converter are depicted in Figures 5 and 6, respectively. The converter operation is considered to be in boost mode, during which the switch (S_1) is pulse-modulated and the diode D_2 freewheels. The boost mode operation is further categorized into two sub-modes of operation over a switching period, and the equivalent circuit of each sub-mode is depicted as shown in Figure 7.

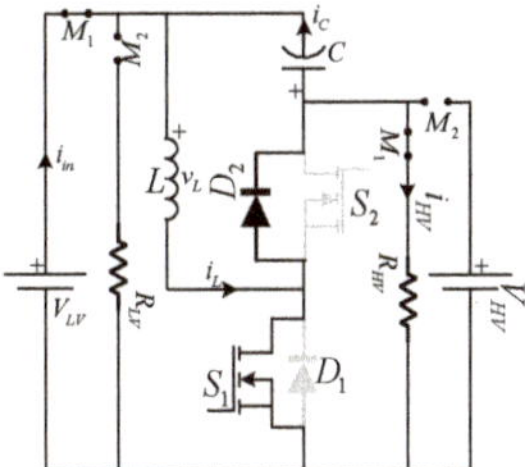

Figure 5. LC bi-directional DC–DC converter in boost mode.

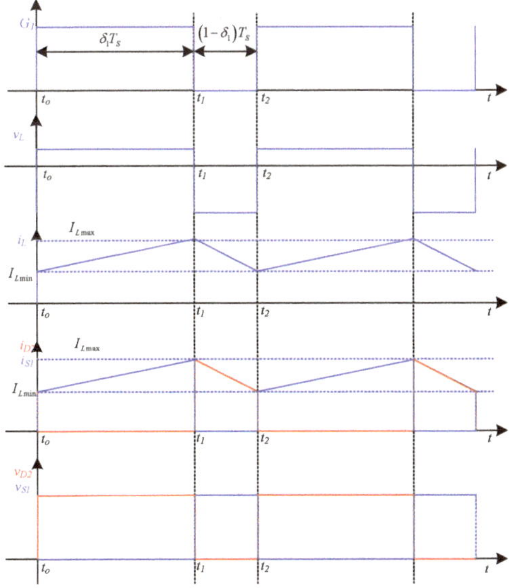

Figure 6. Characteristic waveforms during various boost modes of operation.

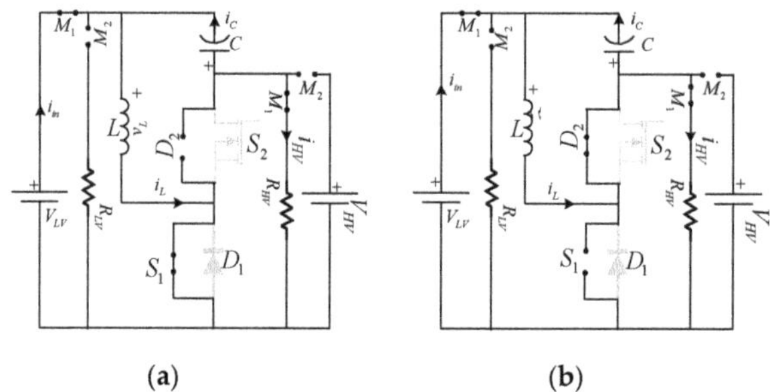

Figure 7. Equivalent circuit of boost operation (**a**) in mode 1 and (**b**) in mode 2.

2.1.1. Mode 1 ($t_0 < t < t_1$): (S_1 ON, D_2 OFF)

In this mode, switch S_1 is turned on by applying a gate signal. The inductor L starts charging linearly through switch S_1, and the capacitor C will discharge through load R_{HV}. Hence, the diode D_2 goes into the "off" state. The equivalent circuit during this mode of operation is shown in Figure 7a. The current through the inductor $L(i_L)$ and the voltage across the capacitor C are given by

$$V_L = V_{LV} = L\frac{di_L}{dt} \tag{1}$$

$$i_L(t) = \frac{V_{LV}}{L}(t - t_0) + i_L(t_0) \tag{2}$$

$$i_C = -\frac{v_{HV}}{R_{HV}} \tag{3}$$

$$v_C(t) = i_{HV}(t)R_{HV} - V_{LV} \tag{4}$$

This mode of operation ends when the gate pulses to switch S_1 are withdrawn.

2.1.2. Mode 2 ($t_1 < t < t_2$): (S_1 OFF, D_2 ON)

At the instant when the gate pulses of switch S_1 are removed, the switch S_1 goes into the "off" state due to which the voltage across the inductor brings the diode D_2 into the forward-biased state. The equivalent circuit during this mode is shown in Figure 7b. In this mode, both inductor and source feed power to the load, and the inductor charges the capacitor. Hence, there is a formation of the LC tank in this mode, which can offer zero voltage switching to the upper switch with the proper selection of the snubber capacitor. In this mode, the current through $L(i_L)$ reaches its minimum value. The current flowing through the inductor $L(i_L)$ and the voltage across the capacitor C are given by

$$V_L = V_{LV} - V_{HV} = L\frac{di_L}{dt} \tag{5}$$

$$i_L(t) = \frac{V_{LV} - V_{HV}}{L}(t - t_1) + i_L(t_1) \tag{6}$$

$$v_C(t) = i_{HV}(t)R_{HV} - V_{LV} \tag{7}$$

$$i_{Cc}(t) = i_L(t) - i_{HV}(t) \tag{8}$$

This mode ends at $t = Ts$ when the gate signal is provided to S_1 in the next switching cycle. Similar operation (Mode 1 and Mode 2) continues for several switching cycles until a power flow is required in the forward direction

2.2. Buck Mode of Operation of LC-BDC Converter

BDC operates in buck mode when there is a requirement of power flow in the reverse direction, and its equivalent circuit is shown in Figure 8. The converter operation is considered to be in reverse buck mode, during which the switch S_2 is pulse-modulated and the diode D_1 in a freewheeling mode. The buck mode of operation is further categorized into two sub-modes (i.e., mode 3 and mode 4) of operation over a switching period. The operating mode from mode 3 to mode 4 in buck mode is similar to the mode 2 to mode 1 of the boost mode of operation, respectively. Figure 9 illustrates the characteristic waveforms of the converter in buck mode, and its equivalent circuits in each sub-mode are depicted as shown in Figure 10.

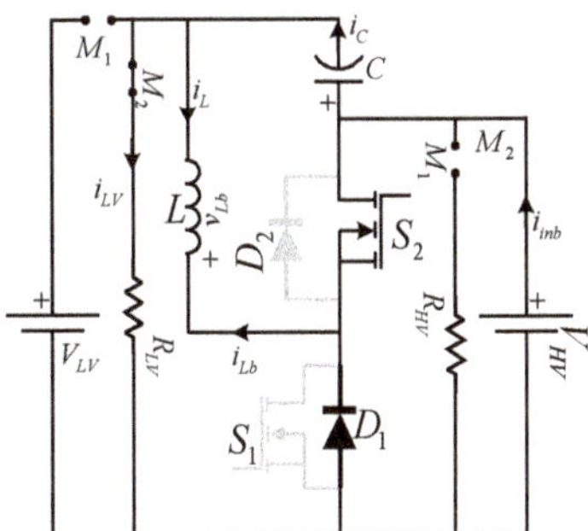

Figure 8. Equivalent circuit in buck mode.

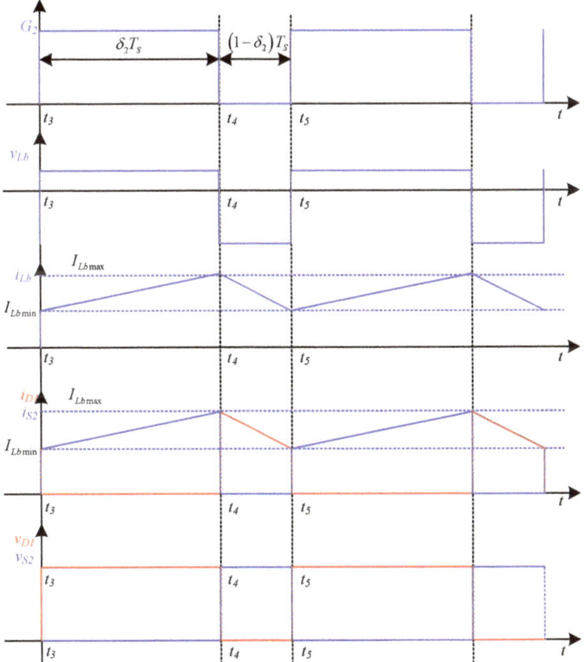

Figure 9. Characteristic waveforms during various modes in buck operation.

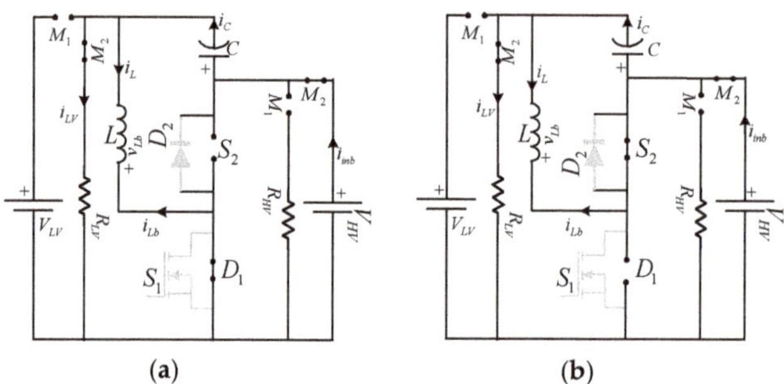

Figure 10. Equivalent circuit of buck operation (**a**) in mode 3 and (**b**) in mode 4.

2.2.1. Mode 3 ($t_3 < t < t_4$): (S_2 ON, D_1 OFF)

This mode starts at $t = t_3$ when the gate signal is given to S_2. At this instant, the main switch S_2 comes into conduction, and the diode D_1 goes into the "off" state. The supply V_{HV} then directly energizes the inductor L. The capacitor is also discharged through the inductor. It leads to the formation of the LC tank, as shown in Figure 10a, similar to mode 2. This feature offers the resonating switching functionality to the upper switch. The current flowing through $L(i_L)$ and the voltage across capacitor C (v_C) are given as

$$v_{Lb}(t) = V_{HV} - v_{LV} \tag{9}$$

$$i_{Lb}(t) = \frac{V_{HV} - v_{LV}}{L}(t - t_3) + i_{Lb}(t_3) \tag{10}$$

$$v_C(t) = V_{HV} - i_{LV}(t)R_{LV} \tag{11}$$

$$i_C(t) = i_{inb}(t) - I_{Lb}(t) \tag{12}$$

This mode continues until the gate pulse of S_2 is withdrawn at $t = t_4$.

2.2.2. Mode 4 ($t_4 < t < t_5$): (S_2 OFF, D_1 ON)

This mode starts at $t = t_4$ when the gate pulses to the main switch are removed. Hence S_2 goes into the "off" state. The voltage across the inductor brings diode D_1 into "on" state, and it continues until $t = t_5$. The energy stored in inductor L discharges through the load. The capacitor charges from the source. During this mode, the current flowing through the inductor L and voltage across the capacitor C can be expressed as

$$v_{Lb}(t) = -i_{LV}(t)R_{HV} \tag{13}$$

$$i_{Lb}(t) = \frac{-v_{LV}}{L}(t - t_4) + i_{Lb}(t_4) \tag{14}$$

$$v_C(t) = V_{HV} - i_{LV}(t)R_{LV} \tag{15}$$

$$i_C(t) = i_{inb}(t) \tag{16}$$

This mode ends at $t = T_s$ when the gate signal is given to S_2 in the next switching cycle. Similar operation of mode 3 and mode 4 continues, for several switching cycles, until power flow is required in the reverse direction.

2.3. State Space Analysis

This section presents the development of a small-signal AC model followed by the derivation of the state-space model equations for one complete switching cycle. For this analysis, few assumptions are considered; (i) the converter is operating in continuous conduction mode, and (ii) there is no trace resistance. For the proposed converter, the state variables are the current through the inductor i_L

and the voltage across the coupling capacitor V_C. A complete derivation of the state-space model and small-signal analysis for boost mode is presented. A similar derivation method can also be used for buck mode. With the inclusion of the parasitic components during both "on" and "off" states, the system can be represented with the help of the state-space model as follows.

$$\begin{bmatrix} \frac{di_L}{dt} \\ \frac{dv_C}{dt} \end{bmatrix} = \begin{bmatrix} -\frac{(r_{on}+r_L)}{L} & 0 \\ 0 & -\frac{1}{C(R+r_C)} \end{bmatrix} \begin{bmatrix} i_L \\ v_C \end{bmatrix} + \begin{bmatrix} \frac{1}{L} & 0 \\ 0 & \frac{-R}{(R+r_C)C} \end{bmatrix} \begin{bmatrix} V_{LV} \\ i_{Load} \end{bmatrix} \tag{17}$$

during "off" state:

$$\begin{bmatrix} \frac{di_L}{dt} \\ \frac{dv_C}{dt} \end{bmatrix} = \begin{bmatrix} \left(\frac{-Rr_c}{(R+r_C)} - r_L\right)\frac{1}{L} & \frac{-R}{(R+r_C)}\frac{1}{L} \\ \frac{R}{(R+r_C)C} & \frac{1}{(R+r_C)C} \end{bmatrix} \begin{bmatrix} i_L \\ v_C \end{bmatrix} + \begin{bmatrix} \frac{1}{L} & \frac{Rr_c}{(R+r_C)} \\ 0 & \frac{-R}{(R+r_C)} \end{bmatrix} \begin{bmatrix} V_{LV} \\ i_{Load} \end{bmatrix} \tag{18}$$

Here, r_{on}—on-state resistance of switching device, r_L—the equivalent series resistance of the inductor, and r_C—equivalent series resistance of the capacitor.

The state-space average model of the converter can be written as follows.

$$\dot{x} = [A_1\delta_1 + A_2(1-\delta_1)]x + [B_1\delta_1 + B_2(1-\delta_1)]U \tag{19}$$

Here, $x = \begin{bmatrix} i_L \\ v_{cc} \end{bmatrix}$, $A_1 = \begin{bmatrix} -\frac{(r_{on}+r_L)}{L} & 0 \\ 0 & -\frac{1}{C(R+r_C)} \end{bmatrix}$, $B_1 = \begin{bmatrix} \frac{1}{L} & 0 \\ 0 & \frac{-R}{(R+r_C)C} \end{bmatrix}$, $A_2 =$

$\begin{bmatrix} \left(\frac{-Rr_c}{(R+r_C)} - r_L\right)\frac{1}{L} & \frac{-R}{(R+r_C)}\frac{1}{L} \\ \frac{R}{(R+r_C)C} & \frac{1}{(R+r_C)C} \end{bmatrix}$, $B_2 = \begin{bmatrix} \frac{1}{L} & \frac{Rr_c}{(R+r_C)} \\ 0 & \frac{-R}{(R+r_C)} \end{bmatrix}$, $u = \begin{bmatrix} V_{LV} \\ i_{Load} \end{bmatrix}$

Define:

$$\delta_1 T_s = t_1 - t_0 \; \& \; t_2 - t_1 = (1-\delta_1)T_s \tag{20}$$

The duty ratio of the main switch S_1 is defined as

$$\delta_1 = \frac{t_1 - t_0}{t_2} \tag{21}$$

The turn-off duty cycle of the main switch S_1 is

$$\delta'_1 = \frac{t_2 - t_1}{t_2} \tag{22}$$

Substituting the duty ratio values from (20)–(22) in Equations (17)–(19) and then incorporating the perturbation effect into the state variables and other variables around the steady-state values gives

$$i_L = I_L + \hat{i}_L, v_{LV} = V_{LV} + \hat{v}_{LV}, v_C = V_C + \hat{v}_C, v_{HV} = V_{HV} + \hat{v}_{HV}, \delta_1 = D_1 + d_1 \tag{23}$$

where D_1 is the duty ratio of the main switch under steady-state condition. After solving the above state-space equation, the steady-state gains of the converter can be obtained as

$$V_C = \frac{D_1 V_{LV}}{1 - D_1} \& I_L = \frac{I_{HV}}{1 - D_1} \tag{24}$$

Comparing small-signal AC parameters while ignoring the considerably very small second-order quantities, and then solving the equations gives the following two transfer functions.

2.3.1. Control-to-Output Transfer Function

From the small-signal AC model, the control-to-output (output voltage to duty ratio) transfer function can be found under the condition of $\hat{v}_{in} = 0 \& \hat{i}_L = 0$, which is shown in Equation (25).

$$\frac{\hat{v}_o}{1-\hat{d}_1} = \frac{[-R(R+r_c)\times(1+CSr_c)\times] \left(\begin{matrix} RV_{LV}r_L - R^2V_{LV}D_1{}^2 - R^2V_{LV} + RV_{LV}r_{on} + V_{LV}r_Lr_c + V_{LV}r_cr_{on} + 2R^2V_{LV}D_1 + 2I_{load}R^2r_L \\ +I_{load}R^2r_c + I_{load}R^2r_{on} - 2I_{load}R^2D_1(r_L+r_c) + I_{load}R^2D_1{}^2r_c - I_{load}R^2D_1{}^2r_{on} + LRSV_{LV} \\ +LSV_{LV}r_c + 2I_{load}Rr_Lr_c + I_{load}Rr_cr_{on} + I_{load}LR^2S + I_{load}LRSr_c - 2I_{load}RD_1r_Lr_c - I_{load}LR^2SD_1 \\ -I_{load}RD_1{}^2r_cr_{on} - I_{load}LRSD_1r_c \end{matrix} \right)}{\left[\left(\begin{matrix} Rr_L + Rr_c + r_Lr_c - 2R^2D_1 + R^2 \\ +R^2D_1{}^2 - RD_1r_c + RD_1r_{on} + D_1r_cr_{on} \end{matrix} \right) \times \left(\begin{matrix} CLR^2S^2 - CR^2SD_1r_c + Cr_{on}R^2SD_1 + CR^2Sr_c + Cr_LR^2S + R^2D_1{}^2 - 2R^2D_1 + R^2 \\ +2CLRS^2r_c - CRSD_1r_c^2 + 2Cr_{on}RSD_1r_c + CRSr_c^2 + 2Cr_LRSr_c + LRS - RD_1r_c \\ +r_{on}RD_1 + Rr_c + r_LR + CLS^2r_c^2 + Cr_{on}SD_1r_c^2 + Cr_LSr_c^2 + LSr_c + r_{on}D_1r_c + r_Lr_c \end{matrix} \right) \right]}$$
(25)

By neglecting parasitic components, it can be simplified as

$$\frac{\hat{v}_{HV}}{1-\hat{d}_1} = \frac{RV_{LV} - LSV_{LV} - 2RV_{LV}D_1 + RV_{LV}D_1{}^2 - I_{load}LRS + I_{load}LRSD_1}{(D_1-1)^2(CLRS^2 + LS + RD_1{}^2 - 2RD_1 + R)}$$
$$\Rightarrow \frac{v_{HV}}{1-\hat{d}_1} = \frac{RV_{LV}(1-D_1)^2 - LSV_{LV} - (1-D_1)I_{load}LRS}{(D_1-1)^2(CLRS^2 + LS + R(1-D_1)^2)}$$
(26)

2.3.2. Control-to-Input Transfer Function

From the small-signal AC model, the inductor current-to-control (input current to duty ratio) transfer function can be found under the condition of $\hat{v}_{LV} = 0 \& \hat{i}_{Load} = 0$, which is shown in Equation (27).

$$\frac{\hat{i}_{in}}{1-\hat{d}_1} = \frac{-(R+r_c) \left(\begin{matrix} 2R^2V_{LV} - I_{load}R^3D_1{}^2 - RV_{LV}r_c - I_{load}R^3 + V_{LV}r_cr_{on} + 2I_{load}R^3D_1 + 2R^2V_{LV}D_1 + I_{laod}R^2r_L + I_{load}R^2r_{on} + I_{laod}Rr_Lr_c \\ +I_{load}Rr_cr_{on} - CR^3SV_{LV}D_1 - CI_{load}R^3Sr_L - CI_{laod}R^3Sr_C - CR^2SV_{LV}r_L - CRSV_{LV}r_c^2 - 2CR^2SV_{LV}r_c + CSV_{LV}r_c^2r_{on} \\ -CI_{laod}LR^3S^2 - CLR^2S^2V_{LV} + CR^3SV_{LV}VD_1{}^2 - CI_{load}LR^2Sr_C - CI_{load}R^3SD_1{}^2r_C + CI_{load}R^3SD_1{}^2r_{on} - CRSV_{LV}r_Lr_C \\ +CRSV_{LV}r_cr_{on} - CLRS^2V_{LV}r_c + 2CI_{load}R^3SD_1r_L + 2CI_{load}R^3SD_1r_C + CR^2SV_{LV}D_1r_c + CI_{load}RSr_Lr_c^2 + CI_{load}RSr_c^2r_{on} \\ +CI_{load}R^2Sr_cr_{on} + CI_{load}LR^3S^2D_1 + 2CI_{load}R^2SD_1r_Lr_c + CI_{load}LR^2S^2D_1r_c + CI_{load}R^2SD_1{}^2r_{on}r + RV_{LV}r_{on} \end{matrix} \right)}{\left[\left(\begin{matrix} Rr_L + Rr_c + r_Lr_c - 2R^2D_1 + R^2 \\ +R^2D_1{}^2 - RD_1r_c + RD_1r_{on} + D_1r_cr_{on} \end{matrix} \right) \times \left(\begin{matrix} CLR^2S^2 - CR^2SD_1r_c + Cr_{on}R^2SD_1 + CR^2Sr_c + Cr_LR^2S + R^2D_1{}^2 - 2R^2D_1 + R^2 \\ +2CLRS^2r_c - CRSD_1r_c^2 + 2Cr_{on}RSD_1r_c + CRSr_c^2 + 2Cr_LRSr_c + LRS - RD_1r_c + r_{on}RD_1 \\ +Rr_c + r_LR + CLS^2r_c^2 + Cr_{on}SD_1r_c^2 + Cr_LSr_c^2 + LSr_c + r_{on}D_1r_c + r_Lr_c \end{matrix} \right) \right]}$$
(27)

By neglecting parasitic components, it can be simplified as

$$\frac{\hat{i}_{in}}{1-\hat{d}_1} = \frac{2V_{LV} - I_{load}R(1-D_1) + CRV_{LV}S}{(1-D_1)(CLRS^2 + LS + RD_1{}^2 - 2RD_1 + R)}$$
$$\Rightarrow \frac{i_{in}}{1-\hat{d}_1} = \frac{RV_{LV}(1-D_1)^2 - LSV_{LV} - I_{load}(1-D_1)}{(1-D_1)(CLRS^2 + LS + R(1-D_1)^2)}$$
(28)

2.3.3. Step and Bode Responses of LC-BDC

From the derived transfer functions, the step responses of various variables are presented in Figure 11. From these results, it can be understood that the variations in input currents and capacitor voltages for both line and load disturbance are low in the case of the proposed converter compared to the existing converter. Moreover, the step responses reveal that the proposed LC-BDC converter response is the same for inductor current and capacitor current and capacitor voltage against the duty ratio, whereas input current transients against duty ratio variations are reduced in LC-BDC. It can also be noted that in the case of supply variations, the transient responses of the inductor current, output voltage, load current, and capacitor voltage are improved.

2.3.4. Ripple Capacitor Voltage

From the charge balance equation and further simplification of the above Equation (25) in the steady-state, the capacitor ripple voltage can be calculated as

$$\Delta v_{Ccboost} = \frac{D_1 V_{HV}}{CRF_S}$$
(29)

From (29), the capacitor value can be sized to minimize the voltage ripple across the capacitor.

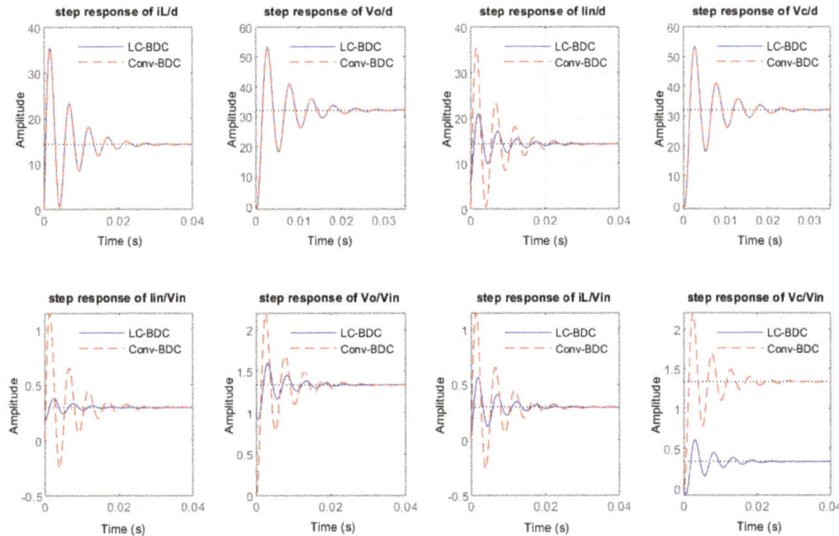

Figure 11. Step responses of the inductor current, output voltage, load current, and capacitor voltage transfer functions with respect to duty ratio and input voltage of LC-BDC and conventional converter.

3. Experimental Results and Discussion

The proposed LC impedance bi-directional dc-dc converter has been successfully validated through experiments in both boost and buck modes. The parameters considered for the experimental validations are summarized in Table 1. The experimental setup of the proposed converter is shown in Figure 12. The system performance is evaluated in both steady-state and transient conditions while feeding power to two series-connected 12 V, 50 W lamp load under various test conditions for the 18 V DC to 24 V DC conversion in forwarding boost mode, and 24 V DC to 18 V DC conversion in reverse buck mode. The inductor current, load current, and capacitor voltage waveforms are captured in both boost and buck modes for both conventional and proposed converters. Comparative analysis through experimental results was carried out, as explained below. In the case of the conventional converter, there is a need for two capacitors (C_{HV} plays a vital role in boost mode, and C_{LV} plays a vital role in buck mode), whereas, in the case of the proposed converter, there is a need for only one capacitor C, which can take care of the functionality of the above mentioned two capacitors in the respective modes. It can be observed that two capacitors are used in the realization of the conventional converter, whereas only one capacitor is used for the realization of the proposed LC-BDC, as shown in Table 1.

Table 1. Parameters of the proposed converter.

	Proposed		Conventional	
Parameter name	Boost Mode (LV to HV)	Buck Mode (HV to LV)	Boost Mode (LV to HV)	Buck Mode (HV to LV)
Input voltage (V)	18	24	18	24
Output voltage (V)	24	18	24	18
Output voltage ripple, %	≤0.50	≤0.50	≤0.50	≤0.50
Load	12 V, 50 W of 2 Lamps in series	12 V, 50 W of 2 Lamps in series	12 V, 50 W of 2 Lamps in series	12 V, 50 W of 2 Lamps in series
Output current (A)	4	3.20	4	3.20
Switching frequency (kHz)	10	10	10	10
Inductor	0.5 mH		0.5 mH	
Filter capacitors	500 μF/500 V(HV Side)	500 μF/500 V (LV Side)	500 μF/500 V (intermediate stage)	

Figure 12. Experimental Setup of Power Converter.

The gate signal of the lower switch, inductor current, load current, and capacitor voltage for four switching cycles are captured and presented in Figures 13–16. From Figure 14, it can be seen that the peak value of the inductor current is 6.17 A in the conventional converter, whereas it is 6.07 A in the proposed converter for the same load current, as shown in Figure 15. From Figure 16, it can be seen that the voltage across the capacitor is 23.20 V in the case of the conventional converter, whereas it is 5.10 V in the case of the proposed converter. Hence, there is 78.02% of capacitor voltage reduction in the proposed converter as compared to the conventional converter for the same input/output voltage conversion.

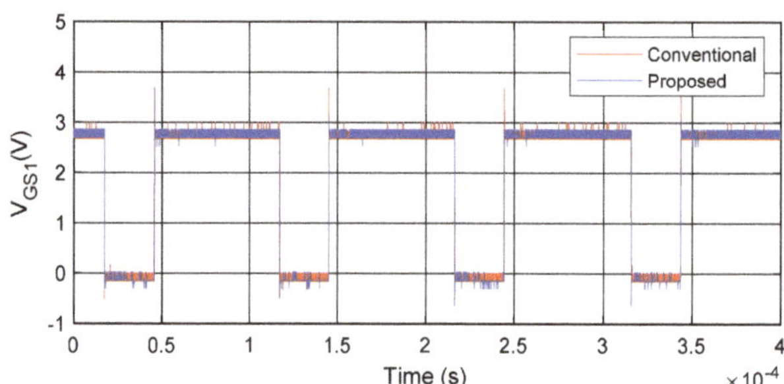

Figure 13. Gate voltage of lower switch (S_1) in LC-BDC (Blue), and conventional BDC (red).

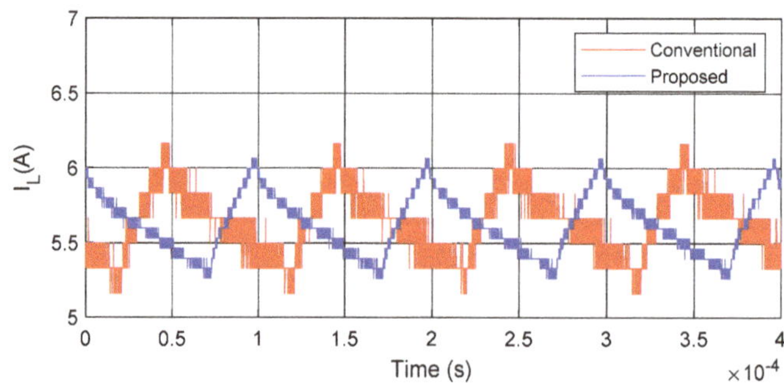

Figure 14. Inductor current in LC-BDC (Blue), and conventional BDC (red).

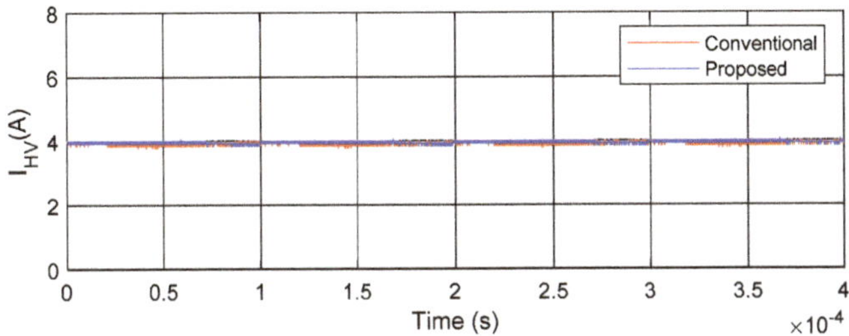

Figure 15. Load current in LC-BDC (Blue), and conventional BDC (red).

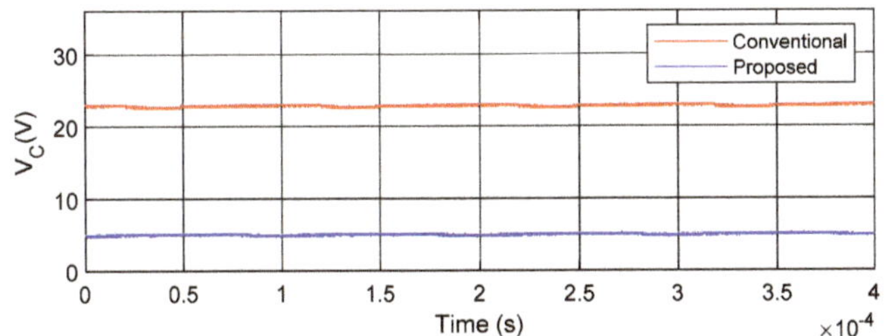

Figure 16. Capacitor voltages in LC-BDC (Blue), and conventional BDC (Red).

In Figures 17–19, respectively, a zoomed view of respective parameters is presented during both "on" and "off" state. From these figures, peak values during both transient and steady-state can be measured as listed in Table 2.

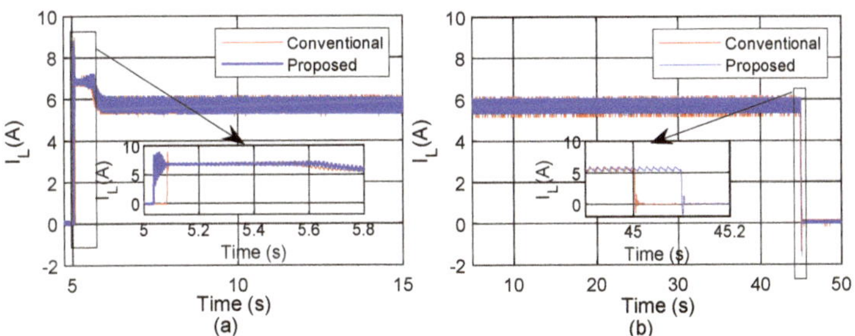

Figure 17. Inductor current of LC-BDC (blue) and conventional BDC (red) converter during (a) "on" and (b) "off" states.

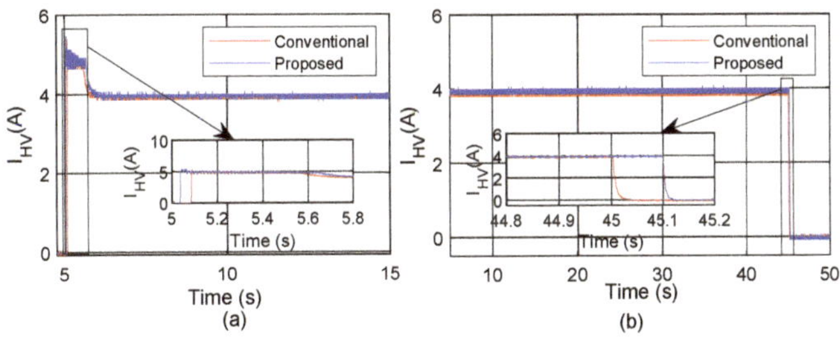

Figure 18. Load current of LC-BDC (blue) and conventional BDC (red) converter during (a) "on" and (b) "off" states.

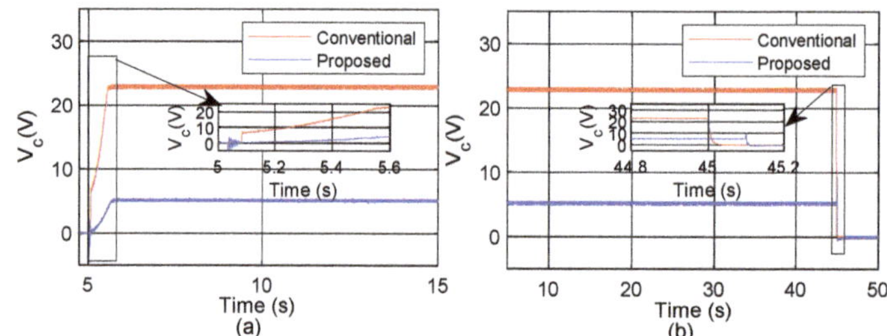

Figure 19. Capacitor voltage of LC-BDC (blue) and conventional BDC (red) converter during (a) "on" and (b) "off" states.

Table 2. Comparison of the various parameter during both transient and steady state for the proposed and conventional converters.

Peak Values	Conventional		Proposed	
	Transient	**Steady State**	**Transient**	**Steady State**
Capacitor voltage	23.20 V	23.20 V	4.80 V	5.20 V
Inductor current	8.80 A	6.00 A	8.80 A	6.00 A
Load Current	5.20 A	4.00 A	5.20 A	4.00 A

Form Table 3, it can be observed that the proposed converter not only offer its best performance during steady-state conditions but also the exhibits same best performance during transient conditions in terms of capacitor voltage stresses.

Table 3. Comparison of ripple values of capacitor voltage, inductor current and load current for the proposed and conventional converters.

Ripples	Conventional	Proposed
Capacitor voltage	0.50 V	0.34 V
Input current	0.99 A	0.79 A
Load Current	0.13 A	0.13 A

For the ripple content investigation, a zoomed view of the inductor current, load current, and capacitor voltages are presented in Figures 20–23, respectively. The summary of the ripple content for both converter topologies is tabulated in Table 3. From this table, it can be understood that there is a 0.4% reduction of ripple content in capacitor voltages and inductor current for the same content of load current ripples.

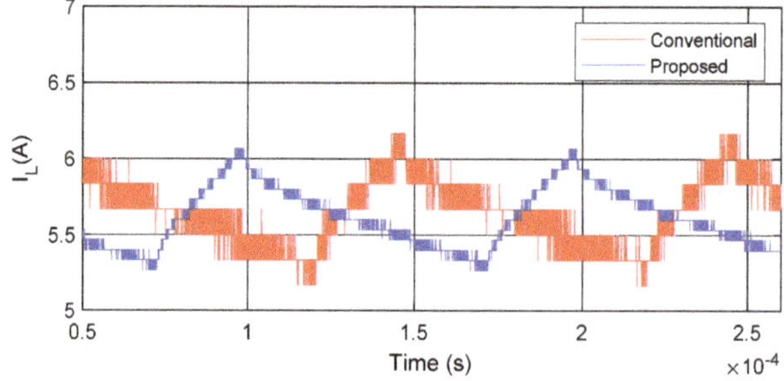

Figure 20. Zoomed view of inductor current for ripple analysis LC-BDC (blue), and conventional converter (red).

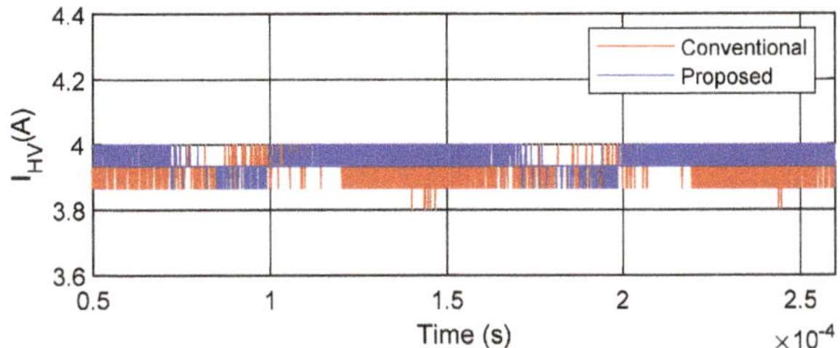

Figure 21. Zoomed view of load current for ripple analysis in LC-BDC (blue), and conventional converter (red).

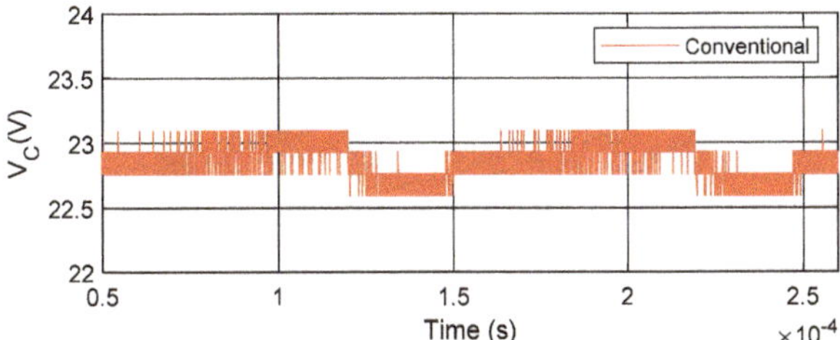

Figure 22. Zoomed view of capacitor voltage in the conventional converter for ripple analysis.

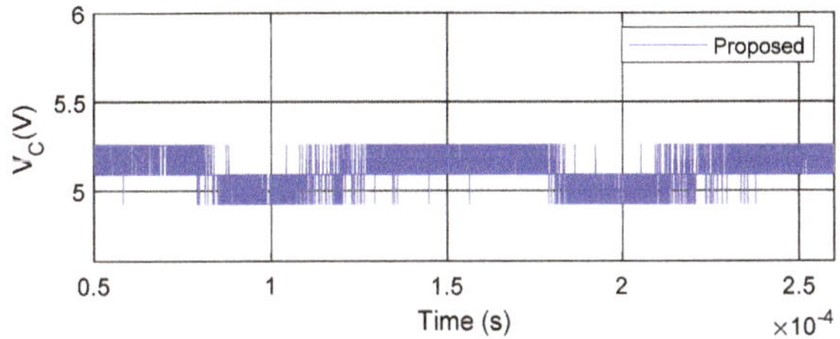

Figure 23. Zoomed view of capacitor voltage in the proposed converter for ripple analysis.

For the reverse buck mode of operation, the gating signal of the upper switch, inductor current, load current, and capacitor voltage for four switching cycles have been captured, as presented in Figures 24–27. From Figure 25, it can be seen that the peak value of the inductor current is 3.61 A in the case of conventional converter, whereas it is 3.62 A in the case of the proposed converter for the same load current as shown in Figure 26. From Figure 27, it can be seen that the voltage across the capacitor is 16.40 V in the case of conventional converter, whereas it is 7.80 V in the case of the proposed converter. A capacitor voltage reduction of 35.80% can be witnessed in this mode of operation.

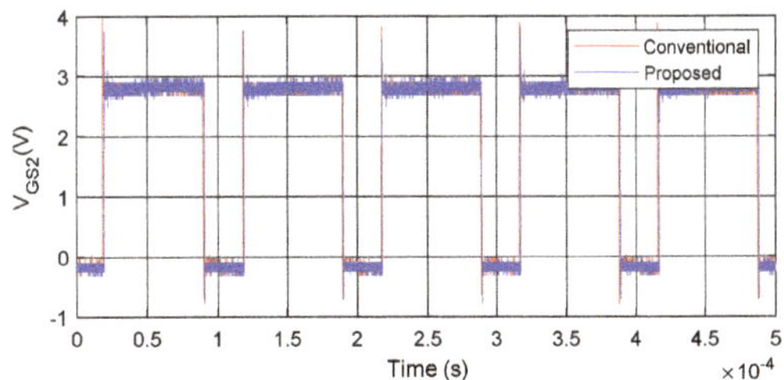

Figure 24. Gate voltage of upper switch (S_2) in LC-BDC (Blue), and conventional BDC (red).

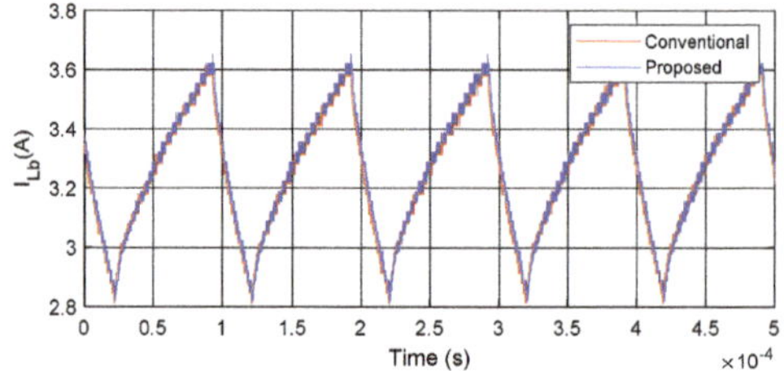

Figure 25. Inductor current in LC-BDC (Blue), and conventional BDC (red).

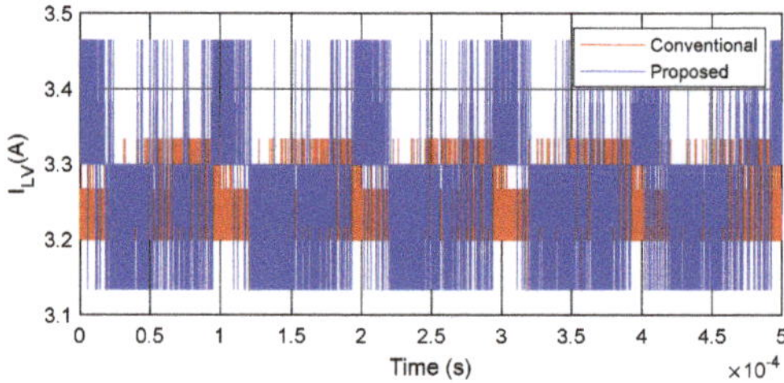

Figure 26. Load current in LC-BDC (Blue), and conventional BDC (red).

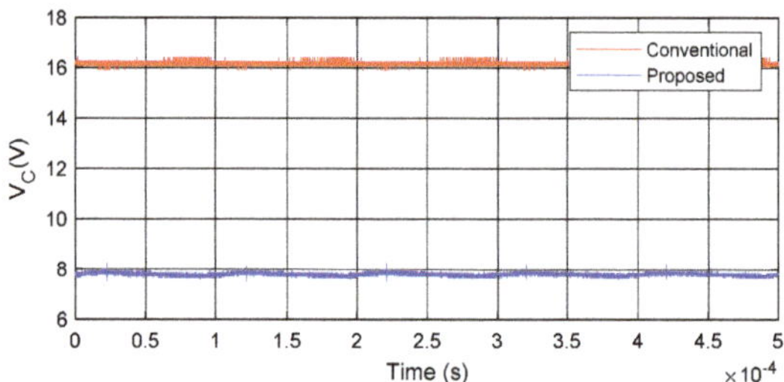

Figure 27. Capacitor voltages in LC-BDC (Blue), and conventional BDC (red).

For the critical investigation, results have been captured under various test conditions to assess the proposed converter suitability for various applications like smooth turn-on, faster load turn-off, and converter on- and off-switching with variable duty. During these conditions, captured inductor current, load current, and capacitor voltage are shown in Figures 28–30, respectively. Moreover, in these figures, a zoomed view of respective parameters is presented during both turn-on and turn-off. From these figures, peak values during both transient and steady states can be measured as listed in Table 4. Form the Table 4, and it can be observed that the proposed converter not only offers its best performance during steady-state conditions but also exhibits the same best performance during transient conditions in terms of capacitor voltage stresses.

Table 4. Comparison of the various parameter during both transient and steady states for the proposed and conventional converters.

Peak Values	Conventional		Proposed	
	Transient	**Steady State**	**Transient**	**Steady State**
Capacitor voltage	5.60 V	16.40 V	6.20 V	7.80 V
Inductor current	4.54 A	3.65 A	4.62 A	3.68 A
Load Current	4.54 A	3.65 A	4.62 A	3.68 A

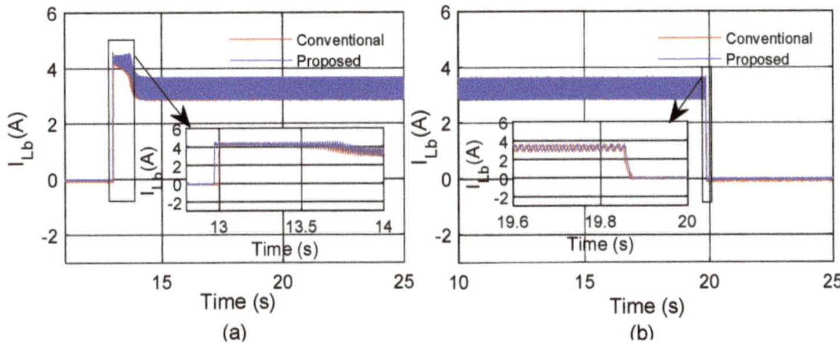

Figure 28. Inductor current of LC-BDC (blue) and conventional BDC (red) converter during (**a**) turn on (**b**) turn off.

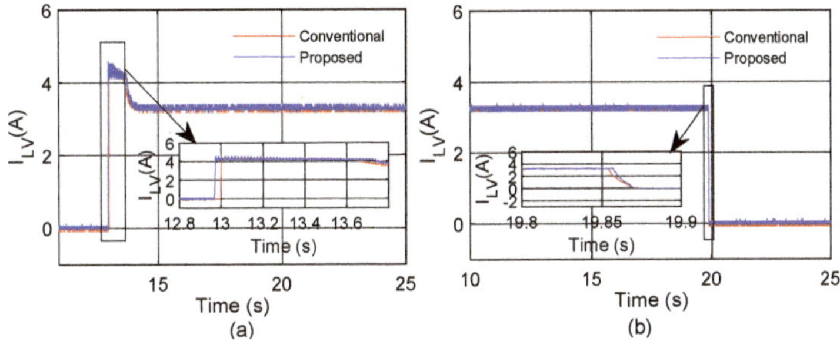

Figure 29. Load current of LC-BDC (blue) and conventional BDC (red) converter during (**a**) turn on and (**b**) turn off.

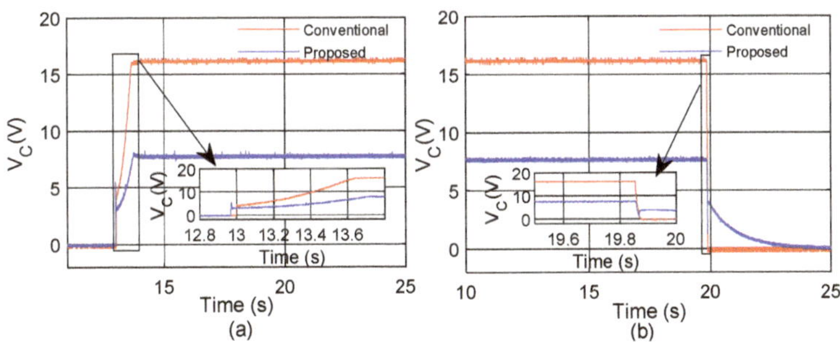

Figure 30. Capacitor voltage of LC-BDC (blue) and conventional BDC (red) converter during (**a**) turn on and (**b**) turn off.

For the ripple content investigation, a zoomed view of the inductor current, load current, and capacitor voltages are presented in Figures 31–34, respectively. The summary of the ripple content in both the converters is tabulated in Table 5. From this table, it can be extracted that there is a 0.3% reduction of ripple content in capacitor voltages ripples for the same load current.

Table 5. Comparison of Ripple Values for The Proposed and Conventional Converters in Buck Mode.

Ripples	Conventional	Proposed
Capacitor voltage	1.40%	1.10%
Inductor current	16.65%	16.65%
Load Current	4.70%	5.00%

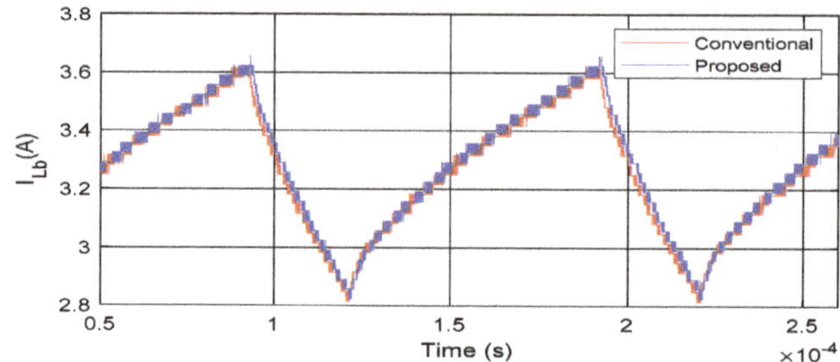

Figure 31. Zoomed view of inductor current for ripple analysis LC-BDC (blue), and conventional converter (red).

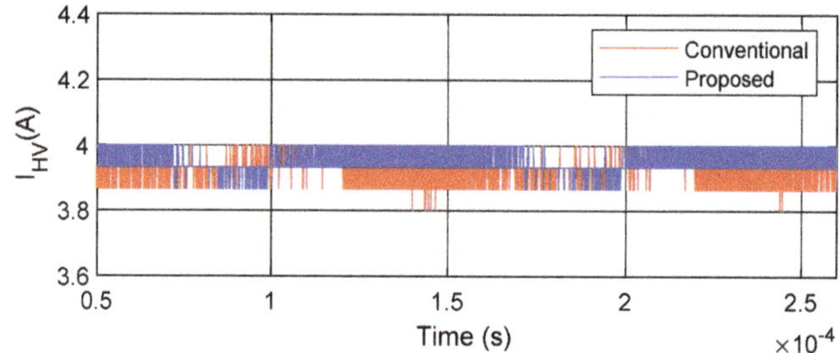

Figure 32. Zoomed view of load current for ripple analysis in LC-BDC (blue), and conventional converter (red).

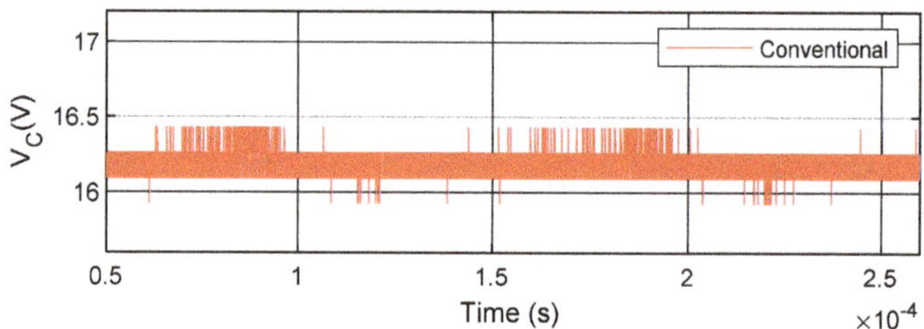

Figure 33. Zoomed view of capacitor voltage in the conventional converter for ripple analysis.

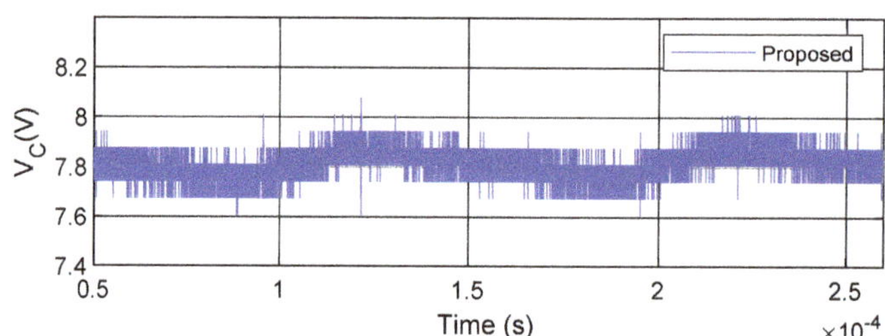

Figure 34. Zoomed view of capacitor voltage in the proposed converter for ripple analysis.

4. Comparative Analysis

Another set of simulations is performed to investigate the performance of the above topologies, and simulation parameters are listed in Table 6 as mentioned below. By using these sets of simulations, various performance parameters such as voltage gains, capacitor voltages, and losses were investigated.

These simulations are carried out for various output voltages ranging from 36 V to 108 V with the dc input voltage of 24 V, i.e., voltage gain ranging from 1.5 to 4.5. With the help of this data, a comparative analysis is presented in the following subsections.

Table 6. Parameters considered for comparative analysis.

Parameter	Value
Input DC Voltage	24 V
Switching Frequency	10 kHz
Output line voltage (RMS)	36 V–108 V
Load Power	500 W

4.1. No. of Components

As mentioned earlier, by changing the impedance network configurations, various topologies are proposed, and hence each topology has a different number of components. The number of components used for different topologies are listed and presented as a bar chart shown in Figure 35. From this chart, it is clear that the proposed converter and conventional BDC require fewer components compared to other topologies.

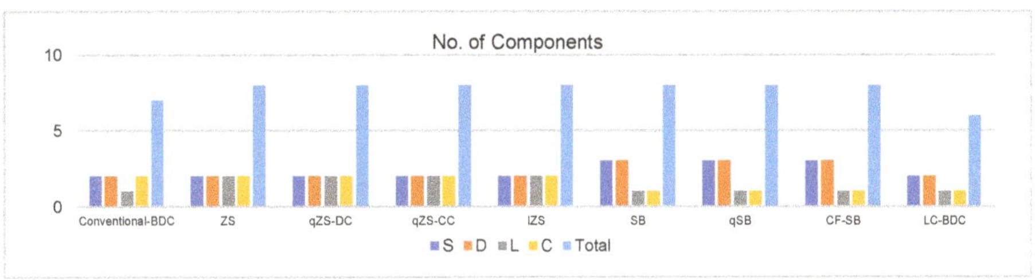

Figure 35. Bar chart of no. of components used in different topology.

4.2. Capacitor Voltage Stress

Since capacitors used in various topologies are different, total voltage stresses in all capacitors are calculated for comparison purposes. Here in Figure 36, the total capacitor stresses are plotted, while 24 V DC is converted into a range of DC voltage ranging from 36 V to 108 V. From this figure, it is clear that capacitor stresses are low in the case of the proposed converter.

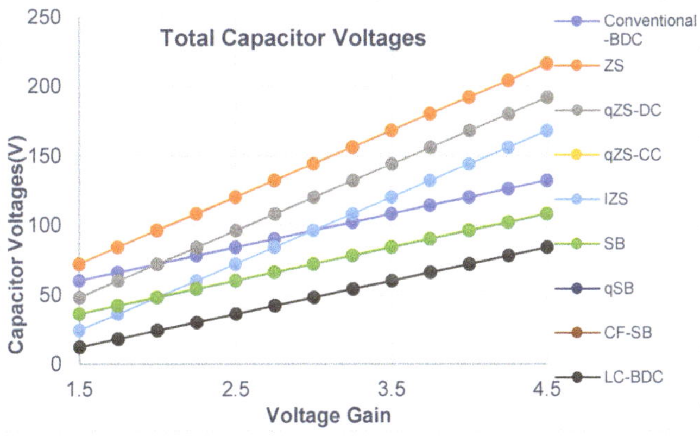

Figure 36. Total capacitor stress in different topology.

4.3. Efficiency Analysis and Loss Comparison

To perform the converter efficiency analysis, the parasitic resistance of inductors and capacitors, and the diode forward conduction losses are considered in this paper. The parasitic resistance of inductor and capacitor are r_L and r_C, respectively, and forward conduction loss of diode due to forward voltage (V_F) was assumed to be the same in all topologies for comparative analysis. The impact of the parasitic resistances and the forward voltage drop of the main power devices (MOSFETs) are also considered in this manuscript. Equivalent circuits of all the considered buck-boost bi-directional converters with the inclusion of various parasitic components are presented in Figure 37 for the efficiency calculations. Formulas derived for losses and efficiencies are presented in Table 7.

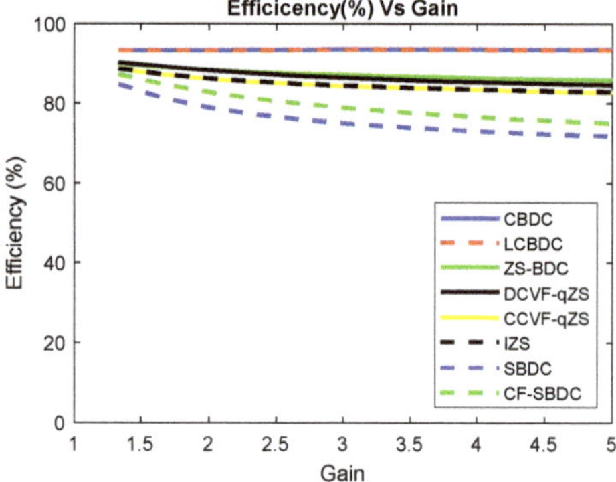

Figure 37. Efficiency comparison of various topologies.

Table 7. Comparison of various parameters (device, inductor and capacitor RMS currents, overall losses, and efficiency) of existing and proposed topologies.

Topology	Equivalent Circuit Diagram	RMS Currents of Various Components	Parameters
Conv-BDC		$I_{Srms} = \frac{I_o\sqrt{D}}{1-D}$ $I_{Drms} = \frac{I_o}{\sqrt{1-D}}$ $I_{Lrms} = \frac{I_o}{1-D}$ $I_{Crms} = I_o\sqrt{\frac{D}{1-D}}$	$\eta = \dfrac{1}{1+\left\{\frac{r_L+Dr_{DS}}{(1-D)^2 R_L}+\frac{R_F+Dr_c}{(1-D)R_L}+\frac{V_f}{V_o}+F_s C_o R_L\right\}}$ $M_{VDS} = \dfrac{1}{(1-D)\left\{\frac{r_L+Dr_{DS}}{(1-D)^2 R_L}+\frac{R_F+Dr_c}{(1-D)R_L}+\frac{V_f}{V_o}+F_s C_o R_L\right\}}$
LC-BDC		$I_{Srms} = \frac{I_o\sqrt{D}}{1-D}$ $I_{Drms} = \frac{I_o}{\sqrt{1-D}}$ $I_{Lrms} = \frac{I_o}{1-D}$ $I_{Crms} = I_o\sqrt{\frac{D}{1-D}}$	$\eta = \dfrac{1}{1+\left\{\frac{r_L+Dr_{DS}}{(1-D)^2 R_L}+\frac{R_F+Dr_c}{(1-D)R_L}+\frac{V_f}{V_o}+F_s C_o R_L\right\}}$ $M_{VDS} = \dfrac{1}{(1-D)\left\{\frac{r_L+Dr_{DS}}{(1-D)^2 R_L}+\frac{R_F+Dr_c}{(1-D)R_L}+\frac{V_f}{V_o}+F_s C_o R_L\right\}}$
ZS-BDC		$I_{Srms} = \frac{I_o\sqrt{D}}{1-2D}$ $I_{Drms} = \frac{(1-D)^{3/2}}{(1-2D)}I_o$ $I_{Lrms} = \frac{1-D}{1-2D}I_o$ $I_{Crms} = \frac{\sqrt{D(1-D)}}{(1-2D)}I_o$	$\eta = \dfrac{1}{1+\left\{\begin{array}{l}\frac{Dr_{Ds}}{(1-2D)^2 R_L}+\frac{(1-D)^2}{(1-2D)}\frac{V_f}{V_o}+\frac{(1-D)^3}{(1-2D)^2}\frac{R_F}{R_L}++2\left(\frac{1-D}{1-2D}\right)^2\frac{r_L}{R_L}\\ +\frac{F_s C_o R_L}{2}+\frac{r_{LF}}{R_L}+2\frac{D(1-D)}{(1-2D)^2}\frac{r_c}{R_L}+\frac{(M_{Vdc}-1)^2}{12(L_f F_s)^2}(1-2D)^2 r_{CF} R_L\end{array}\right\}}$ $M_{VDS} = \dfrac{(1-D)}{(1-2D)\left\{\begin{array}{l}\frac{Dr_{Ds}}{(1-2D)^2 R_L}+\frac{(1-D)^2}{(1-2D)}\frac{V_f}{V_o}+\frac{(1-D)^3}{(1-2D)^2}\frac{R_F}{R_L}++2\left(\frac{1-D}{1-2D}\right)^2\frac{r_L}{R_L}\\ +\frac{F_s C_o R_L}{2}+\frac{r_{LF}}{R_L}+2\frac{D(1-D)}{(1-2D)^2}\frac{r_c}{R_L}+\frac{(M_{Vdc}-1)^2}{12(L_f F_s)^2}(1-2D)^2 r_{CF} R_L\end{array}\right\}}$

Table 7. *Cont.*

Topology	Equivalent Circuit Diagram	RMS Currents of Various Components	Parameters
DCVF-qZS BDC	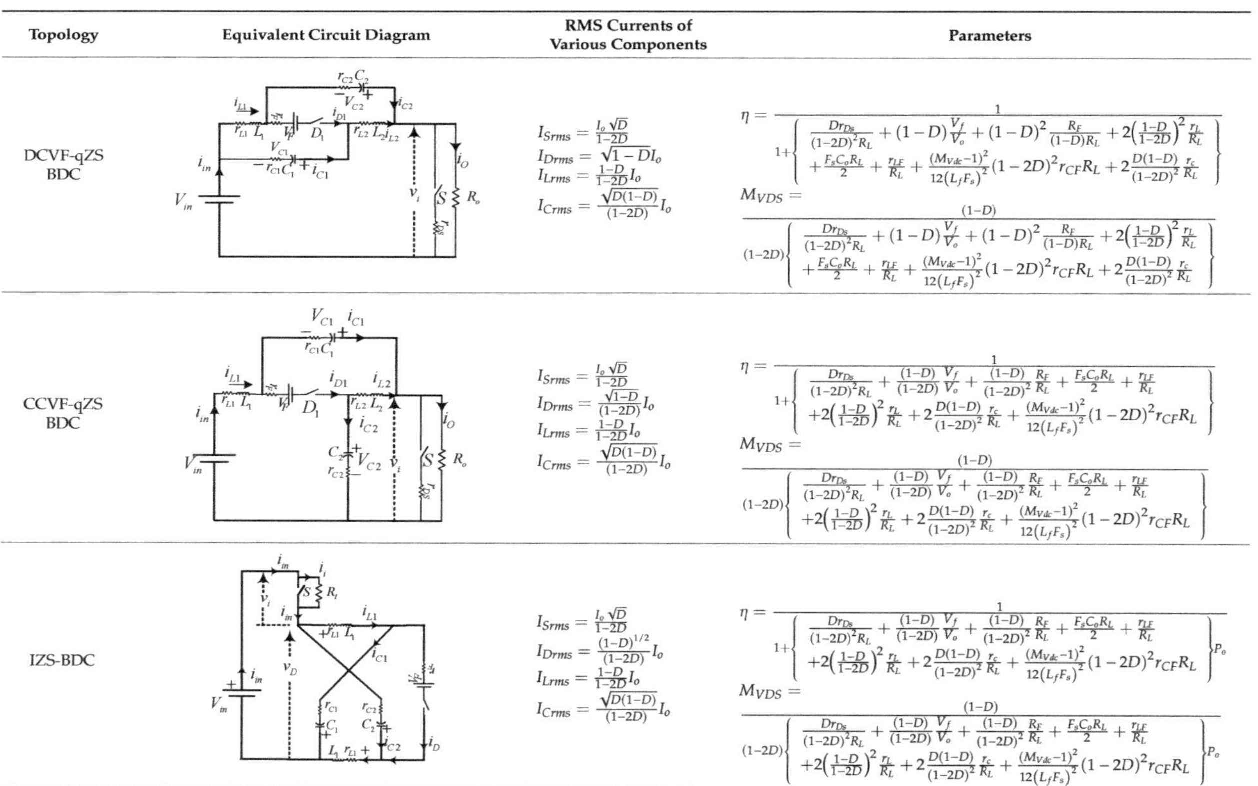	$I_{Srms} = \frac{I_o \sqrt{D}}{1-2D}$ $I_{Drms} = \sqrt{1-D}I_o$ $I_{Lrms} = \frac{1-D}{1-2D}I_o$ $I_{Crms} = \frac{\sqrt{D(1-D)}}{(1-2D)}I_o$	$\eta = \dfrac{1}{1+\left\{\begin{array}{l}\frac{Dr_{Ds}}{(1-2D)^2 R_L} + (1-D)\frac{V_f}{V_o} + (1-D)^2\frac{R_F}{(1-D)R_L} + 2\left(\frac{1-D}{1-2D}\right)^2\frac{r_L}{R_L} \\ + \frac{F_s C_o R_L}{2} + \frac{r_{LF}}{R_L} + \frac{(M_{Vdc}-1)^2}{12(L_f F_s)^2}(1-2D)^2 r_{CF}R_L + 2\frac{D(1-D)}{(1-2D)^2}\frac{r_c}{R_L}\end{array}\right\}}$ $M_{VDS} = \dfrac{(1-D)}{(1-2D)\left\{\begin{array}{l}\frac{Dr_{Ds}}{(1-2D)^2 R_L} + (1-D)\frac{V_f}{V_o} + (1-D)^2\frac{R_F}{(1-D)R_L} + 2\left(\frac{1-D}{1-2D}\right)^2\frac{r_L}{R_L} \\ + \frac{F_s C_o R_L}{2} + \frac{r_{LF}}{R_L} + \frac{(M_{Vdc}-1)^2}{12(L_f F_s)^2}(1-2D)^2 r_{CF}R_L + 2\frac{D(1-D)}{(1-2D)^2}\frac{r_c}{R_L}\end{array}\right\}}$
CCVF-qZS BDC		$I_{Srms} = \frac{I_o \sqrt{D}}{1-2D}$ $I_{Drms} = \frac{\sqrt{1-D}}{(1-2D)}I_o$ $I_{Lrms} = \frac{1-D}{1-2D}I_o$ $I_{Crms} = \frac{\sqrt{D(1-D)}}{(1-2D)}I_o$	$\eta = \dfrac{1}{1+\left\{\begin{array}{l}\frac{Dr_{Ds}}{(1-2D)^2 R_L} + \frac{(1-D)}{(1-2D)}\frac{V_f}{V_o} + \frac{(1-D)}{(1-2D)^2}\frac{R_F}{R_L} + \frac{F_s C_o R_L}{2} + \frac{r_{LF}}{R_L} \\ + 2\left(\frac{1-D}{1-2D}\right)^2\frac{r_L}{R_L} + 2\frac{D(1-D)}{(1-2D)^2}\frac{r_c}{R_L} + \frac{(M_{Vdc}-1)^2}{12(L_f F_s)^2}(1-2D)^2 r_{CF}R_L\end{array}\right\}}$ $M_{VDS} = \dfrac{(1-D)}{(1-2D)\left\{\begin{array}{l}\frac{Dr_{Ds}}{(1-2D)^2 R_L} + \frac{(1-D)}{(1-2D)}\frac{V_f}{V_o} + \frac{(1-D)}{(1-2D)^2}\frac{R_F}{R_L} + \frac{F_s C_o R_L}{2} + \frac{r_{LF}}{R_L} \\ + 2\left(\frac{1-D}{1-2D}\right)^2\frac{r_L}{R_L} + 2\frac{D(1-D)}{(1-2D)^2}\frac{r_c}{R_L} + \frac{(M_{Vdc}-1)^2}{12(L_f F_s)^2}(1-2D)^2 r_{CF}R_L\end{array}\right\}}$
IZS-BDC		$I_{Srms} = \frac{I_o \sqrt{D}}{1-2D}$ $I_{Drms} = \frac{(1-D)^{1/2}}{(1-2D)}I_o$ $I_{Lrms} = \frac{1-D}{1-2D}I_o$ $I_{Crms} = \frac{\sqrt{D(1-D)}}{(1-2D)}I_o$	$\eta = \dfrac{1}{1+\left\{\begin{array}{l}\frac{Dr_{Ds}}{(1-2D)^2 R_L} + \frac{(1-D)}{(1-2D)}\frac{V_f}{V_o} + \frac{(1-D)}{(1-2D)^2}\frac{R_F}{R_L} + \frac{F_s C_o R_L}{2} + \frac{r_{LF}}{R_L} \\ + 2\left(\frac{1-D}{1-2D}\right)^2\frac{r_L}{R_L} + 2\frac{D(1-D)}{(1-2D)^2}\frac{r_c}{R_L} + \frac{(M_{Vdc}-1)^2}{12(L_f F_s)^2}(1-2D)^2 r_{CF}R_L\end{array}\right\}} P_o$ $M_{VDS} = \dfrac{(1-D)}{(1-2D)\left\{\begin{array}{l}\frac{Dr_{Ds}}{(1-2D)^2 R_L} + \frac{(1-D)}{(1-2D)}\frac{V_f}{V_o} + \frac{(1-D)}{(1-2D)^2}\frac{R_F}{R_L} + \frac{F_s C_o R_L}{2} + \frac{r_{LF}}{R_L} \\ + 2\left(\frac{1-D}{1-2D}\right)^2\frac{r_L}{R_L} + 2\frac{D(1-D)}{(1-2D)^2}\frac{r_c}{R_L} + \frac{(M_{Vdc}-1)^2}{12(L_f F_s)^2}(1-2D)^2 r_{CF}R_L\end{array}\right\}} P_o$

Table 7. *Cont.*

Topology	Equivalent Circuit Diagram	RMS Currents of Various Components	Parameters
SB-BDC		$I_{Srms} = \frac{I_o \sqrt{D}}{1-2D}$ $I_{Drms} = \frac{(1-D)^{1/2}}{(1-2D)} I_o$ $I_{Lrms} = \frac{1}{1-2D} I_o$ $I_{Crms} = \frac{2\sqrt{D(1-D)}}{(1-2D)} I_o$	$\eta = \dfrac{1}{1 + \left\{ \dfrac{2Dr_{Ds}}{(1-2D)^2 R_L} + \dfrac{(1-D)}{(1-2D)} \dfrac{2V_f}{V_o} + \dfrac{(1-D)}{(1-2D)^2} \dfrac{2R_F}{R_L} + F_s C_o R_L + \left(\dfrac{1}{1-2D}\right)^2 \dfrac{r_L}{R_L} + \dfrac{4D(1-D)}{(1-2D)^2} \dfrac{r_c}{R_L} \right\} P_o}$ $M_{VDS} = \dfrac{(1-D)}{(1-2D)\left\{ \dfrac{2Dr_{Ds}}{(1-2D)^2 R_L} + \dfrac{(1-D)}{(1-2D)} \dfrac{2V_f}{V_o} + \dfrac{(1-D)}{(1-2D)^2} \dfrac{2R_F}{R_L} + F_s C_o R_L + \left(\dfrac{1}{1-2D}\right)^2 \dfrac{r_L}{R_L} + \dfrac{4D(1-D)}{(1-2D)^2} \dfrac{r_c}{R_L} \right\} P_o}$
CFSB-BDC		$I_{Srms} = \frac{I_o \sqrt{D}}{1-2D}$ $I_{Drms} = \frac{(1-D)^{1/2}}{(1-2D)} I_o$ $I_{Lrms} = \frac{1}{1-2D} I_o$ $I_{Crms} = \frac{2\sqrt{D(1-D)}}{(1-2D)} I_o$	$\eta = \dfrac{1}{1 + \left\{ \dfrac{2Dr_{Ds}}{(1-2D)^2 R_L} + \dfrac{(1-D)}{(1-2D)} \dfrac{2V_f}{V_o} + \dfrac{(1-D)}{(1-2D)^2} \dfrac{2R_F}{R_L} + F_s C_o R_L + \left(\dfrac{1}{1-2D}\right)^2 \dfrac{r_L}{R_L} + \dfrac{4D(1-D)}{(1-2D)^2} \dfrac{r_c}{R_L} \right\} P_o}$ $M_{VDS} = \dfrac{(1-D)}{(1-2D)\left\{ \dfrac{2Dr_{Ds}}{(1-2D)^2 R_L} + \dfrac{(1-D)}{(1-2D)} \dfrac{2V_f}{V_o} + \dfrac{(1-D)}{(1-2D)^2} \dfrac{2R_F}{R_L} + F_s C_o R_L + \left(\dfrac{1}{1-2D}\right)^2 \dfrac{r_L}{R_L} + \dfrac{4D(1-D)}{(1-2D)^2} \dfrac{r_c}{R_L} \right\} P_o}$

By using the above-derived formulas, the efficiencies and non-ideal voltage conversion ratios of each topology with respect to gain are presented in Figures 37 and 38, respectively. From these results, it can be observed that the efficiency is higher in conventional BDC and LC-BDC compared to other existing topologies. Moreover, it can be seen that the voltage conversion ratio is more linear in the case of conventional BDC, and the proposed converter compared to other existing topologies. In all existing topologies (except conventional BDC), it can be noted that the performance of the converter is becoming poor as the gain is increased further from the designed gain value.

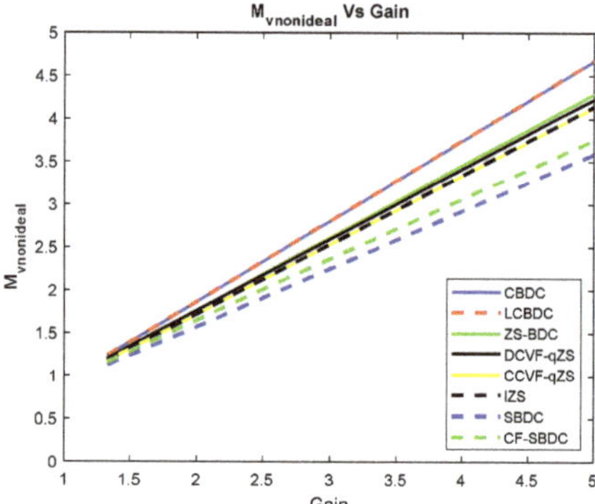

Figure 38. Voltage conversion ratio comparison of various topologies.

5. Conclusions

In this paper, a new LC bi-directional DC–DC converter utilizing small passive components has been proposed and successfully validated. Experimental results proved that there is a reduction of 75% and 35.8% in capacitor voltage for 24 V to 18 V conversion in boost mode, and 18 V to 24 V conversion in buck mode, respectively. Moreover, there is a reduction of one capacitor compared to conventional BDC for the same conversion. In this paper, the proposed converter performance in both transient and steady-state conditions is investigated and presented. This investigation reveals that the proposed converter is able to offer superior performance in both transient and steady-state conditions. Moreover, a comparative analysis of the proposed converter with the conventional BDC, Z-source converter, discontinues current quasi Z-source converter, continues current quasi Z-source converter, improved Z-source converter, switched boost converter, current-fed switched boost converter, and quasi switched boost converter is presented. This comparative analysis proved that the proposed converter offers superior performance compared to existing converters for the same conversion ratio.

Author Contributions: Idea, Conceptualization, Formal Analysis, Investigation, Methodology, Software, Validation, and Writing—Original Draft are the main contributions of D.R.; Funding acquisition, Project administration, Resources, Validation, Visualization and Writing—review & editing are the main contributions of R.D.; H.R. and S.M. contributed in terms of Validation and Writing—review & editing. All authors have read and agreed to the published version of the manuscript.

Funding: This work was supported by the Petrofac Research Chair in Renewable Energy Endowment Fund at the American University of Sharjah.

Acknowledgments: The authors are particularly grateful to the Zunik Energies Pvt. Ltd. and authors of the Indian Patent application published on 30 December 2016 (application number: 201641038705) for providing approval to use the patent data for this research work and publications.

Conflicts of Interest: The authors declare no conflict of interest. The funders had no role in the design of the study; in the collection, analyses, or interpretation of data; in the writing of the manuscript, or in the decision to publish the results.

References

1. Onar, O.C.; Kobayashi, J.; Erb, D.C.; Khaligh, A. A bidirectional high-power-quality grid interface with a novel bidirectional non-inverted buck-boost converter for PHEVs. *IEEE Trans. Veh. Technol.* **2012**, *61*, 2018–2032. [CrossRef]
2. Zhang, Z.; Chau, K.-T. Pulse-width-modulation-based electromagnetic interference mitigation of bidirectional grid-connected converters for electric vehicles. *IEEE Trans. Smart Grid* **2017**, *8*, 2803–2812. [CrossRef]
3. Naden, M.; Bax, R. Generator with DC Boost and Split Bus Bidirectional DC-to-DC Converter for Uninterruptible Power Supply System or for Enhanced Load Pickup. U.S. Patent 7,786,616, 7 February 2003.
4. Naayagi, R.T.; Forsyth, A.J.; Shuttleworth, R. High-power bidirectional DC-DC converter for aerospace applications. *IEEE Trans. Power Electron.* **2012**, *27*, 4366–4379. [CrossRef]
5. Forouzesh, M.; Siwakoti, Y.P.; Gorji, S.A.; Blaabjerg, F.; Lehman, B. Step-up DCDC converters: A comprehensive review of voltage boosting techniques, topologies, and applications. *IEEE Trans. Power Electron.* **2017**, *32*, 9143–9178. [CrossRef]
6. Elasser, A.; Chow, T.P. Silicon carbide benefits and advantages for power electronics circuits and systems. *Proc. IEEE* **2002**, *90*, 969–986. [CrossRef]
7. Wai, R.-J.; Liu, L.-W.; Duan, R.-Y. High-efficiency Voltage-clamped DC-DC converter with reduced reverse-recovery current and Switch-Voltage stress. *IEEE Trans. Ind. Electron.* **2006**, *53*, 272–280. [CrossRef]
8. Calderon-Lopez, G.; Scoltock, J.; Wang, Y.; Laird, I.; Yuan, X.; Forsyth, A.J. Power-Dense Bi-Directional DC–DC Converters with High-Performance Inductors. *IEEE Trans. Veh. Technol.* **2019**, *68*, 11439–11448. [CrossRef]
9. Zhang, S.; Yu, X. A Unified analytical modeling of the interleaved Pulse Width Modulation (PWM) DC–DC converter and its applications. *IEEE Trans. Power Electron.* **2013**, *28*, 5147–5158. [CrossRef]
10. Ahmadi, M.; Mohammadi, M.R.; Adib, E.; Farzanehfard, H. Family of non-isolated zero current transition bi-directional converters with one auxiliary switch. *IET Power Electron.* **2012**, *5*, 158–165. [CrossRef]
11. Chen, X.; Pise, A.A.; Elmes, J.; Batarseh, I. Ultra-Highly efficient low-power bidirectional cascaded buck-boost converter for portable PV-Battery-Devices Applications. *IEEE Trans. Ind. Appl.* **2019**, *55*, 3989–4000. [CrossRef]
12. Narashimharaju, B.L.; Dubey, S.P.; Singh, S.P. Design and analysis of coupled inductor bidirectional DC-DC converter for high-voltage diversity applications. *IET Power Electron.* **2012**, *5*, 998–1007. [CrossRef]
13. Fang, X.; Ji, X. Bidirectional power flow Z-Source DC-DC converter. In Proceedings of the IEEE Vehicle Power and Propulsion Conference (VPPC), Harbin, China, 3–5 September 2008.
14. Siwakoti, Y.P.; Blaabjerg, F.; Loh, P.C.; Town, G.E. High-voltage boost quasi-Z-source isolated dc/dc converter. *IET Power Electron.* **2014**, *7*, 2387–2395. [CrossRef]
15. Vinnikov, D.; Roasto, I. Quasi-Z-source-based isolated DC/DC converters for distributed power generation. *IEEE Trans. Ind. Electron.* **2011**, *58*, 192–201. [CrossRef]
16. Zhang, F.; Peng, F.Z.; Qian, Z. Z-H converter. In Proceedings of the 2008 IEEE Power Electronics Specialists Conference (PESC), Rhodes, Greece, 15–19 June 2008; pp. 1004–1007.
17. Tang, Y.; Xie, S.; Zhang, C.; Xu, Z. Improved Z-Source inverter with reduced Z-source capacitor voltage stress and soft-start capability. *IEEE Trans. Power Electron.* **2009**, *24*, 409–415. [CrossRef]
18. Adda, R.; Ray, O.; Mishra, S.K.; Joshi, A. Synchronous-Reference-Frame-Based control of switched boost inverter for standalone DC nanogrid applications. *IEEE Trans. Power Electron.* **2013**, *28*, 1219–1233. [CrossRef]
19. Siwakoti, Y.P.; Blaabjerg, F.; Loh, P.C. Z-source converters. In *Power Electronic Converters and Systems: Frontiers and Applications*; Institution of Engineering and Technology: London, UK, 2016; pp. 205–243.
20. Siwakoti, Y.; Forouzesh, M.; Pham, H. Power Electronics Converters—An overview. In *Control of Power Electronic Converters and Systems*; Academic Press: Cambridge, MA, USA, 2018.
21. Zeng, Z.; Wang, X.; Wei, Y.; Yu, Y. Research of bi-directional DC/DC converter topology based on supercapacitor energy storage system in IP transmitter. *J. Eng.* **2019**, *2019*, 1962–1967. [CrossRef]
22. Li, W.; He, X. Review of non-isolated high-step-up dc/dc converters in photovoltaic grid-connected applications. *IEEE Trans. Ind. Electron.* **2011**, *58*, 1239–1250. [CrossRef]
23. Cao, D.; Peng, F.Z. A family of Z-source and quasi-Z-source dc-dc converters. In Proceedings of the Twenty-Eighth Annual IEEE Applied Power Electronics Conference and Exposition (APEC), Long Beach, CA, USA, 20–24 September 2013; pp. 1097–1101.

24. Mehrnami, S.; Mazumder, S.K.; Soni, H. Modulation scheme for three-phase differential-mode Ćuk inverter. *IEEE Trans. Power Electron.* **2016**, *31*, 2654–2668. [CrossRef]

25. Meraj, M.; Iqbal, A.; Al-emadi, N.; Rahman, S.; Bhaskar, M.S. Novel PWM Technique for Quasi Switched Boost Converter for the Nano-grid Applications. In Proceedings of the 2019 IEEE 28th International Symposium on Industrial Electronics (ISIE), Vancouver, BC, Canada, 12–14 June 2019; pp. 2659–2664.

26. Nguyen, M.K.; Phan, Q.D.; Nguyen, V.N.; Lim, Y.C.; Park, J.K. Trans-Z-source-based isolated DC-DC converters. In Proceedings of the 2012 IEEE International Symposium on Industrial Electronics (ISIE), Taipei, Taiwan, 28–31 May 2013; pp. 1–6.

27. Lee, H.; Kim, H.G.; Cha, H. Parallel operation of trans-Z-source network full bridge dc/dc converter for wide input voltage range. In Proceedings of the International Conference on Power Electronics and ECCE Asia (ICPE-ECCE Asia), Aalborg, Denmark, 25–28 June 2012; pp. 1707–1712.

28. Flicker, J.; Kaplar, R.; Marinella, M.; Granata, J. PV inverter performance and reliability: What is the role of the bus capacitor? In Proceedings of the Photovoltaic Specialists Conference (PVSC), Austin, TX, USA, 3–8 June 2012; pp. 1–3.

29. Ma, Z.J.; Thomas, S. Reliability and maintainability in photovoltaic inverter design. In Proceedings of the Reliability and Maintainability Symposium (RAMS), 2011 Proceedings-Annual, Lake Buena Vista, FL, USA, 24–27 January 2011; pp. 1–5.

30. Wang, H.; Blaabjerg, F. Reliability of capacitors for DC-Link applications in power electronic Converters—An overview. *Ind. Appl. IEEE Trans.* **2014**, *50*, 3569–3578. [CrossRef]

31. Ravi Kumar, R.D.; Nagesh Kumar, K.S.; Sastry Vedula, V.V. Bidirectional DC-DC converters. *Pat. Off. J.* **2016**, *54*, 51–52.

Article

Nonlinear Voltage Control for Three-Phase DC-AC Converters in Hybrid Systems: An Application of the PI-PBC Method

Federico M. Serra [1,*], Lucas M. Fernández [1], Oscar D. Montoya [2,3], Walter Gil-González [3] and Jesus C. Hernández [4]

1 Automatic Control Laboratory (LCA), Facultad de Ingeniería y Ciencias Agropecuarias, Universidad Nacional de San Luis, Villa Mercedes, San Luis 5730, Argentina; lucasfern@unsl.edu.ar
2 Facultad de Ingeniería, Universidad Distrital Francisco José de Caldas, Carrera 7 No. 40B-53, Bogotá D.C. 11021, Colombia; o.d.montoyagiraldo@ieee.org
3 Laboratorio Inteligente de Energía, Universidad Tecnológica de Bolívar, km 1 vía Turbaco, Cartagena 131001, Colombia; wjgil@utp.edu.co
4 Department of Electrical Engineering, University of Jaén, Campus Lagunillas s/n, Edificio A3, 23071 Jaén, Spain; jcasa@ujaen.es
* Correspondence: fmserra@unsl.edu.ar

Received: 17 April 2020; Accepted: 18 May 2020; Published: 20 May 2020

Abstract: In this paper, a proportional-integral passivity-based controller (PI-PBC) is proposed to regulate the amplitude and frequency of the three-phase output voltage in a direct-current alternating-current (DC-AC) converter with an LC filter. This converter is used to supply energy to AC loads in hybrid renewable based systems. The proposed strategy uses the well-known proportional-integral (PI) actions and guarantees the stability of the system by means of the Lyapunov theory. The proposed controller continues to maintain the simplicity and robustness of the PI controls using the Hamiltonian representation of the system, thereby ensuring stability and producing improvements in the performance. The performance of the proposed controller was validated based on simulation and experimental results after considering parametric variations and comparing them with classical approaches.

Keywords: hybrid system; voltage source converter; passivity-based control; proportional-integral control; voltage regulation

---

1. Introduction

1.1. General Context

Hybrid Renewable Based Systems (HRS) are promising alternatives for electricity supply in remote areas and are also known as stand-alone microgrid systems [1]. The objective of the stand-alone microgrids is to provide energies based on green technologies to people in remote areas, permitting them to augment their productive capabilities and enhance their quality of life [2]. This is possible to implement, thanks to the advances in renewable energy technologies that have allowed the installation of power generations in remote areas, which in turn benefit and cover non-interconnected areas.

Stand-alone microgrid systems can include different types of energy sources (photovoltaic and wind) [3], storage systems (battery banks and supercapacitors) [4], and loads. These elements can be connected through alternating current (AC) or direct current (DC) grids. In Reference [5], a comprehensive review of AC and DC microgrids was presented. DC grids are preferred because they have a higher power density than AC grids; in addition, they do not require synchronization and incur only minor losses due to the skin effect [6]. Figure 1 shows a representative scheme for an HRS where the elements are connected through a DC grid.

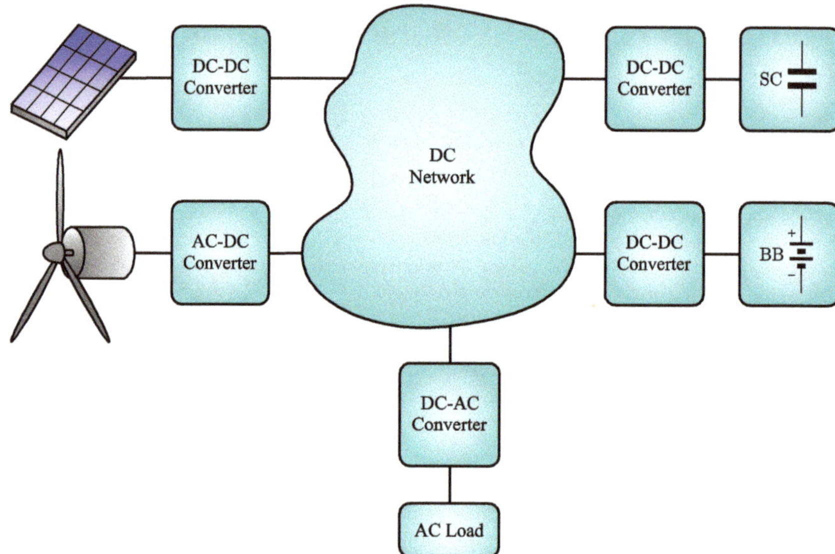

Figure 1. Renewable-based hybrid DC system.

Note that the DC network concept comprehends an extensive range of applications, from high-voltage (examples are given in [7,8]) to low-voltage levels (examples are provided in [9,10]).

In addition, DC networks are especially attractive in control applications since the droop controls of reactive power and frequency disappear in these networks, making it easier for power flow control through lines in high-voltage levels or voltage regulation in low-voltage usages [11]. Another important aspect of the DC network is the possibility of providing service to rural or remote areas with renewable source and energy storage devices, as depicted in Figure 1; this helps improve the living conditions in those areas. In Reference [10], a nonlinear controller for a typical configuration of a rural microgrid was presented and in [12] a comparison between DC and AC microgrids implementations for rural social-economic development was performed.

1.2. Motivation

The interconnection of each part of the system with the DC grid is achieved using power electronics converters, which are responsible for managing the power flow among the sources, storage systems, and loads. In Reference [13,14], examples of the use of the power electronics converters for such interconnections were described. The objective of the power flow control in a hybrid system is to satisfy the energy demand on the loads, maximize the energy extracted from renewable energy sources, and use storage systems efficiently.

The converter is entrusted with controlling the AC voltage applied to the loads, which is usually a DC-AC converter with an LC output filter [15]. This converter can be single-phase or three-phase, depending on the load type. Meanwhile, its control regulates the amplitude and frequency of the output voltage based on a DC voltage applied on the input, which can then be controlled for the remaining HRS [16–18].

The control strategy proposed in this research is motivated by the necessity to have robust and stable control methods for providing sinusoidal voltages in remote areas where conventional power systems are nonexistent [19]. This entails that the opportunity to provide electrical service is by interfacing renewable energy resources (mainly wind turbines and photovoltaic plants) with power electronic converters that can regulate voltage and frequency by tracking sinusoidal references [20]. The approach that uses sinusoidal references is different from a conventional emulation of synchronous generators via virtual inertia control [21] since it is recommended for weak grids with large frequency variations. In these control schemes, active and reactive power measurements are used to define the frequency and voltage references [20]. Nevertheless, the proposed controller in this paper is focusing on supplying electrical service to linear and non-linear loads directly interfaced with VSCs, which implies that the measures of active and reactive power are not efficient in regulating the output voltage. For this reason, our aim is to have a direct voltage control strategy based on trajectory tracking via passivity-based control approach with experimental validations, allowing supporting three-phase balanced voltages in passive and switched loads.

1.3. Brief State-of-the-Art

In specialized literature, several strategies have been developed for the control of DC-AC converters. Due to their simplicity, the most widely used approaches are based on classic linear controllers, which are proportional-integral (PI) controllers [22]. Even though these strategies are the most used, they cannot guarantee the stability of the system. Additionally, they do not perform well away from the point of operation as in the case of non-linear loads. [16]. Therefore, advanced strategies have been developed to address the poor performance of classic controllers. In Reference [23], a feedback linearization control method was proposed based on a power-balance model between the converter and the load. This method improves the performance of linear and nonlinear loads, but the selection of gains is critical. In Reference [24], a current control algorithm for uninterruptible power supplies based on PID compensator was presented. [25] showed the reduction of voltage distortion caused as a result of slowly varying harmonic currents that use synchronous-frame harmonic regulators.Reference [26] describes an integral resonant controller of the output voltage management arrangement in a three-phase VSI. Reference [27] presented a model predictive control for output voltage regulation of a three-phase inverter with output LC filter feeding linear and nonlinear loads. In addition, authors in [28] use the same control strategy for a single-phase voltage source with linear and nonlinear loads. Despite previous works demonstrating good performance of their objective controls, none of them can guarantee the stability of the system.

On the other hand, the application of passivity-based control (PBC) techniques to power converters has the advantage of providing stable closed-loop controllers with good dynamic behavior. In Reference [16], an interconnection and damping assignment (IDA-PBC) approach was proposed to regulate the output voltage from a DC-AC converter and a comparison was also made with classic controllers. The results of [16] demonstrated the good performance of the proposed controller, even when a nonlinear load was considered. An adaptive robust control method for a DC-AC converter with high dynamic performance under nonlinear and unbalanced loads was also proposed by [29]. In both of these methods, the stability is ensured by the passive properties of the controlled system [30]. The problem with the PBC controllers applied to power converters is that the control laws depend on the system parameters and, so, stable-state errors occur when these parameters vary. The errors caused by variations in the system parameters can be eliminated with different techniques; for e.g., a dynamic extension with an integral action was proposed by [16] but at the expense of increasing the complexity of the system.

PI-PBC controllers have been proposed to combine the advantages (simplicity and robustness) of PI-based designs with the typical stability analysis based on the Lyapunov theorem employed in passive strategies. These controllers have been used in power converters for several applications [31–34].

Authors in [31] have presented the general basis of the PI-PBC theory applied to power electronic converters (switched systems). These authors demonstrate that with PI gains in a Hamiltonian representation of the averaged dynamic of the converter is possible to provide constant direct current–voltage to linear loads. Simulation and experimental results demonstrated that when VSCs are used in conversion mode (sinusoidal input to DC constant output), the PI-PBC method guarantees asymptotic stability if the load is completely linear (i.e., resistive). Observe that the VSC was operated with sinusoidal voltage imposed on the AC side to generate constant DC voltage. This is a different case, compared to the approach presented in this paper as we work with constant DC voltage provided by a combination of batteries and renewables to support three-phase balanced voltage in linear and nonlinear loads, guaranteeing stability conditions in the sense of Lyapunov. In reference [32], a general design using the PI-PBC method was presented for tracking trajectories in power electronic converters (sinusoidal or constant references) if they are bounded and differentiable (i.e., admissible trajectories). The stability in closed-loop is ensured via Barbalat's lemma. The authors of this paper validate their control design in an interleaved boost and the modular multilevel converter, including simulation and experimental validations. Note that the first converter works with AC input to provide a constant DC output, while the second one generates single-phase voltages in linear loads considering a constant DC input. This implies that the application of the developed PI-PBC method is different from our approach since we work with the isolated network applications to generate three-phase voltage signals in linear and nonlinear loads. The authors in [33] presented a methodology based on dynamic power compensation of active and reactive power in transmission systems considering superconducting coils integrated via a cascade connection between DC-DC chopper converter and the VSC. The control for this system is developed with PI-PBC, guaranteeing stability in closed-loop. The main aim of this paper is to compensate subsynchronous oscillations in power systems when faults occur in the power grid. Note that the proposal of these authors works with the VSC connected to the grid by controlling active and reactive power flow; while in our approach, the power grid is non-existent and the objective is to provide voltage service to isolated loads, i.e., we generate the power system node with constant voltage and frequency via PI-PBC design. In [34], standard passivity-based control design for integrating renewable energy resources in power systems was developed. The main idea of [34] is to provide a stable control design via PBC, which is made via energy functions using a Lagrangian formulation. Additionally, it is assumed that the wind generator would be connected to the power grid. This implies that the electrical network supports the voltage on the AC side of the converter. For this reason, the authors of this study focused on active and reactive power control and not on the three-voltage generation for isolated power applications as the case studied in our contribution.

In Reference [35], a general control design of controllers for single-phase network applications was presented via interconnection and damping assignment PBC and PI-PBC approaches. In this work, the authors considered isolated power grids composed of batteries, wind turbines, photovoltaic plants, and energy storage devices composed of superconducting coils and supercapacitors. The main contribution in [35] was to demonstrate stability in single-phase networks under well-defined load conditions. Even if this research uses isolated systems by applying PI-PBC control, it is different from our contribution since, in our work, the grid has a three-phase structure and the loads are strong, nonlinear loads (switched devices), which were not considered in [35]. Note that in [36] the initial design based on PI-PBC and IDA-PBC was complemented with modifications on the controller structure to integrate renewables in single-phase networks. In addition, the difference with our approach is that the authors do not present any experimental test that validates their simulation analysis.

Authors in Reference [37] presented a general stability analysis for single-phase networks feed-through power electronic converters considering constant power load. This analysis was performed assuming a Hamiltonian representation of the system and the perfect operation of the controllers that manage the power flow between the distributed energy resources and the grid. The authors of this work do not mention how this approach is extensible to AC grids with strong nonlinear loads as the case study in our proposal.

Even if controllers based on PI-PBC have been proposed for controlling power, electronic converters in single-phase and three-phase applications. In this paper, we focus on the problem of the voltage generation in three-phase nonlinear loads located in isolated areas by deriving the

PI-PBC approach from the classical IDA-PBC method [16], which has not been reported in the scientific literature yet. In addition, our work contains multiple simulation scenarios and some experimental validations that validate the proposed approach, demonstrating its easy implementation in real-life operative cases that combine renewables, batteries, power converters, and nonlinear loads.

It is important to mention that it is necessary to employ optimal tuning of the PI gains so that PI controllers (including classical PI and PI-PBC approaches) perform excellently [38]. Active/passive tuning methods have been reported in the scientific literature. In Reference [39], it was presented an interactive tool for adjusting PI controls in first-order systems from a graphical point of view for first-order systems with time delays, numerical results confirm the efficiency of the tool developed in comparison with other literature reports. In Reference [40], an algorithm for the PID controller based on the gain margin and phase margin concept was presented. However, the controller parameters depend on a single parameter, these parameters are subjected to the desired phase margin, and a minimum required gain margin constraint. The main advantage of these tuning an approach with respect to previous works is that it is easy to implement applicable to any linear as well non-linear model structures. Authors of [41] have presented a simple method to design PI controllers in the frequency domain by proposing an optimization model with constraints. This method uses a single tuning parameter, defined as the quotient between the final crossover frequency and the zero of the controller. This adjusting procedure maximizes the controller gain by considering the equality constraint on the phase margin and an inequality restriction in the gain margin. Numerical results confirm the effectiveness of this proposal in comparison with literature reports. Additional methods for tuning PI controllers have been reported in specialized literature, some of them are particle swarm optimization [42], ant-lion optimizer [43], genetic algorithms [44], and so on. The main feature of these metaheuristic optimization methods is that they work with the minimization of integral indices to find the optimal set of control gains by using sequential programming methods [45].

Remark 1. *The selection of control gains is an important task in the design of PI controllers in power converter applications. These methods can be passive or active approaches that work with optimization models or desired performances [46]. Nevertheless, in this research, our focus is on presenting a simple controller based on the properties of the passivity theory combined with classical and well-known PI actions to generate ideal three-phase voltages for non-linear loads in isolated areas. This implies that the focus in the grid performance with load variations and no optimal adjusting of the control gains. In this sense, we employ a basic tuning method based on the root locus design approach [47].*

1.4. Contribution and Scope

In the present study, a PI-PBC controller is proposed for regulating the amplitude and frequency of the output voltage in a three-phase DC-AC converter with an LC filter, providing a well-defined sinusoidal service to linear and nonlinear loads by transforming the DC signal from the transmission/distribution network to local loads [48].

The main contribution of this research in the literature reports about the control of VSCs for feeding isolated three-phase loads can be summarized as follows:

✓   A passivity-based control design that is easily implementable with the main advantages of the classical PI controllers that allows tracking a sinusoidal trajectory by transforming this into a regulation problem. The proposed PI-PBC design also allows guaranteeing stability conditions based on the Lyapunov theory by applying the properties of the Hamiltonian energy models.

✓   The proposed controller can maintain objective controls, which are to regulate constant voltage amplitude and constant frequency although the test system feeds a non-linear load, demonstrating the generation of a robust three-phase balanced signal. This is achieved by avoiding the use of classical phase-locked loops embedded in virtual synchronous emulations that emulates inertia properties in converters.

✓ The experimental validation in a laboratory prototype with a realistic model of the system include switching effects, losses, and a detailed transistor model to feed passive loads and nonlinear ones.

In addition, the performance of the controller under parametric variation is shown, and a comparison with the classic PI controller demonstrates the superiority of the proposal to mitigate the harmonic content produced by non-linear loads.

Regarding the scope of this research, it is important to mention that in the control design as well as in the simulation, experimental validations will be considered unique voltage source converters that are forced to work as an ideal voltage source to provide sinusoidal voltages to linear and nonlinear loads. For doing so, we consider that the DC side of the converter is fed by a strong DC network (transmission/distribution DC grid) or by a combination of renewable energy resources and batteries [19,28]. In addition, to determine the amount of instantaneous power absorbed by the load (linear and nonlinear), it is considered that there is a current measure at the load side which is important since the amount of current provided by the converter is a linear function of the load consumption. This implies that the existence of this measure is indispensable when no load estimators are implemented, as in the case studied in this research. Note that the implementation of load estimators could be considered for future work since only a few studies have been reported in the scientific literature with experimental validations.

1.5. Organization of the Document

The remainder of this paper is organized as follows. In Section 2, we describe the configuration of the DC-AC converter and its dynamical modeling using a Hamiltonian representation. In Section 3, we explain the proposed control design based on the PI-PBC approach, which guarantees asymptotic stability according to Lyapunov. In Section 4, we demonstrate the numerical performance of the proposed method based on simulations and experimental validations using a laboratory prototype. Finally, the main conclusions derived based on this study are presented in Section 5.

2. System Configuration and Dynamical Model

The DC-AC converter comprises of a three-phase voltage source converter with IGBTs ($S_1, \ldots, S_6$) and an LC output filter. Figure 2 shows the structure of the DC-AC converter considered in this study for supplying an arbitrary load (i.e., linear or nonlinear consumption) [49].

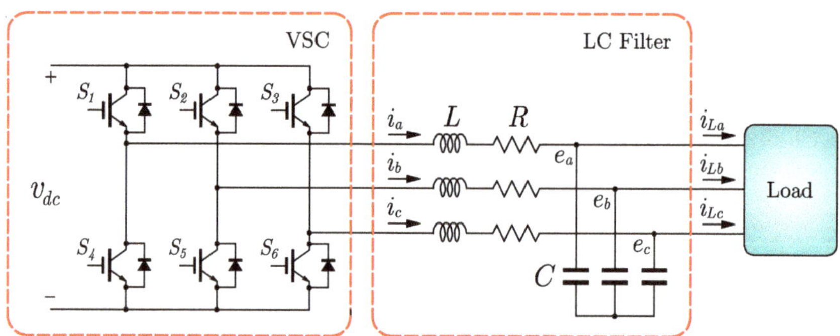

Figure 2. Three-phase DC-AC converter structure.

The DC-link voltage, v_{dc}, is considered to be approximately constant and controlled by other converters involved in the HRS shown in Figure 1, and thus, its dynamics are not considered in our model [16,49].

Dynamical Model

The DC-AC converter model in dq coordinates is as follows [3],

$$L\dot{i}_d = m_d v_{dc} - Ri_d - \omega_{dq} Li_q - e_d, \tag{1}$$

$$L\dot{i}_q = m_q v_{dc} - Ri_q + \omega_{dq} Li_d - e_q, \tag{2}$$

$$C\dot{e}_d = i_d - \omega_{dq} Ce_q - i_{Ld}, \tag{3}$$

$$C\dot{e}_q = i_q + \omega_{dq} Ce_d - i_{Lq}, \tag{4}$$

where ω_{dq} is the angular frequency of the dq reference frame, which is set equal to the desired output-voltage frequency; i_d and i_q represent the currents in the dq frame; e_d and e_q are the output voltages; i_{Ld} and i_{Lq} correspond to the load currents; and m_d and m_q are the inverter modulation indexes. All the variables represented in the dq reference frame are obtained through Park's transformation from the abc variables. The parameters L, C, and R represent the inductance, capacitance of the output filter, and equivalent output resistance, which model the filter inductance losses and converter losses, respectively.

The port-Hamiltonian (pH) model of the system can be written as follows:

$$\dot{\mathbf{x}} = [\mathbf{J} - \mathbf{R}] \frac{\partial H(\mathbf{x})}{\partial \mathbf{x}} + \mathbf{g}\mathbf{u} + \zeta, \tag{5}$$

where the state vector is $\mathbf{x}$, $\mathbf{J}$, and $\mathbf{R}$ are the interconnection and damping matrices, respectively, $H(\mathbf{x})$ represents the total energy stored in the system, $\mathbf{g}$ is the input matrix, $\mathbf{u}$ is the control input vector, and ζ represents the external input.

The pH model of the DC-AC converter is represented by Equation (6).

$$
\begin{bmatrix} L\dot{i}_d \\ L\dot{i}_q \\ C\dot{e}_d \\ C\dot{e}_q \end{bmatrix} =
\begin{bmatrix} -R & -\omega_{dq}L & -1 & 0 \\ \omega_{dq}L & -R & 0 & -1 \\ 1 & 0 & 0 & -\omega_{dq}C \\ 0 & 1 & \omega_{dq}C & 0 \end{bmatrix}
\begin{bmatrix} i_d \\ i_q \\ e_d \\ e_q \end{bmatrix} +
\begin{bmatrix} v_{dc} & 0 \\ 0 & v_{dc} \\ 0 & 0 \\ 0 & 0 \end{bmatrix}
\begin{bmatrix} m_d \\ m_q \end{bmatrix} +
\begin{bmatrix} 0 \\ 0 \\ -i_{Ld} \\ -i_{Lq} \end{bmatrix}. \tag{6}
$$

In this case, the interconnection and damping matrices are defined based on Equation (6) as

$$
\mathbf{J} = \begin{bmatrix} 0 & -\omega_{dq}L & -1 & 0 \\ \omega_{dq}L & 0 & 0 & -1 \\ 1 & 0 & 0 & -\omega_{dq}C \\ 0 & 1 & \omega_{dq}C & 0 \end{bmatrix}, \tag{7}
$$

$$
\mathbf{R} = \begin{bmatrix} R & 0 & 0 & 0 \\ 0 & R & 0 & 0 \\ 0 & 0 & 0 & 0 \\ 0 & 0 & 0 & 0 \end{bmatrix}, \tag{8}
$$

where $\mathbf{J} = -\mathbf{J}^T$ is antisymmetric and $\mathbf{R} = \mathbf{R}^T \geq 0$ is symmetric positive semidefinite.

The total energy stored in the system, $H(\mathbf{x})$, is shown by the sum of the energy stored on the output-filter inductors and capacitors,

$$H(\mathbf{x}) = \frac{1}{2} \left(Li_d^2 + Li_q^2 + Ce_d^2 + Ce_q^2 \right). \tag{9}$$

It should be noted that $H(\mathbf{x})$ is a hyperboloidal function that is convex and positive definite.

3. PI-PBC Approach

This section presents three main aspects of power electronic converters' control using passivity-based control theory. (1) The design of the proposed PI-PBC approach by transforming the trajectory tracking problem into a regulation problem using the incremental representation. (2) The control objective of the problem, i.e., the definition of the desired sinusoidal trajectory. (3) The stability analysis of the proposed control scheme via Lyapunov's stability theorem for autonomous dynamical systems. In the next subsections, each of these aspects will be discussed.

3.1. Control Design

The proposed PI-PBC approach depends on the existence of an admissible trajectory, defined as [34],

$$\mathbf{x}^* = \begin{bmatrix} Li_d^* & Li_q^* & Ce_d^* & Ce_q^* \end{bmatrix}^T, \tag{10}$$

such that the dynamical system in Equation (6) is continuous, differentiable, and bounded, which implies that

$$\dot{\mathbf{x}}^* = [\mathbf{J} - \mathbf{R}]\frac{\partial H(\mathbf{x}^*)}{\partial \mathbf{x}^*} + \mathbf{g}\mathbf{u}^* + \zeta, \tag{11}$$

with some $\mathbf{u}^*$ bounded.

If we define $\tilde{\mathbf{x}} = \mathbf{x} - \mathbf{x}^*$ and $\tilde{\mathbf{u}} = \mathbf{u} - \mathbf{u}^*$, the system in Equation (5) can be written as follows:

$$\dot{\tilde{\mathbf{x}}} + \dot{\mathbf{x}}^* = [\mathbf{J} - \mathbf{R}]\frac{\partial H(\tilde{\mathbf{x}} + \mathbf{x}^*)}{\partial(\tilde{\mathbf{x}} + \mathbf{x}^*)} + \mathbf{g}(\tilde{\mathbf{u}} + \mathbf{u}^*) + \zeta. \tag{12}$$

The energy function of the system given by Equation (9) can be represented as:

$$H(\mathbf{x}) = H(\tilde{\mathbf{x}} + \mathbf{x}^*) = \frac{1}{2}\mathbf{x}^T\mathbf{P}^{-1}\mathbf{x} = \frac{1}{2}(\tilde{\mathbf{x}} + \mathbf{x}^*)^T\mathbf{P}^{-1}(\tilde{\mathbf{x}} + \mathbf{x}^*), \tag{13}$$

and the gradient of this function is,

$$\frac{\partial H(\mathbf{x})}{\partial \mathbf{x}} = \mathbf{P}^{-1}\mathbf{x}, \tag{14}$$

then,

$$\begin{aligned}\frac{\partial H(\tilde{\mathbf{x}} + \mathbf{x}^*)}{\partial(\tilde{\mathbf{x}} + \mathbf{x}^*)} &= \mathbf{P}^{-1}(\tilde{\mathbf{x}} + \mathbf{x}^*) = \mathbf{P}^{-1}\tilde{\mathbf{x}} + \mathbf{P}^{-1}\mathbf{x}^* \\ &= \frac{\partial H(\tilde{\mathbf{x}})}{\partial\tilde{\mathbf{x}}} + \frac{\partial H(\mathbf{x}^*)}{\partial \mathbf{x}^*}.\end{aligned} \tag{15}$$

Using Equation (15), the system defined by Equation (12) can be written as follows:

$$\dot{\tilde{\mathbf{x}}} + \dot{\mathbf{x}}^* = [\mathbf{J} - \mathbf{R}]\frac{\partial H(\tilde{\mathbf{x}})}{\partial\tilde{\mathbf{x}}} + [\mathbf{J} - \mathbf{R}]\frac{\partial H(\mathbf{x}^*)}{\partial \mathbf{x}^*} + \mathbf{g}\tilde{\mathbf{u}} + \mathbf{g}\mathbf{u}^* + \zeta, \tag{16}$$

Thus, the following error dynamics are reached.

$$\dot{\tilde{\mathbf{x}}} = [\mathbf{J} - \mathbf{R}]\frac{\partial H(\tilde{\mathbf{x}})}{\partial\tilde{\mathbf{x}}} + \mathbf{g}\tilde{\mathbf{u}}. \tag{17}$$

In addition, the output of the system can be calculated as follows:

$$\tilde{\mathbf{y}} = \mathbf{g}^T\frac{\partial H(\tilde{\mathbf{x}})}{\partial(\tilde{\mathbf{x}})}. \tag{18}$$

The system in Equation (17) is passive if $\dot{H}(\tilde{\mathbf{x}}) \leq \tilde{\mathbf{y}}^T \tilde{\mathbf{u}}$ [33,34]. Therefore, to obtain a closed-loop dynamic based on Lyapunov theory, the energy function is shown as:

$$H(\tilde{\mathbf{x}}) = \frac{1}{2}\tilde{\mathbf{x}}^T \mathbf{P}^{-1}\mathbf{x}, \tag{19}$$

where $H(0) = 0$ and $H(\tilde{\mathbf{x}}) > 0, \forall \tilde{\mathbf{x}} \neq 0$, which satisfy the first and second conditions of the Lyapunov theorem. The time derivative of the proposed energy function is

$$\begin{aligned} \dot{H}(\tilde{\mathbf{x}}) &= \tilde{\mathbf{x}}^T \mathbf{P}^{-1}\dot{\tilde{\mathbf{x}}} = -\tilde{\mathbf{x}}^T \mathbf{P}^{-1}\mathbf{R}\mathbf{P}^{-1}\tilde{\mathbf{x}} + \tilde{\mathbf{x}}^T \mathbf{P}^{-1}\mathbf{g}\tilde{\mathbf{u}}, \\ &= -\tilde{\mathbf{x}}^T \mathbf{P}^{-1}\mathbf{R}\mathbf{P}^{-1}\tilde{\mathbf{x}} + \tilde{\mathbf{y}}^T \tilde{\mathbf{u}} \leq \tilde{\mathbf{y}}^T \tilde{\mathbf{u}}, \end{aligned} \tag{20}$$

which proves that the dynamics of the error are also passive.

The proposed controller can be written as follows:

$$\begin{aligned} \dot{\mathbf{z}} &= -\tilde{\mathbf{y}}, &&(21) \\ \tilde{\mathbf{u}} &= -K_P \tilde{\mathbf{y}} + K_I \mathbf{z}, &&(22) \end{aligned}$$

and it has a PI structure, where K_P and K_I are the proportional and integral gain matrices, respectively, which are diagonal and positive definite, and $\tilde{\mathbf{y}}$ is

$$\tilde{\mathbf{y}} = \mathbf{g}^T \frac{\partial H(\tilde{\mathbf{x}})}{\partial \tilde{\mathbf{x}}} = \begin{bmatrix} v_{dc}\left(i_d - i_d^*\right) \\ v_{dc}\left(i_q - i_q^*\right) \end{bmatrix}. \tag{23}$$

Note that $\mathbf{z}$ represents a set of auxiliary variables that allow the passive output feedback and the inclusion of an integral action to minimize steady-state errors on the control objective. For more details of the PI-PBC approach for non-affine dynamic systems, refer to [32].

Using Equation (11), the references of the dq axis currents and the modulation index can be obtained as follows:

$$\begin{aligned} i_d^* &= \omega_{dq} C e_q^* + i_{Ld}, &&(24) \\ i_q^* &= -\omega_{dq} C e_d^* + i_{Lq}, &&(25) \\ m_d^* &= \frac{1}{v_{dc}}\left(L \dot{i}_d^* + R i_d^* + \omega_{dq} L i_q^* + e_d^*\right), &&(26) \\ m_q^* &= \frac{1}{v_{dc}}\left(L \dot{i}_q^* + R i_q^* - \omega_{dq} L i_d^* + e_q^*\right), &&(27) \end{aligned}$$

where we have considered that $\dot{e}_d^* = \dot{e}_q^* = 0$ because e_d^* and e_q^* are constant values in the Park's reference frame.

Figure 3 shows the proposed control scheme. It can be observed that the modulation indexes are obtained using Equations (22) and (23) with Equation (27), and the references e_d^*, e_q^*, and ω_{dq} are selected by the users. The angle θ_{dq} used in the Park's transformations is obtained from $\frac{d\omega_{dq}}{dt}$.

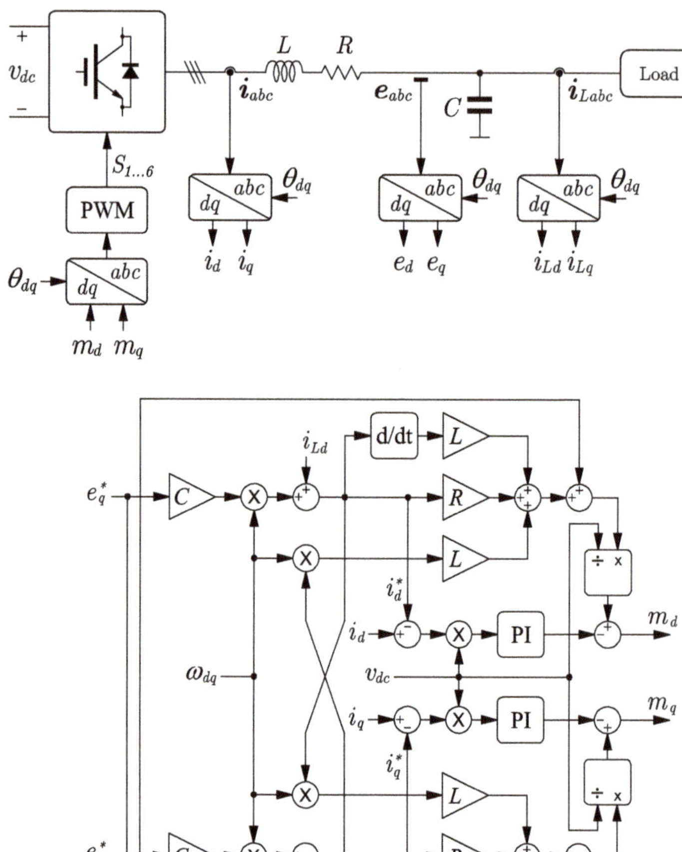

Figure 3. Block diagram of the proposed controller.

3.2. Control Objective

The main idea behind implementing VSCs for the integration of non-linear loads in the AC side using the energy provided by the DC network (see Figure 1) is to generate a three-phase sinusoidal signal with constant amplitude and frequency [19]. This implies that the references for the voltage signals in the capacitors in parallel to the load can be defined as follows:

$$
\begin{aligned}
e_a^*(t) &= \sqrt{2}E_{rms}\sin(\omega t), \\
e_b^*(t) &= \sqrt{2}E_{rms}\sin\left(\omega t - 2\frac{\pi}{3}\right), \\
e_c^*(t) &= \sqrt{2}E_{rms}\sin\left(\omega t + 2\frac{\pi}{3}\right),
\end{aligned}
\tag{28}
$$

where E_{rms} is the desired voltage magnitude and ω represents the angular frequency, i.e., for this study $\omega = 2\pi f$, with $f = 50\,\text{Hz}$.

From the desired three-phase voltage reference defined in Equation (28), we can observe that the electrical frequency is considered as an input for the Park's transformation (see Figure 3). This implies that if the controller makes its task, the voltage outputs in the load will be sinusoidal waves regardless of the load variations (linear or non-linear behavior) [50].

Note that if we send a constant frequency as input to the Parks' transformation, then the sinusoidal references in Equation (28) become constant values in the dq reference frame, i.e.,

$$
\begin{aligned}
e_d^*(t) &= E_{rms}, \\
e_q^*(t) &= e_0^*(t) = 0,
\end{aligned}
\tag{29}
$$

which implies that if the controller reaches these values in the dq reference frame, then the voltage output will be perfectly sinusoidal in the abc reference frame sinusoidal [19].

Remark 2. *Note that the voltage and frequency and regulation reported in this paper allows supporting a three-phase sinusoidal signal in load coupling based on the properties of the controller for following constant references in the dq frame [19]. This is different from classical approaches that regulate voltage and frequency by measuring active and reactive power by emulating synchronous machines using the virtual inertia concept [20,21].*

3.3. Stability Analysis

The stability analysis in modern control applications is an important aspect and should be demonstrated to show that they are suitable for implementation without fails [8,51]; one of the main attractive characteristics of the PBC design is that in a large class of linear and nonlinear system, stability can be guaranteed using Hamiltonian or Lagrangian representations [14,32]. Here, we present basic proof for analyzing the proposed PI-PBC approach, since it was extensively studied in [32,35].

We will enunciate the necessary conditions for guaranteeing stability of a nonlinear autonomous dynamical $\dot{x} = f(x, u)$ system in the sense of Lyapunov around the equilibrium point $x = x^*$ as follows [52]:

✓   If there is a Lyapunov candidate function $V(x)$ that is positive definite for all $x \neq x^*$, and zero only for $x = x^*$, i.e., $V(x) > 0$, $x \neq x^*$ & $V(x) = 0$, $x = x^*$,

✓   and the derivative of the Lyapunov function with respect to the time $(\dot{V}(x))$ is negative semidefinite, i.e., $(\dot{V}(x)) \leq 0$, $\forall x \neq x^*$, & $(\dot{V}(x^*)) = 0$.

Note that to prove these two necessary conditions for the proposed closed-loop dynamical system with state variables $\tilde{x}$ and z, we employed a quadratic function as recommended in [32] which fulfills the positive definiteness in all the solution space which is zero only in the equilibrium point, as follows:

$$
V(\tilde{x}, z) = H(\tilde{x}) + \frac{1}{2} z^T K_I z,
\tag{30}
$$

where $H(\tilde{x})$ is the desired Hamiltonian function. Note that the second term is a quadratic expression as function of the integral variables that deal with these new stable variables introduced by the PI-PBC method [33].

Now, if we take the time derivative of the candidate Lyapunov function in Equation (30), then the following result is yielded:

$$
\dot{V}(\tilde{x}, z) = \dot{H}(\tilde{x}) + z^T K_I \dot{z}
\tag{31}
$$

where, if it substitutes the expression in Equation (20) by considering the PI-PBC design in Equation (22), then, the following result is reached:

$$
\begin{aligned}
\dot{V}(\tilde{x}, z) &= -\tilde{x}^T P^{-1} R P^{-1} \tilde{x} + \tilde{x}^T P^{-1} g \tilde{u} + z^T K_I (-\tilde{y}) \\
&\leq -\tilde{y}^T K_P \tilde{y} \leq 0,
\end{aligned}
\tag{32}
$$

which clearly fulfills the second condition of the Lyapunov's theorem [34].

Remark 3. *As the derivative of the Lyapunov function with respect to time does not directly contain the state variables of interest, i.e., $\tilde{x}$ and z, based on the Lyapunov's stability theorem, we can affirm that the dynamical system $\dot{\tilde{x}}$ (see Equation (17)) is stable in a closed-ball that contain the equilibrium point $x = \tilde{x}$, as demonstrated in [35].*

Regarding the range of application of the aforementioned stability analysis of the real systems and its relation to the parameters of the system, it is important to highlight the following facts:

- The PI-PBC design can guarantee stability independently of the value of the parameters of the filter LC since its demonstration is based on the positive definiteness of the matrix $\mathbf{P}$ in Equation (14) [13]. This matrix is constant and contains at its diagonal the parameters of the filter, which of course are positive in real physical systems. This implies that if there exists variations between the parameters assigned to the controller and the real parameters, it will not compromise the stability of the system in closed-loop [32].
- The control gains assigned to the PI-PBC controller plays an important role in the stability analysis [33]; nevertheless, these need to fulfill an important condition related to the positiveness of their values. Since these are design parameters, we can ensure that they will be positive and the system will remain stable during the closed-loop operation.
- Some unmodeled dynamics such as parasitic resistances in the capacitors connected in parallel to the load or power losses in the converter will help the stable behavior of the system since these parameters introduce additional dampings in the dynamical response of the physical system that we no longer observe in the simulation environment [53].

4. Results

The proposed PI-PBC controller was validated based on simulations and experimental results. The parameters used in the system in both the simulations and experimental tests are listed in Table 1.

Table 1. System parameters.

Parameter	Value	Unit	Parameter	Value	Unit
Inductance (L)	1.25	mH	Switching frequency (f_s)	20	kHz
Resistance (R)	0.2	Ω	Output frequency (f)	50	Hz
Capacitance (C)	45	μF	DC-link voltage (v_{dc})	311	V
Internal IGBT resistance (R_{on})	10	mΩ	DC-link capacitance (C_{dc})	5400	μF

4.1. Simulation Results

The simulation results were obtained using a realistic model of the system, including the switching effects, losses, and a detailed transistor model. The simulations were implemented using the SimPowerSystem in MATLAB.

To validate the performance of the proposed controller, a test was performed by changing the load (linear and nonlinear). Figure 4a shows the dq axis output voltage and its reference, and Figure 4b shows the output voltage and inductor current for phase a. Initially, the system operated with a resistive load of 10 Ω, and the reference of the output voltage was $e_d^* = 100$ V and $e_q^* = 0$ V. When the load was changed from 10 Ω to 5 Ω (at 0.1 s), the d-axis voltage exhibited a transient variation; this also occurred for the voltage on the q-axis, but within a short time (approximately one cycle of the voltage), the controller was able to regulate the amplitude and frequency of the output voltage in the expected time and with the desired dynamic behavior. At 0.2 s, the nonlinear load was connected to the system. Again, the controller regulated the dq axis output voltages in an acceptable time and with suitable behavior. In addition, as shown in Figure 4b, the waveform of the voltage was sinusoidal (free of significant harmonics) even under a nonlinear load.

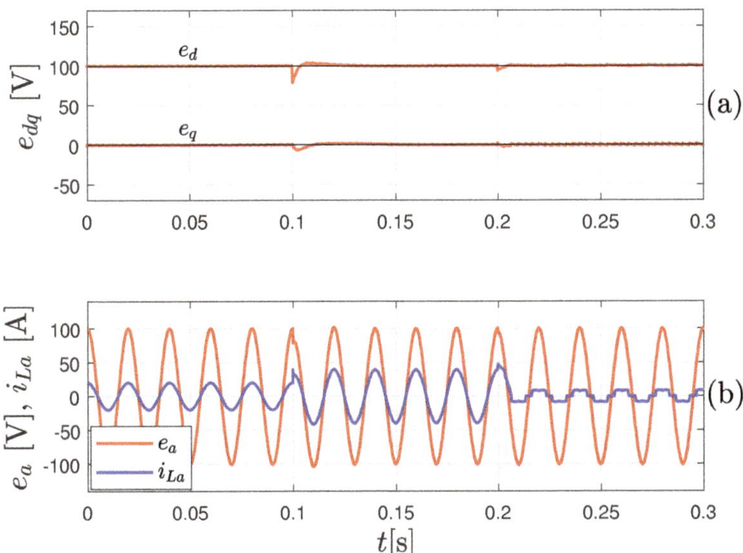

Figure 4. Voltage behavior in the Park's reference frame, and voltage and load current per-phase.
(**a**) e_{dq} profile. (**b**) Voltage and load current in phase a.

Figure 5a shows the three-phase output voltage and Figure 5b depicts the currents in the filter inductors. The amplitude of the three-phase output voltage was regulated in advance of the load changes as expected.

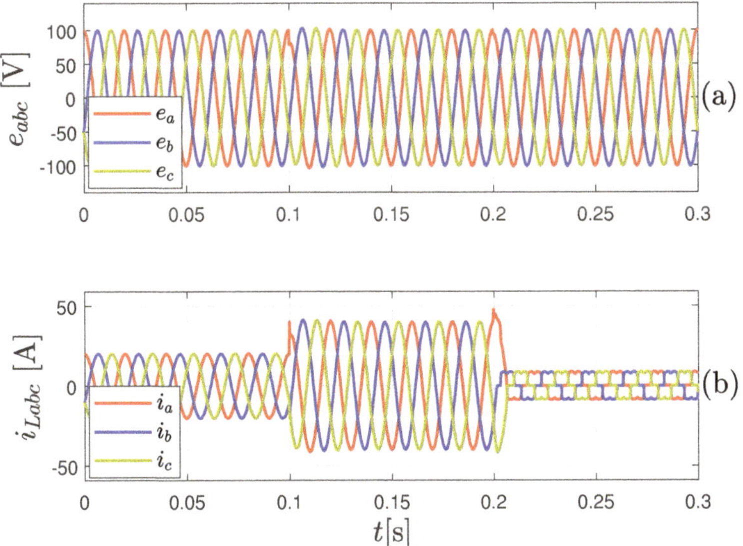

Figure 5. Dynamical performance of the voltage and currents at the point of load connection.
(**a**) Three-phase voltage profile. (**b**) Current profile for linear and nonlinear consumption.

With the objective to show the performance of the proposed controller in the presence of parametric variations, a test with varying inductance and resistance values of the LC filter was performed. Figure 6 shows the d-axis output voltage for the same test of Figure 4 when the value of the inductance and resistance of the filter are different and the same are then considered in the controller. The parameter variation consists of a change of 50% in the inductance and resistance of the filter. It can be observed that the d-axis voltage exhibited transient differences between the situations with and without parameter variation, while in regimen, the voltage is regulated in the reference value as the integral action of the PI eliminates possible steady-state errors (see [54]).

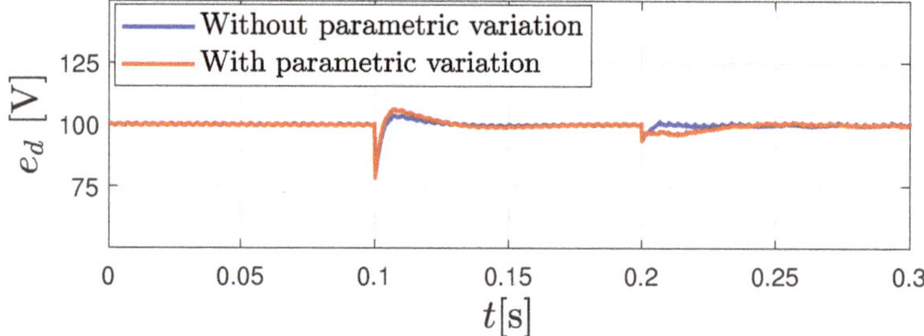

Figure 6. Behavior of d-axis voltage for parametric variation.

For comparison purposes with classic approaches (PI controllers), Figure 7 shows the output voltage behavior in dq-coordinates for the same test of Figure 4, when a PI controller (in blue) and the proposed PI-PBC controller (in red) are working. It can be observed that the classic PI controller performs similar to PI-PBC when a linear load is considered, but when a non-linear load is connected, the system with the PI controller presents high harmonic content, and therefore the performance is reduced. This test shows the superiority of the proposed strategy to mitigate the harmonics due to non-linear loads.

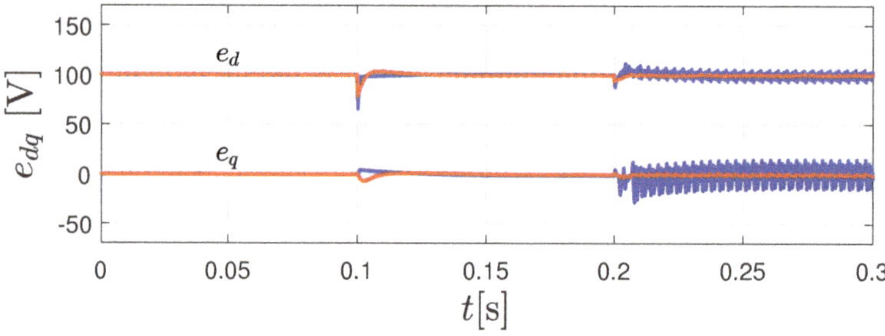

Figure 7. Voltage behavior in the Park's reference frame, comparison between the proposed controller (in red) and the classic PI approach (in blue).

With the aim to demonstrate the performance of the proposed controller under highly demanding conditions, Figure 8 presents a test with a change of the amplitude reference of the output voltage and a change in load when the converter supplies the nonlinear load, as considered in Figure 4. In Figure 8a, the dq axis output voltage and its reference is illustrated and in Figure 8b, the output voltage and inductor current for phase a is shown. It can be seen that the proposed PI-PBC controller (in red) regulates the amplitude of the output voltage in the presence of changes of the amplitude reference (0.1 s) and load changes (0.2 s) even when the power of the nonlinear load is increased. Furthermore, in this test, a comparison between the proposed controller (in red) and the classic PI (in yellow) approach was performed. The performance of the classic PI presents high harmonic content as the value of the power of the nonlinear charge increases while the oscillations presented by the PI-PBC are small. It should be noted that the elimination of these oscillations can be accomplished using a multiple reference controller but it will increase the complexity of the control algorithm [24].

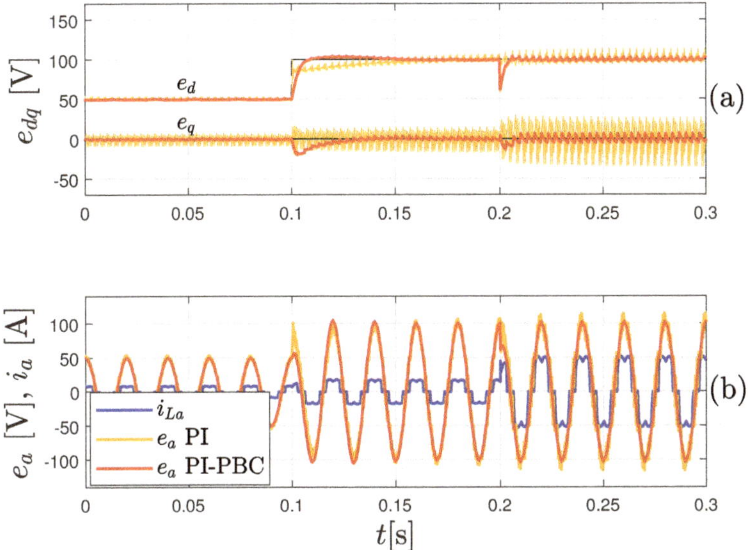

Figure 8. Test system behavior under very demanding conditions for the proposed controller (in red) and the classic PI approach (in yellow). (**a**) Voltage behavior in the Park's reference frame. (**b**) Phase *a* voltage profile.

Figure 9 depicts the harmonic spectrum of the output voltage using the proposed PI-PBC controller. It can be seen that the amplitude of individual harmonics remains below the required limits by the standards and the total harmonic distortion (THD) is 2.17%. This value is lower than the values established by IEC 62040-3 (less than 8%) [16]. Figure 10 shows the harmonic spectrum of the output voltage for the classic PI controller, the THD is 7.37%, which is very close to the allowed limits. The harmonic spectra were obtained for the system feeding the nonlinear load corresponding to Figure 8 between 0.2 s to 0.3 s.

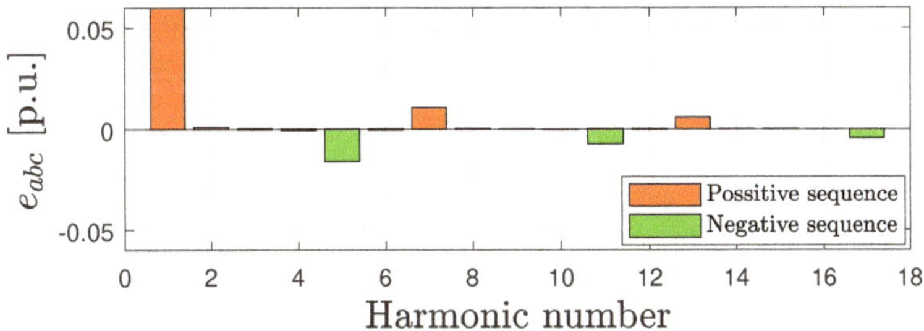

Figure 9. Voltage harmonic spectrums of proportional-integral passivity-based controller (PI-PBC).

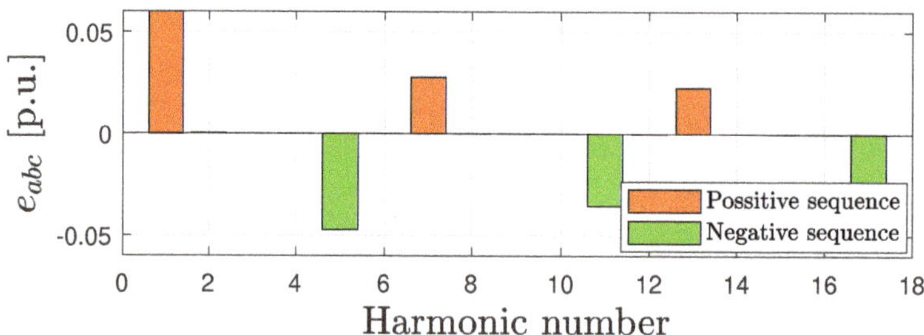

Figure 10. Voltage harmonic spectrums of Classical PI controller.

Finally, Figure 11 shows a test for an unbalanced load. It can be seen that the three-phase voltage is maintained with constant amplitude and frequency, and without significative harmonics when considering these types of loads.

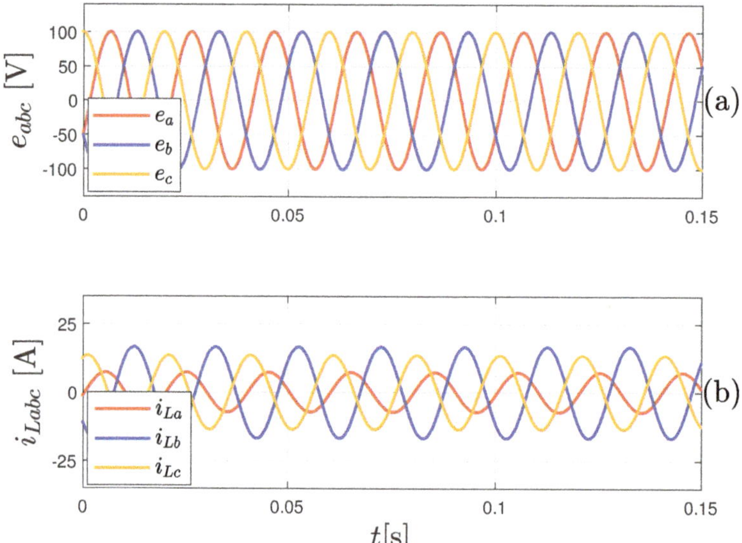

Figure 11. Test system behavior under very demanding conditions for the proposed controller (in red) and the classic PI approach (in yellow). (**a**) Voltage behavior in the Park's reference frame. (**b**) Phase *a* voltage profile.

4.2. Experimental Results

The experimental results were obtained using a laboratory prototype with the same parameters employed in the simulation tests (see Table 1). The DC-AC converter was constructed using an IGBT module (SEMiX101GD12E4s), and the controller was implemented in a TMS320F28335 floating-point DSP (Texas Instruments).

Figure 12 shows the three-phase voltage generated by the system under a resistive load change (50 Ω to 25 Ω). The controller regulated the amplitude and frequency of the output voltage, and the time response required for the load change was approximately one cycle, thereby agreeing with the previously described simulation results, and this time is acceptable in practical applications as well.

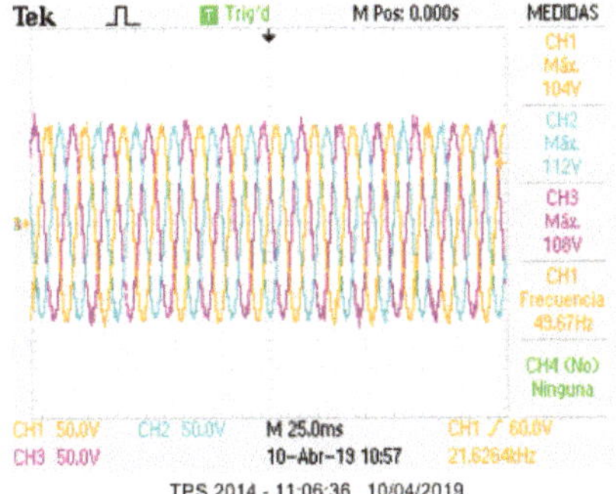

Figure 12. Experimental results: three-phase generated voltage for a resistive load change.

The output voltage and inductor current for phase *a* are shown in Figure 13 for the resistive load change and in Figure 14 for a nonlinear load. The results show that dynamic behavior was acceptable in both cases. In addition, when a nonlinear load was considered (Figure 14), the controller generated an output voltage that was free of significant harmonics. The voltage total harmonic distortion is 3.6%. This value is lower than the values established by IEC 62040-3 (less than 8%) [16].

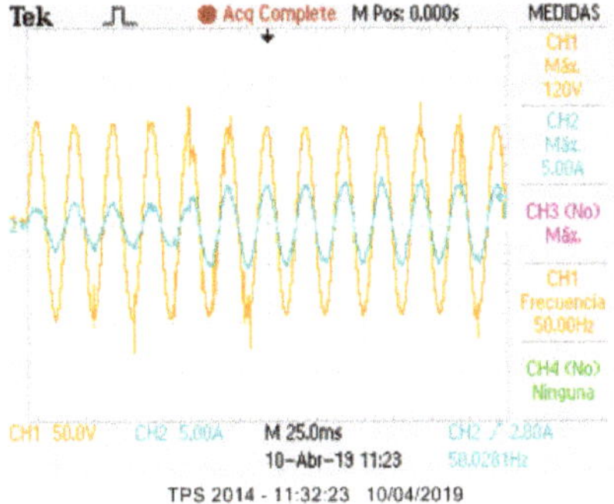

Figure 13. Experimental results: output voltage and inductor current for phase *a* with a resistive load change.

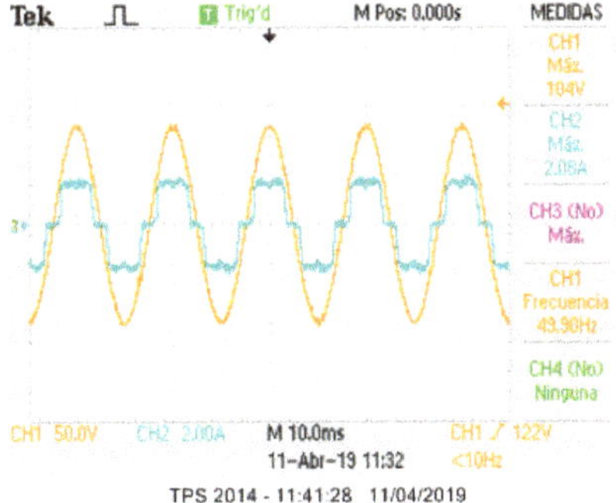

Figure 14. Experimental results: output voltage and inductor current for phase *a* under a nonlinear load.

5. Conclusions

In this article, we present a PI-PBC controller to regulate the amplitude and frequency of a three-phase output voltage in a DC-AC converter with an LC filter. Simulations and experimental results show that the proposed controller allowed regulating the amplitude and frequency of the output voltage of the converter under both linear and nonlinear load changes. Additionally, the PI-PBC allowed generating well-defined sinusoidal signals to linear and nonlinear loads. The results of the performance analysis showed that the proposed design combines the simplicity and robustness of PI-based controllers with the stability and performance characteristics of PBC as a powerful tool for the design of power converter controllers used in hybrid power systems.

Derived from this research, some future recommendations have been made as follows: (1) to develop a direct power control approach to operate DC-AC converters in constant power load applications; (2) to combine the advantages of the passivity-based control designs with virtual inertia emulators to develop robust controllers for weak three-phase distribution networks with unbalanced and nonlinear loads; and (3) to employ advanced methodologies for optimal adjustment of PI control gains using active and passive strategies in order to improve the dynamical performance of the proposed PI-PBC method in applications with large variations in the load terminals.

Author Contributions: Conceptualization and writing—review and editing, F.M.S., L.M.F., O.D.M., W.G.-G., and J.C.H. All authors have read and agreed to the published version of the manuscript.

Funding: This work was partially supported by the Universidad Nacional de San Luis under project PROICO: 142318 "Control de convertidores de potencia para la integración de fuentes de energía renovables y vehículos eléctricos a la red"and by Fondo para la Investigación Científica y Tecnológica (FONCyT) under project PICT-2017-0794 "Cargadores de baterías para vehículos eléctricos: integración con la red y fuentes de energía renovables" and in part by the Universidad Tecnológica de Bolívar under grant CP2019P011 associated with the project: "Operación eficiente de redes eléctricas con alta penetración de recursos energéticos distribuidos considerando variaciones en el recurso energético primario".

Acknowledgments: The authors want to express to thanks to Universidad Nacional de San Luis and Consejo Nacional de Investigaciones Científicas y Técnicas (CONICET) in Argentina, Universidad Distrital Francisco José de Caldas, Universidad Tecbológica de Bolivar in Colombia and University of Jaén in Spain.

Abbreviations

The following abbreviations and nomenclature are used in this manuscript:

Acronyms

AC	Alternating current
BB	Battery
IDA-PBC	Interconnection and damping assignment passivity-based controller
pH	port-Hamiltonian
PI-PBC	Proportional-integral passivity-based controller
DC	Direct current
SC	Supercapacitor
RHS	Renewable-based hybrid systems

Subscripts and superscripts

$*$	Admissible trajectory
dq	Direct-quadrature reference frame

Parameters

R	Resistance filter
L	Inductance filter
C	Capacitance filter

Variables

i_{dq}	Output currents of VSC in the dq frame
i_{Ldq}	Load currents in the dq frame
e_{dq}	Output voltages of VSC in the dq frame
m_{dq}	modulation indexes in the dq frame
$\mathbf{x} \in R^n$	State vector
$\tilde{x} \in R^n$	State vector with incremental variables
$\mathbf{u} \in R^m$	Control input vector
$\mathbf{z} \in R^m$	auxiliary variable vector
$\mathbf{R} \in R^{n \times n}$	Inertia matrix
$\mathbf{R} \in R^{n \times n}$	Damping matrix
$\mathbf{J} \in R^{n \times n}$	Interconnection matrix
$\mathbf{g} \in R^{n \times m}$	Input matrix
$H(\mathbf{x})$	Energy storage function
$V(\mathbf{x})$	Candidate Lyapunov function
$K_p \in R^{m \times m}$	Proportional gain matrix
$K_I \in R^{m \times m}$	Integral gain matrix

References

1. Ellabban, O.; Abu-Rub, H.; Blaabjerg, F. Renewable energy resources: Current status, future prospects and their enabling technology. *Renew. Sustain. Energy Rev.* **2014**, *39*, 748–764. [CrossRef]
2. Bueno-Lopez, M.; Lemos, S.G. Electrification in non-interconnected areas: Towards a new vision of rurality in Colombia. *IEEE Technol. Soc. Mag.* **2017**, *36*, 73–79. [CrossRef]
3. Montoya, O.D.; Garcés, A.; Serra, F.M. DERs integration in microgrids using VSCs via proportional feedback linearization control: Supercapacitors and distributed generators. *J. Energy Storage* **2018**, *16*, 250–258. [CrossRef]
4. Zou, P.; Chen, Q.; Xia, Q.; He, G.; Kang, C. Evaluating the Contribution of Energy Storages to Support Large-Scale Renewable Generation in Joint Energy and Ancillary Service Markets. *IEEE Trans. Sustain. Energy* **2016**, *7*, 808–818. [CrossRef]
5. Justo, J.J.; Mwasilu, F.; Lee, J.; Jung, J.W. AC-microgrids versus DC-microgrids with distributed energy resources: A review. *Renew. Sustain. Energy Rev.* **2013**, *24*, 387–405. [CrossRef]
6. Kumar, D.; Zare, F.; Ghosh, A. DC Microgrid Technology: System Architectures, AC Grid Interfaces, Grounding Schemes, Power Quality, Communication Networks, Applications, and Standardizations Aspects. *IEEE Access* **2017**, *5*, 12230–12256. [CrossRef]
7. Gavriluta, C.; Candela, I.; Citro, C.; Luna, A.; Rodriguez, P. Design considerations for primary control in multi-terminal VSC-HVDC grids. *Electr. Power Syst. Res.* **2015**, *122*, 33–41. [CrossRef]
8. Gil-González, W.; Montoya, O.D.; Garces, A. Direct power control for VSC-HVDC systems: An application of the global tracking passivity-based PI approach. *Int. J. Electr. Power Energy Syst.* **2019**, *110*, 588–597. [CrossRef]
9. Ornelas-Tellez, F.; Rico-Melgoza, J.J.; Espinosa-Juarez, E.; Sanchez, E.N. Optimal and Robust Control in DC Microgrids. *IEEE Trans. Smart Grid* **2018**, *9*, 5543–5553. [CrossRef]
10. Roy, T.K.; Mahmud, M.A.; Oo, A.M.T.; Haque, M.E.; Muttaqi, K.M.; Mendis, N. Nonlinear Adaptive Backstepping Controller Design for Islanded DC Microgrids. *IEEE Trans. Ind. Appl.* **2018**, *54*, 2857–2873. [CrossRef]
11. Garces, A. Uniqueness of the power flow solutions in low voltage direct current grids. *Electr. Power Syst. Res.* **2017**, *151*, 149–153. [CrossRef]
12. Opiyo, N.N. A comparison of DC- versus AC-based minigrids for cost-effective electrification of rural developing communities. *Energy Rep.* **2019**, *5*, 398–408. [CrossRef]
13. Serra, F.M.; Angelo, C.H.D. IDA-PBC controller design for grid connected Front End Converters under non-ideal grid conditions. *Electr. Power Syst. Res.* **2017**, *142*, 12–19. [CrossRef]
14. Montoya, O.D.; Gil-González, W.; Serra, F.M. PBC Approach for SMES Devices in Electric Distribution Networks. *IEEE Trans. Circuits Syst. II* **2018**, *65*, 2003–2007. [CrossRef]

15. Han, J.; Liu, Z.; Liang, N.; Song, Q.; Li, P. An Autonomous Power-Frequency Control Strategy Based on Load Virtual Synchronous Generator. *Processes* **2020**, *8*, 433. [CrossRef]

16. Serra, F.M.; Angelo, C.H.D.; Forchetti, D.G. IDA-PBC control of a DC–AC converter for sinusoidal three-phase voltage generation. *Int. J. Electron.* **2017**, *104*, 93–110. [CrossRef]

17. Ye, Y.; Zhou, K.; Zhang, B.; Wang, D.; Wang, J. High-Performance Repetitive Control of PWM DC-AC Converters With Real-Time Phase-Lead FIR Filter. *IEEE Trans. Circuits Syst. II* **2006**, *53*, 768–772. [CrossRef]

18. Khefifi, N.; Houari, A.; Machmoum, M.; Ghanes, M.; Ait-Ahmed, M. Control of grid forming inverter based on robust IDA-PBC for power quality enhancement. *Sustain. Energy Grids Netw.* **2019**, *20*, 100276. [CrossRef]

19. Amin, W.T.; Montoya, O.D.; Garrido, V.M.; Gil-González, W.; Garces, A. Voltage and Frequency Regulation on Isolated AC Three-phase Microgrids via s-DERs. In Proceedings of the 2019 IEEE Green Technologies Conference (GreenTech), Lafayette, LA, USA, 3–6 April 2019; pp. 1–6.

20. Huang, X.; Wang, K.; Li, G.; Zhang, H. Virtual Inertia-Based Control Strategy of Two-Stage Photovoltaic Inverters for Frequency Support in Islanded Micro-Grid. *Electronics* **2018**, *7*, 340. [CrossRef]

21. Fang, J.; Lin, P.; Li, H.; Yang, Y.; Tang, Y. An Improved Virtual Inertia Control for Three-Phase Voltage Source Converters Connected to a Weak Grid. *IEEE Trans. Power Electron.* **2019**, *34*, 8660–8670. [CrossRef]

22. Haque, M.E.; Negnevitsky, M.; Muttaqi, K.M. A Novel Control Strategy for a Variable-Speed Wind Turbine With a Permanent-Magnet Synchronous Generator. *IEEE Trans. Ind. Appl.* **2010**, *46*, 331–339. [CrossRef]

23. Kim, D.; Lee, D. Feedback Linearization Control of Three-Phase UPS Inverter Systems. *IEEE Trans. Ind. Electron.* **2010**, *57*, 963–968. [CrossRef]

24. Loh, P.C.; Newman, M.J.; Zmood, D.N.; Holmes, D.G. A comparative analysis of multiloop voltage regulation strategies for single and three-phase UPS systems. *IEEE Trans. Power Electron.* **2003**, *18*, 1176–1185. [CrossRef]

25. Mattavelli, P. Synchronous-frame harmonic control for high-performance AC power supplies. *IEEE Trans. Ind. Appl.* **2001**, *37*, 864–872. [CrossRef]

26. Lidozzi, A.; Calzo, G.L.; Solero, L.; Crescimbini, F. Integral-resonant control for stand-alone voltage source inverters. *IET Power Electron.* **2013**, *7*, 271–278. [CrossRef]

27. Mohamed, I.S.; Zaid, S.A.; Elsayed, H.M.; Abu-Elyazeed, M. Three-phase inverter with output LC filter using predictive control for UPS applications. In Proceedings of the 2013 International Conference on Control, Decision and Information Technologies (CoDIT), Hammamet, Tunisia, 6–8 May 2013; IEEE: Piscataway, NJ, USA, 2013, pp. 489–494.

28. Talbi, B.; Krim, F.; Laib, A.; Sahli, A. Model predictive voltage control of a single-phase inverter with output LC filter for stand-alone renewable energy systems. *Electr. Eng.* **2020**. [CrossRef]

29. Valderrama, G.E.; Stankovic, A.M.; Mattavelli, P. Dissipativity-based adaptive and robust control of UPS in unbalanced operation. *IEEE Trans. Power Electron.* **2003**, *18*, 1056–1062. [CrossRef]

30. Ortega, R.; van der Schaft, A.; Maschke, B.; Escobar, G. Interconnection and damping assignment passivity-based control of port-controlled Hamiltonian systems. *Automatica* **2002**, *38*, 585–596. [CrossRef]

31. Perez, M.; Ortega, R.; Espinoza, J.R. Passivity-based PI control of switched power converters. *IEEE Trans. Control Syst. Technol.* **2004**, *12*, 881–890. [CrossRef]

32. Cisneros, R.; Pirro, M.; Bergna, G.; Ortega, R.; Ippoliti, G.; Molinas, M. Global tracking passivity-based PI control of bilinear systems: Application to the interleaved boost and modular multilevel converters. *Control Eng. Pract.* **2015**, *43*, 109–119. [CrossRef]

33. Gil-González, W.; Montoya, O.D.; Garces, A. Control of a SMES for mitigating subsynchronous oscillations in power systems: A PBC-PI approach. *J. Energy Storage* **2018**, *20*, 163–172. [CrossRef]

34. Cisneros, R.; Mancilla-David, F.; Ortega, R. Passivity-Based Control of a Grid-Connected Small-Scale Windmill With Limited Control Authority. *IEEE J. Emerg. Sel. Top. Power Electron.* **2013**, *1*, 247–259. [CrossRef]

35. Montoya, O.D.; Gil-González, W.; Garces, A. Distributed energy resources integration in single-phase microgrids: An application of IDA-PBC and PI-PBC approaches. *Int. J. Electr. Power Energy Syst.* **2019**, *112*, 221–231. [CrossRef]

36. Montoya, O.D.; Gil-González, W.; Avila-Becerril, S.; Garces, A.; Espinosa-Pérez, G. Integración de REDs en Redes AC: Una Familia de Controladores Basados en Pasividad. *Revista Iberoamericana De Automática E Informática Industrial* **2019**, *16*, 212. [CrossRef]

37. Montoya, O.D.; Garces, A.; Avila-Becerril, S.; Espinosa-Pérez, G.; Serra, F.M. Stability Analysis of Single-Phase Low-Voltage AC Microgrids With Constant Power Terminals. *IEEE Trans. Circuits Syst. II Express Briefs* **2019**, *66*, 1212–1216. [CrossRef]

38. Luyben, W.L. Tuning Proportional-Integral-Derivative Controllers for Integrator/Deadtime Processes. *Ind. Eng. Chem. Res.* **1996**, *35*, 3480–3483. [CrossRef]

39. Ruz, M.; Garrido, J.; Vazquez, F.; Morilla, F. Interactive Tuning Tool of Proportional-Integral Controllers for First Order Plus Time Delay Processes. *Symmetry* **2018**, *10*, 569. [CrossRef]

40. Ahamad, A.; Yadav, C.; Ahuja, S.; Nema, S.; Padhy, P.K. PID tuning procedure based on simplified single parameter optimization. In Proceedings of the 2013 International Conference on Control, Automation, Robotics and Embedded Systems (CARE), Jabalpur, India, 16–18 December 2013; pp. 1–5.

41. Sanchis, R.; Romero, J.A.; Balaguer, P. Tuning of PID controllers based on simplified single parameter optimisation. *Int. J. Control* **2010**, *83*, 1785–1798. [CrossRef]

42. Irshad, M.; Ali, A. Optimal tuning rules for PI/PID controllers for inverse response processes. *IFAC-PapersOnLine* **2018**, *51*, 413–418. [CrossRef]

43. Pradhan, R.; Majhi, S.K.; Pradhan, J.K.; Pati, B.B. Antlion optimizer tuned PID controller based on Bode ideal transfer function for automobile cruise control system. *J. Ind. Inf. Integr.* **2018**, *9*, 45–52. [CrossRef]

44. Meena, D.C.; Devanshu, A. Genetic algorithm tuned PID controller for process control. In Proceedings of the 2017 International Conference on Inventive Systems and Control (ICISC), Coimbatore, India, 19–20 January 2017; pp. 1–6.

45. Barisal, A. Comparative performance analysis of teaching learning based optimization for automatic load frequency control of multi-source power systems. *Int. J. Electr. Power Energy Syst.* **2015**, *66*, 67–77. [CrossRef]

46. Ab. Talib, M.H.; Mat Darns, I.Z. Self-tuning PID controller for active suspension system with hydraulic actuator. In Proceedings of the 2013 IEEE Symposium on Computers Informatics (ISCI), Langkawi, Malaysia, 7–9 April 2013; pp. 86–91.

47. Tejado, I.; Vinagre, B.; Traver, J.; Prieto-Arranz, J.; Nuevo-Gallardo, C. Back to Basics: Meaning of the Parameters of Fractional Order PID Controllers. *Mathematics* **2019**, *7*, 530. [CrossRef]

48. Komurcugil, H. Steady-State Analysis and Passivity-Based Control of Single-Phase PWM Current-Source Inverters. *IEEE Trans. Ind. Electron.* **2010**, *57*, 1026–1030. [CrossRef]

49. He, J.; Li, Y.W. Generalized Closed-Loop Control Schemes with Embedded Virtual Impedances for Voltage Source Converters with LC or LCL Filters. *IEEE Trans. Power Electron.* **2012**, *27*, 1850–1861. [CrossRef]

50. Peña-Asensio, A.; Arnaltes-Gómez, S.; Rodriguez-Amenedo, J.L.; García-Plaza, M.; Carrasco, J.E.G.; de las Morenas, J.M.A.M. A Voltage and Frequency Control Strategy for Stand-Alone Full Converter Wind Energy Conversion Systems. *Energies* **2018**, *11*, 474. [CrossRef]

51. Sivadas, D.; Vasudevan, K. Stability Analysis of Three-Loop Control for Three-Phase Voltage Source Inverter Interfaced to the Grid Based on State Variable Estimation. *IEEE Trans. Ind. Appl.* **2018**, *54*, 6508–6518. [CrossRef]

52. Gil-González, W.; Serra, F.M.; Montoya, O.D.; Ramírez, C.A.; Orozco-Henao, C. Direct Power Compensation in AC Distribution Networks with SCES Systems via PI-PBC Approach. *Symmetry* **2020**, *12*, 666. [CrossRef]

53. Zahid, Z.U.; Lai, J.J.; Huang, X.K.; Madiwale, S.; Hou, J. Damping impact on dynamic analysis of LLC resonant converter. In Proceedings of the 2014 IEEE Applied Power Electronics Conference and Exposition—APEC 2014, Fort Worth, TX, USA, 16–20 March 2014; pp. 2834–2841.

54. Serra, F.M.; Angelo, C.H.D.; Forchetti, D.G. Interconnection and damping assignment control of a three-phase front end converter. *Int. J. Electr. Power Energy Syst.* **2014**, *60*, 317–324. [CrossRef]

 *electronics*

Article

Coordinated Control Scheme of Battery Storage System to Augment LVRT Capability of SCIG-Based Wind Turbines and Frequency Regulation of Hybrid Power System

Md. Rifat Hazari [1,*], Effat Jahan [1], Mohammad Abdul Mannan [1] and Junji Tamura [2]

[1] Department of Electrical and Electronic Engineering, American International University-Bangladesh (AIUB), 408/1, Kuratoli, Khilkhet, Dhaka 1229, Bangladesh; effat@aiub.edu (E.J.); mdmannan@aiub.edu (M.A.M.)

[2] Department of Electrical and Electronic Engineering, Kitami Institute of Technology (KIT), 165 Koen-cho, Kitami, Hokkaido 090-8507, Japan; tamuraj@mail.kitami-it.ac.jp

* Correspondence: rifat@aiub.edu

Received: 12 January 2020; Accepted: 29 January 2020; Published: 1 February 2020

Abstract: Fixed speed wind turbine-squirrel cage induction generator (FSWT-SCIG)-based wind farms (WFs) are increasing significantly. However, FSWT-SCIGs have no low voltage ride-through (LVRT) and frequency control capabilities, which creates a significant problem on power system transient and steady-state stability. This paper presents a new operational strategy to control the voltage and frequency of the entire power system, including large-scale FSWT-SCIG-based WFs, by using a battery storage system (BSS). The proposed cascaded control of the BSS is designed to provide effective quantity of reactive power during transient periods, to augment LVRT capability and real power during steady-state periods in order to damp frequency fluctuations. The cascaded control technique is built on four proportional integral (PI) controllers. The droop control technique is also adopted to ensure frequency control capability. Practical grid code is taken to demonstrate the LVRT capability. To evaluate the validity of the proposed system, simulation studies are executed on a reformed IEEE nine-bus power system with three synchronous generators (SGs) and SCIG-based WF with BSS. Triple-line-to-ground (3LG) and real wind speed data are used to analyze the hybrid power grid's transient and steady-state stability. The simulation results indicate that the proposed system can be an efficient solution to stabilize the power system both in transient and steady-state conditions.

Keywords: FSWT-SCIG; battery storage system; power system stability; synchronous generator

---

1. Introduction

Wind energy is a clean energy, the use of which can avoid 5.6 billion tons of CO_2 by 2050, equivalent to the yearly emissions of the 80 most polluting cities in the world, home to around 720 million people [1]. This would help to save up to four million lives annually by 2030 by reducing pollution, because one in eight deaths in the world is linked to air pollution [1].

The total worldwide wind power capacity in 2015 was 432.9 GW, which is a summative market growth of more than 17% [2–4]. By 2030, wind power could exceed 2110 GW and supply up to 20% of worldwide power demands [4].

1.1. Motivation

This massive penetration of wind energy into the power system, replacing fossil fuel-based power plants, has introduced some burdens to the power grid.

Basically, fixed speed wind turbine-squirrel cage induction generators (FSWT-SCIGs) are mostly used to develop wind farms (WFs). SCIGs have some advantages, such as their low cost, fewer maintenance requirements, good speed regulation, high efficiency in converting mechanical energy to electrical energy, better heat regulation, small size, and light weight. However, they cannot ensure low voltage ride-through (LVRT) capability and frequency stability of entire power systems during transient and steady-state periods [5], respectively.

1.2. Literature Reviews

Many auxiliary devices can be applied to SCIGs to augment their LVRT capability. For example, a dynamic voltage restorer (DVR) [6], thyristor controlled series capacitor (TCSC) [7], magnetic energy recovery switch (MERS) [8], series dynamic braking resistor (SDBR) [9], fault current limiter (FCL) [10], bridge-type fault current limiter (BFCL) [11], static synchronous compensator (STATCOM) [12], static VAR compensator (SVC) [13], superconductor dynamic synchronous condenser (SDSC) [14], unified compensation system (UCS) [15], and a unified power quality conditioner (UPQC) [15], were installed in SCIGs to inject reactive power in order to ensure LVRT capability.

Even though the LVRT was ensured by using the above-mentioned schemes, there are several drawbacks [6–15]. For example: TCSC creates resonance and injects objectionable harmonics, DVR has phase angle jumps and absorbs real power, SDBR and SFCL cannot control reactive power, MERS necessitates mechanical bypass switches, BFCL needs a large-scale coupling transformer, SVC offers voltage oscillations, STATCOM necessities a cut-off in a high-voltage drop, SDSC is less effective for applications of low-voltage drops, UPQC needs a large dc-link capacitor which increases the system cost, and UCS has high losses of conduction in the series bypass switch [16].

On the other hand, the battery storage system (BSS) is well known, and can respond more swiftly and quicker with better performance [17]. Additionally, it is steadily established technology and has appropriate energy density. Thus, BSS is the most prevalent solution in wind power applications [18].

1.3. Contribution

Therefore, based on the above discussion, a BSS for a SCIG-based WF is proposed in this paper to enhance the LVRT capability during transient periods and to damp frequency oscillations during steady-state periods. A suitable PI controller-based cascaded control approach is constructed to guarantee the stable operation of the power system. The proposed control strategy also incorporates droop gain and frequency signals to provide an effective amount of real power from the BSS which will ensure smaller frequency fluctuations. Detailed design procedures and control strategies are adequately presented in this paper.

The transient and steady-state responses of the entire power system including IEEE nine-bus system, WF with BSS is compared with that of SCIG without BSS. Actual wind speed values of Hokkaido Island, Japan are considered for steady-state analysis.

The simulation results clearly indicate that the proposed strategy can confirm the LVRT capability of WFs, and damp the frequency fluctuations of a hybrid power grid.

The paper is structured in six sections. Section 1 presented the introduction, motivation behind this work, literature reviews, and contribution. Sections 2 and 3 present the model of a hybrid power system and an aerodynamic model of wind turbine. Section 4 describes the proposed BSS along with its control strategy. The simulation results and analysis are presented in Section 5. Finally, Section 6 concludes the paper with a brief summary.

2. Model of a Hybrid Power System

The hybrid power system model presented in Figure 1 is used for transient and steady-state analysis. It has a WF and IEEE nine-bus main model. Basically, three different conventional synchronous generators (SGs) are used for the main system. The ratings of the SGs are 150 MVA (SG1), 250 MVA (SG2), and 200 MVA (SG3).

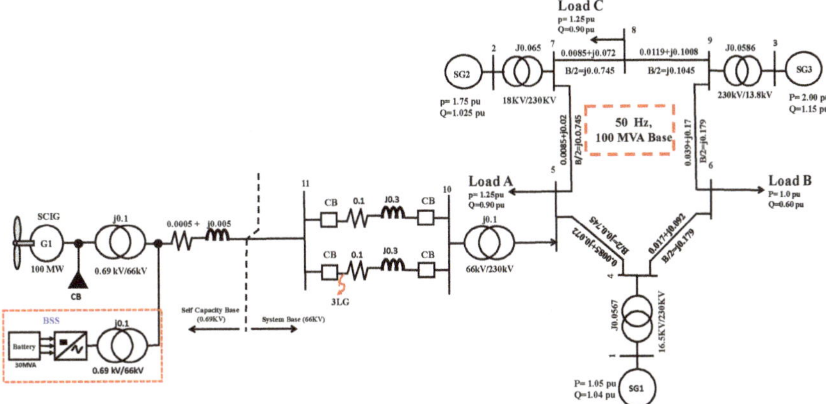

Figure 1. Model of hybrid power system.

An AC4A-type exciter model is used for all SGs as shown in Figure 2 [19]. The thermal-based power station as depicted in Figure 3 is considered for SG1 and SG2, and the hydro-based power station is considered for SG3 as illustrated in Figure 4 [19].

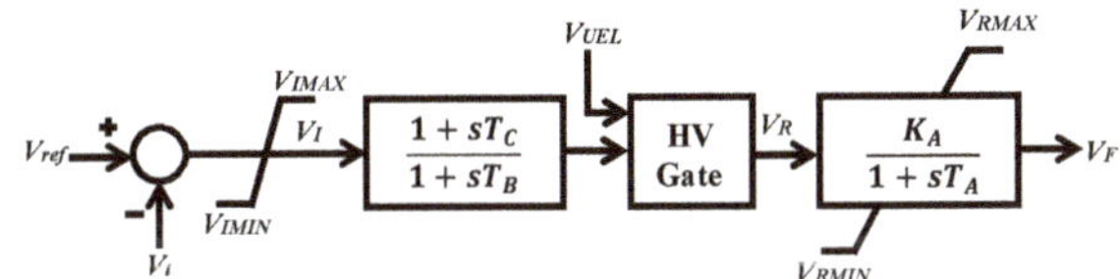

Figure 2. Exciter model of synchronous generators (SGs).

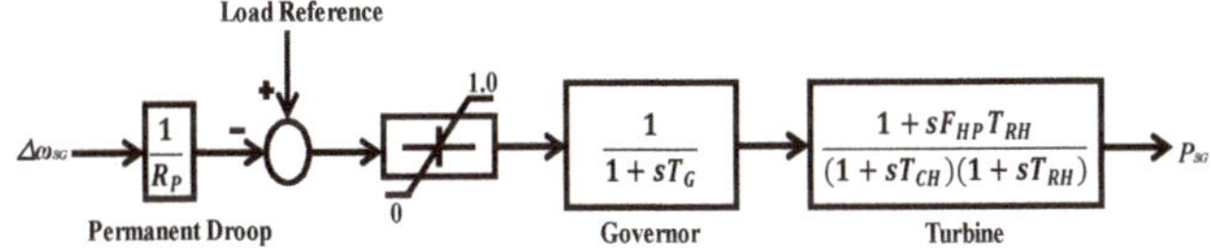

Figure 3. Governor model of thermal turbine.

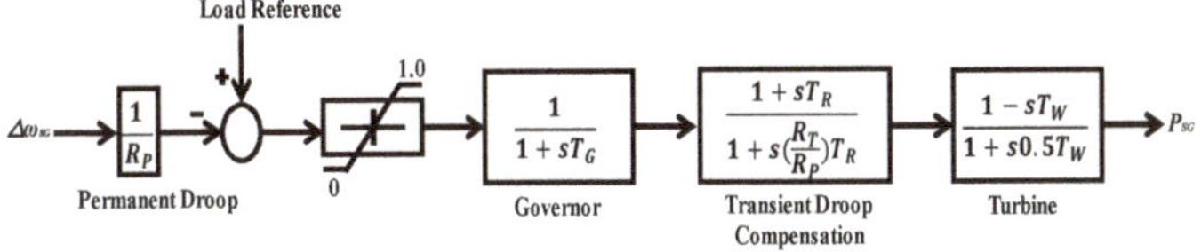

Figure 4. Governor model of hydro turbine.

Here, the FSWT-SCIG is linked to the main system at Bus 5 through 0.69 kV/66 kV, 66 kV/230 kV transformers and a dual transmission line. It has one SCIG (rated capacity: 100 MW) as shown in Figure 1. The reactive power is delivered to the SCIG using a capacitor bank during steady-state periods. Additionally, the BSS is connected next to the capacitor bank. The capacity of the BSS is 30 MVA. The frequency is 50 Hz and the base power of the power system is 100 MVA. The governor systems of SG1 and SG3 are controlled by an integral control to ensure automatic generation control (AGC) as depicted in Figure 5 [19]. The parameters of the SGs and SCIGs are presented in the Appendix A.

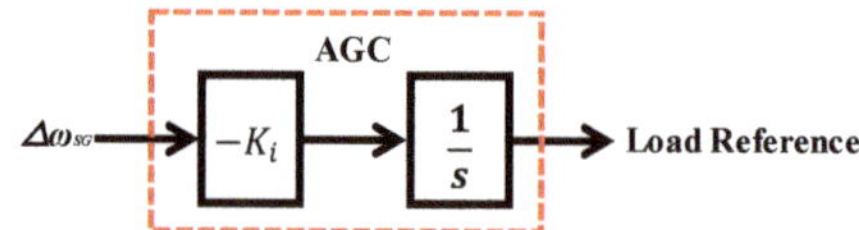

Figure 5. Automatic generation control (AGC).

3. Model of a Wind Turbine

The expression of the wind turbine's mechanical power can be written as follows [20]:

$$P_w = 0.5\rho\pi R^2 V_w{}^3 C_p(\lambda, \beta) \tag{1}$$

Here, P_w = captured wind power, R = rotor blade radius (m), C_p = power coefficient, ρ = air density (kg/m^3), and V_w = wind speed (m/s).

The expression C_p can be written as follows [21]:

$$C_p(\lambda, \beta) = c_1\left(\frac{c_2}{\lambda_i} - c_3\beta - c_4\right)e^{\frac{-c_5}{\lambda_i}} + c_6\lambda \tag{2}$$

$$\frac{1}{\lambda_i} = \frac{1}{\lambda - 0.08\beta} - \frac{0.035}{\beta^3 + 1} \tag{3}$$

$$\lambda = \frac{\omega_r R}{V_w} \tag{4}$$

Here, β = pitch angle (deg), ω_r = wind turbine rotor speed (rad/s), c_1 to c_6 = wind turbine characteristic coefficients [21], and λ = tip speed ratio.

Figure 6 presents the characteristics curve of Cp vs. λ. The curve is found for a different β from Equation (2). From the graph, the optimum λ (λ_{opt}) = 8.1 and the optimum Cp (Cp_{opt}) = 0.48. Figure 7 shows the model for the blade pitch control system of FSWT [22]. In FSWT, the pitch controller is used to control the real power output of the SCIG when it exceeds the rated power.

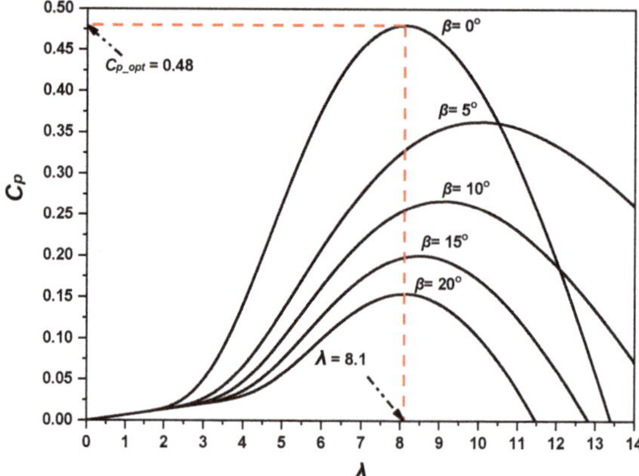

Figure 6. Cp vs. λ characteristics curve.

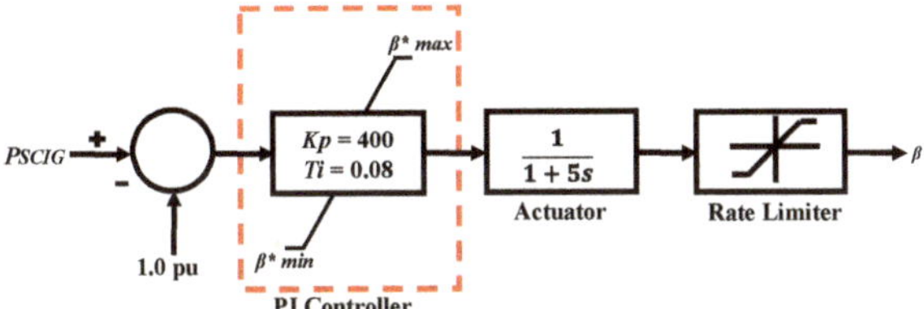

Figure 7. Pitch controller of fixed speed wind turbine-squirrel cage induction generator (FSWT-SCIG).

4. Proposed Coordinated Control of Battery Storage System

The BSS model which is used in this work is presented in Figure 8. It consists of a lead–acid battery unit, a voltage source converter (VSC) based on pulse width modulation (PWM), and a step-up transformer. For simplicity, the battery is symbolized using a constant DC voltage source. The DC voltage is transformed to grid-synchronized three-phase AC voltage using VSC.

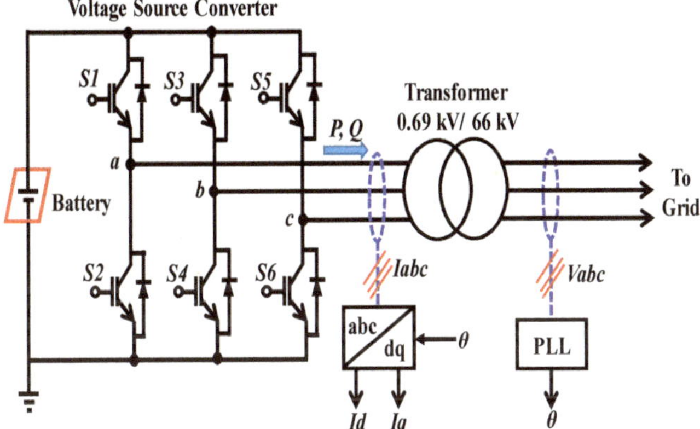

Figure 8. Proposed battery storage system (BSS).

The proposed control technique of the BSS is depicted in Figure 9. The different error signals are compensated using a cascaded control technique based on four PI controllers. The upper portion of the proposed control system controls the real power injected to the power grid system by adjusting the d-axis current (Id), whereas the lower portion is controlling the reactive power injected to the power system by adjusting the q-axis current (Iq). Additionally, in the upper portion the frequency of the grid system (f_{sys}) is taken as feedback. Depending upon the frequency deviation, the upper controller portion will minimize the frequency fluctuations by injecting effective amounts of real power from the BSS during steady-state conditions. The trial and error technique is applied to choose the droop gain (Kp) which ensures optimized results.

In the lower portion, the r.m.s voltage (V) of the grid system is taken as feedback. During transient conditions (e.g., fault conditions), the lower controller portion will inject effective amounts of reactive power until the terminal voltage reaches its pre-fault value.

Finally, the reference voltages (Va^*, Vb^*, and Vc^*) are compared with a high-frequency triangular carrier wave to get the gate drive pulses of the VSC. In this way, the LVRT's capability and minimization of frequency fluctuations can be ensured.

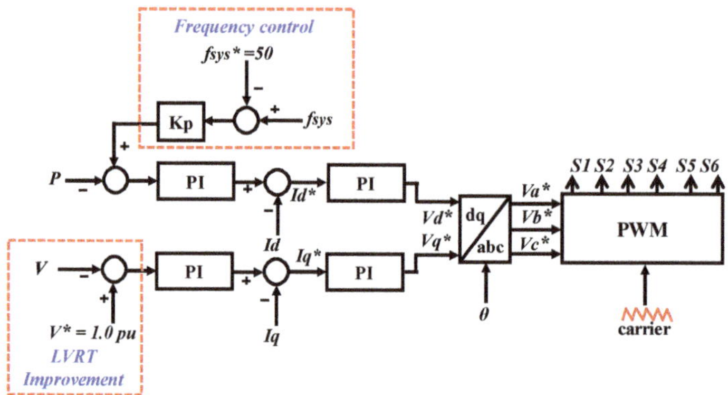

Figure 9. Proposed control strategy of BSS.

5. Simulation Results

In this work, simulation investigation has been completed on the same hybrid power system model presented in Figure 1. The well-known PSCAD/EMTDC software is used for simulation analysis. Due to detailed modeling of the whole system, the simulation time step is taken as 10 μs. The system frequency is 50 Hz. Two case studies are executed in order to authenticate the appropriateness of the proposed BSS. In Case 1, simulation scrutiny is accomplished without BSS, and in Case 2, simulation scrutiny is accomplished by including the proposed BSS.

5.1. Transient Stability Analysis

As shown in Figure 1, the triple-line-to-ground (3LG) fault is considered as a network disturbance near Bus 11 at one of the double circuit transmission lines. The fault conditions are presented in Figure 10. The wind speed data applied to the FSWT-SCIG are sustained constantly at 12 m/s based on the supposition that the wind speed does not change often within this short period.

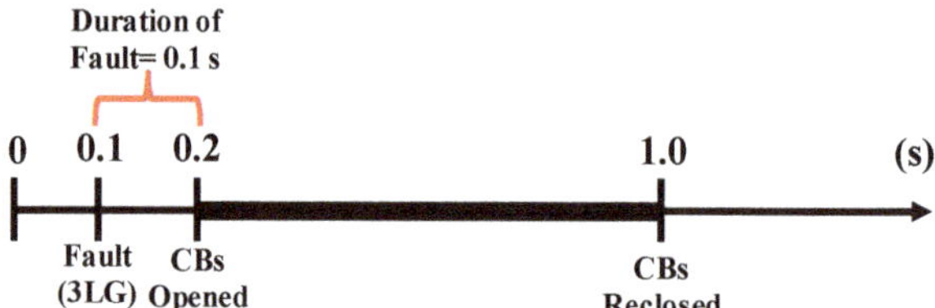

Figure 10. Triple-line-to-ground (3LG) fault conditions.

Figure 11 presents the reactive power response of BSS (Case 2), which indicates that it can provide an effective amount of reactive power to the SCIG during transient periods as depicted in Figure 12. Due to this effective injection of reactive power from the BSS, the terminal voltage of the SCIG goes back to the nominal value more quickly in Case 2, whereas it fails in Case 1, as shown in Figure 13. As the terminal voltage does not reach 90% of the nominal value within 1.5 s based on standard grid code [23], it is disconnected from the main power grid by opening the circuit breaker (CB) at 2.0 s. The rotor speed response of the SCIG, as shown in Figure 14, is unstable in Case 1 but is stable in Case 2 after the fault. This is because the SCIG requires more reactive power during transient periods than steady-state periods to improve the air gap flux. If enough reactive power is not provided, the developed electromagnetic torque of the SCIG declines considerably. Thus, the SCIG's rotor speed increases considerably in Case 1 and makes the whole system unstable. On the other hand, in Case 2 the

SCIG gets enough reactive power from the BSS during a network disturbance situation, and therefore its rotor speed response is stable. The real power output of the SCIG is shown in Figure 15. The real power output goes back to the nominal value more effectively in Case 2 than Case 1. Finally, the rotor speed responses of conventional SGs are presented in Figure 16. The rotor speed of SGs are more stable in Case 2 than Case 1. Thus, from the above analysis it is clear that the LVRT capability can be improved by incorporating the proposed BSS.

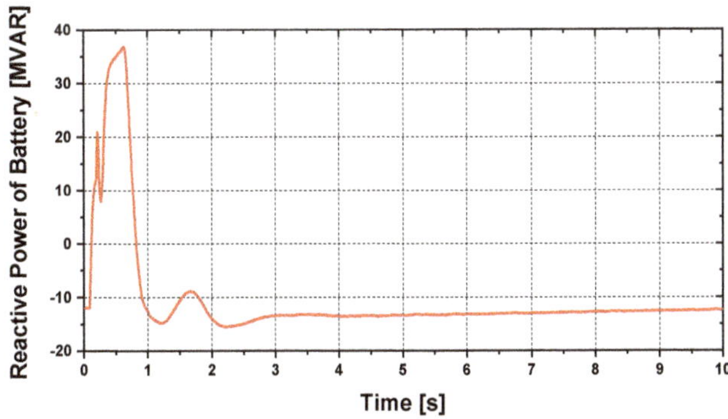

Figure 11. Reactive power response of BSS (Case 2).

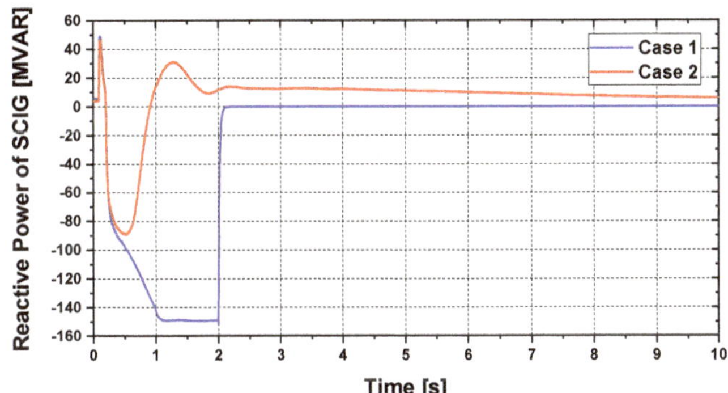

Figure 12. Reactive power response of SCIG.

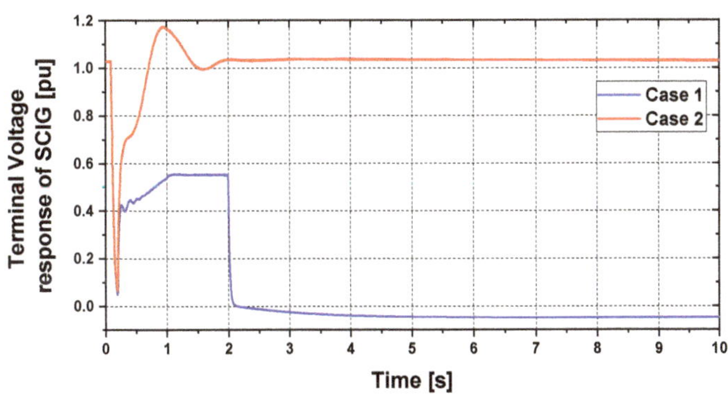

Figure 13. Terminal voltage response of SCIG.

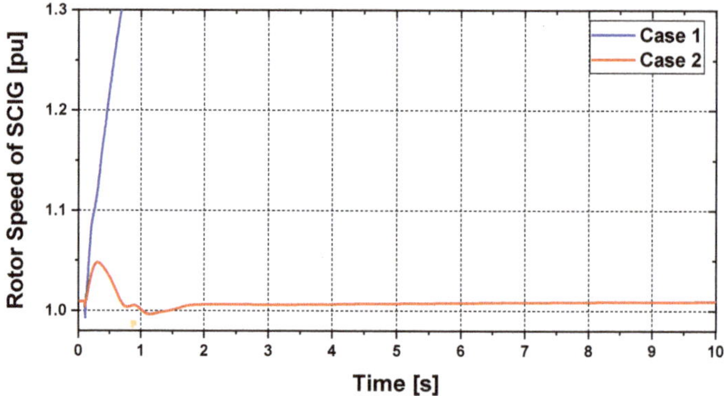

Figure 14. Rotor speed response of SCIG.

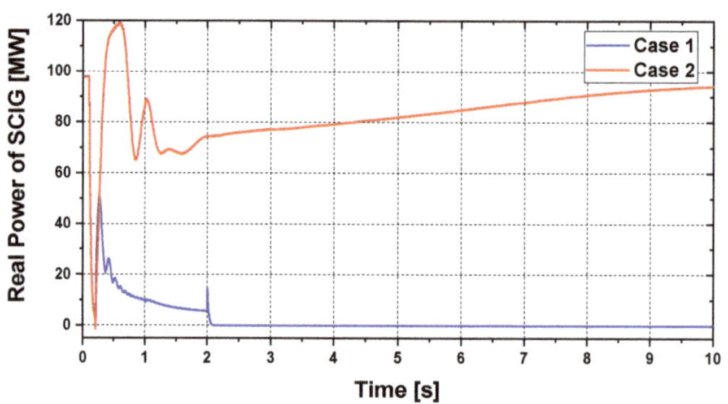

Figure 15. Real power response of SCIG.

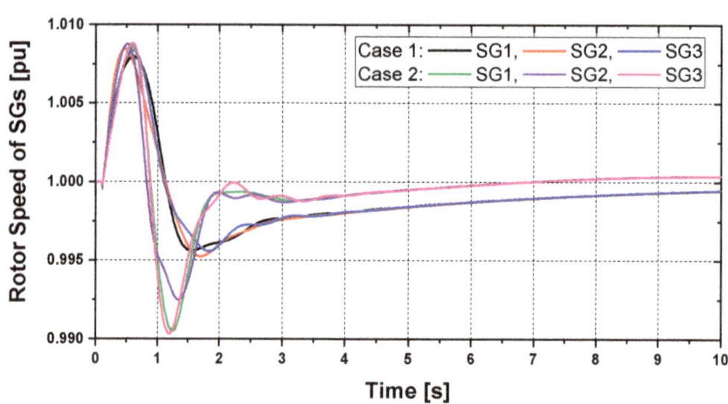

Figure 16. Rotor speed response of SGs.

5.2. Steady-State Stability Analysis

The actual wind speed value of Hokkaido Island, Japan is taken in this steady-state analysis as depicted in Figure 17. The total computational time is considered as 70 s.

Figure 18 shows the real power profile of SCIG-based WFs. The responses are identical for both cases, because no control system is involved in the SCIG. Additionally, the real power output is fluctuating because of the variable wind speed data.

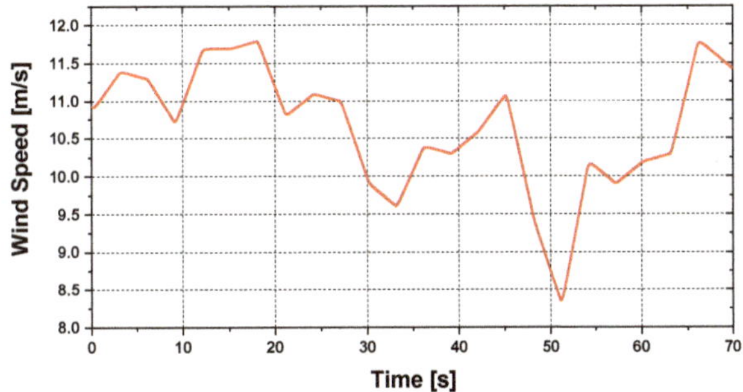

Figure 17. Wind speed applied to FSWT-SCIG.

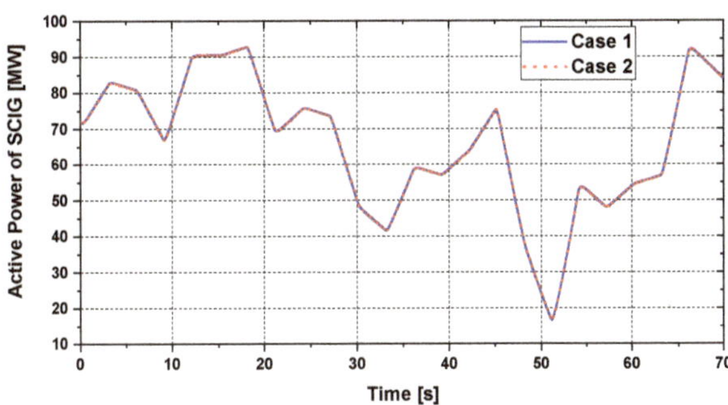

Figure 18. Real power response of FSWT-SCIG.

The BSS real power profile for Case 2 is presented in Figure 19. From the figure, it is clear that the BSS can ensure an effective amount of real power based on the frequency fluctuations. Thus, the frequency variation is smaller in Case 2 compared to Case 1, as depicted in Figure 20, which validates the importance of the proposed BSS. Figure 21 shows the real power profiles of conventional power plants (SGs). The real power outputs of SGs are fluctuating because of the variable outputs of SCIGs. In addition, there is small variance between the Case 1 and Case 2 responses. This is because BSS is providing and taking real power to and from the grid system, based upon frequency variations.

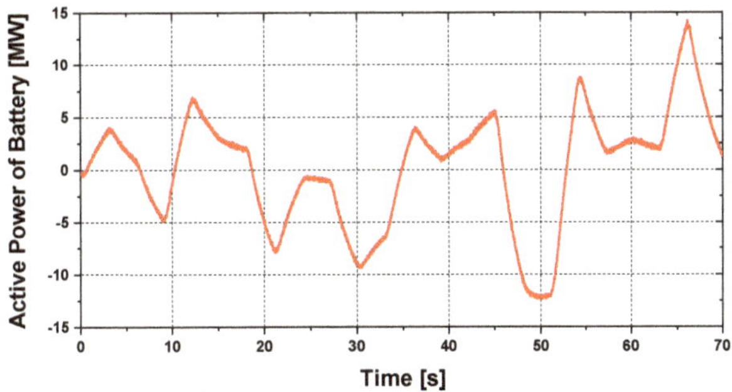

Figure 19. Real power output of BSS (Case 2 only).

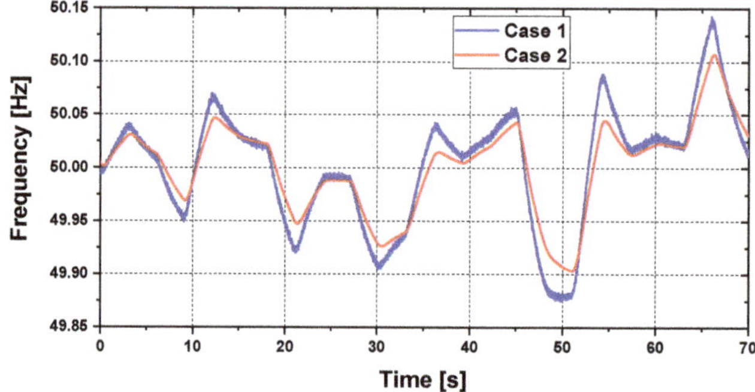

Figure 20. Hybrid power system frequency response.

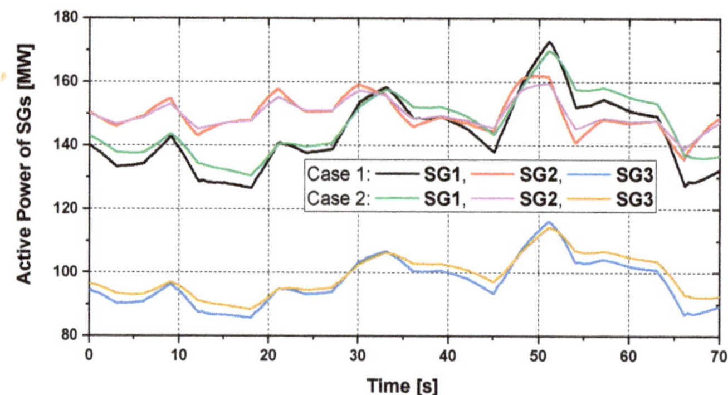

Figure 21. Real power response of SGs.

Finally, Figure 22 presents the mechanical power output of FSWT. The responses are identical for both cases, as no control system is involved in FSWT-SCIGs.

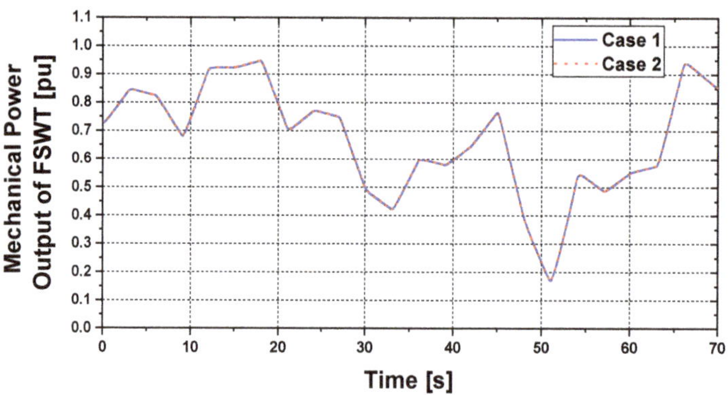

Figure 22. Mechanical power output of FSWT.

Table 1 shows the $+\Delta f$, $-\Delta f$, and σ for both cases, which are calculated from Figure 20. The $+\Delta f$, $-\Delta f$, and σ are smaller in Case 2 compared to Case 1.

Table 1. Comparison of different parameters of the frequency response graph.

Parameters of frequency	Case 1	Case 2
Maximum frequency deviation in positive direction ($+\Delta f$)	0.1428	0.1072
Maximum frequency deviation in negative direction ($-\Delta f$)	-0.1249	-0.0969
Standard deviation (σ)	0.0536	0.0419

6. Conclusions

To augment LVRT aptitude and minimize the frequency oscillations of a hybrid power system during transient and steady-state periods, a novel BSS-based FSWT-SCIG is proposed in this paper. Detailed design procedures of the proposed BSS, WFs, and hybrid power systems are explained adequately. The BSS can provide real and reactive power during steady-state and transient periods, respectively. The proposed BSS is composed of both LVRT enhancement techniques and frequency control algorithms. Different case studies are executed to show the usefulness of the proposed BSS. The LVRT characteristic is determined with respect to the standard grid code, taking a 3LG fault. Steady-state performance is tested using the actual wind speed data of Hokkaido Island, Japan. The simulation results clearly indicate that the LVRT aptitude can be improved, and frequency oscillations can be minimized effectively, by using the proposed BSS. Therefore, this proposed control technique has an encouraging prospective value.

As future work, the variable droop controller techniques of BSS with FSWT-SCIGs will be a strong candidate.

Author Contributions: M.R.H. and E.J. prepared the theoretical conceptions, and designed the proposed BSS and model of hybrid power system. M.R.H. executed the simulation studies. M.R.H. and E.J. wrote the manuscript. All authors have read and agreed to the published version of the manuscript.

Funding: This research received no external funding.

Conflicts of Interest: The authors declare no conflict of interest.

Appendix A

The parameters of conventional SGs and SCIGs are depicted in Tables A1 and A2, respectively.

Table A1. SGs parameters.

Parameter	SG1 (Thermal)	SG2 (Thermal)	SG3 (Hydro)
Voltage	16.5 kV	18 kV	13.8 kV
R_a	0.003 pu	0.003 pu	0.003 pu
X_l	0.1 pu	0.1 pu	0.1 pu
X_d	2.11 pu	2.11 pu	1.20 pu
X_q	2.05 pu	2.05 pu	0.700 pu
X'_d	0.25 pu	0.25 pu	0.24 pu
X''_d	0.21 pu	0.21 pu	0.20 pu
X''_q	0.21 pu	0.21 pu	0.20 pu
T'_{do}	6.8 s	7.4 s	7.2 s
T''_{do}	0.033 s	0.033 s	0.031 s
T''_{qo}	0.030 s	0.030 s	0.030 s
H	4.0 s	4.0 s	4.0 s

Table A2. SCIG parameters.

Squirrel Cage Induction Generator (SCIG)	
R_1	0.01 pu
X_1	0.1 pu
X_m	3.5 pu
R_{21}	0.035 pu
R_{22}	0.014 pu
X_{21}	0.03 pu
X_{22}	0.089 pu
H	1.5 s

References

1. Ula, A.H.M.S. Global warming and electric power generation: What is the connection? *IEEE Trans. Energy Convers.* **1991**, *6*, 599–604. [CrossRef]
2. Global Status Report 2017. Available online: https://www.worldgbc.org/sites/default/files/UNEP%20188_GABC_en%20%28web%29.pdf (accessed on 10 January 2020).
3. Global Wind Energy Council (GWEC). Annual Market Update 2015, Global Wind Report 2015. Available online: http://www.gwec.net/ (accessed on 15 October 2017).
4. Global Wind Energy Council (GWEC). Global Wind Energy Outlook 2016: Wind Power to Dominate Power Sector Growth 2016. Available online: http://www.gwec.net/ (accessed on 25 November 2019).
5. Hazari, M.; Mannan, M.; Umemura, A.; Takahashi, R.; Tamura, J. Stabilization of Wind Farm by Using PMSG Based Wind Generator Taking Grid Codes into Consideration. *J. Power Energy Eng.* **2018**, *6*, 40–52. [CrossRef]
6. Priyavarthini, S.; Nagamani, C.; Ilango, G.S.; Rani, M.A.A. An improved control for simultaneous sag/swell mitigation and reactive power support in a grid-connected wind farm with DVR. *Int. J. Electr. Power Energy Syst.* **2018**, *101*, 38–49. [CrossRef]
7. Mohammadpour, H.A.; Ghaderi, A.; Mohammadpour, H.; Ali, M.H. Low voltage ride-through enhancement of fixed-speed wind farms using series FACTS controllers. *Sustain. Energy Technol. Assess.* **2015**, *9*, 12–21. [CrossRef]
8. Wei, Y.; Kang, L.; Qi, R.; Wen, M.; Cheng, M.M. Optimal control of SVC-MERS and application in SCIG powered micro-grid. In Proceedings of the 2014 IEEE Applied Power Electronics Conference and Exposition—APEC 2014, Fort Worth, TX, USA, 16–20 March 2014; pp. 3367–3373.
9. Gatavi, E.; Hellany, A.; Nagrial, M.; Rizk, J. Improved Low Voltage Ride-Through capability of DFIG-based wind turbine with breaking resistor and converter control. In Proceedings of the 2018 IEEE 12th International Conference on Compatibility, Power Electronics and Power Engineering (CPE-POWERENG 2018), Doha, Qatar, 10–12 April 2018; pp. 1–6.
10. Kamel, R.M. Three fault ride through controllers for wind systems running in isolated micro-grid and Effects of fault type on their performance: A review and comparative study. *Renew. Sustain. Energy Rev.* **2014**, *37*, 698–714. [CrossRef]
11. Marei, M.I.; El-Goharey, H.S.K.; Toukhy, R.M. Fault ride-through enhancement of fixed speed wind turbine using bridge-type fault current limiter. *J. Electr. Syst. Inf. Technol.* **2016**, *3*, 119–126. [CrossRef]
12. Heydari-doostabad, H.; Khalghani, M.R.; Khooban, M.H. A novel control system design to improve LVRT capability of fixed speed wind turbines using STATCOM in presence of voltage fault. *Int. J. Electr. Power Energy Syst.* **2016**, *77*, 280–286. [CrossRef]
13. Ren, J.; Hu, Y.; Wang, J.; Ji, Y. Principle of low voltage ride-through ability realization of fixed speed wind generator using series reactor and SVC. In Proceedings of the 7th International Power Electronics and Motion Control Conference, Harbin, China, 2–5 June 2012; Volume 3, pp. 2173–2177.
14. Kalsi, S.; Madura, D.; Howard, R.; Snitchler, G.; MacDonald, T.; Bradshaw, D.; Grant, I.; Ingram, M. Superconducting dynamic synchronous condenser for improved grid voltage support. In Proceedings of the 2003 IEEE PES Transmission and Distribution Conference and Exposition (IEEE Cat. No.03CH37495), Dallas, TX, USA, 7–12 September 2003; Volume 2, pp. 742–747.
15. Farias, M.F.; Battaiotto, P.E.; Cendoya, M.G.; de Ingeniería, F. Investigation of UPQC for sag compensation in wind farms to weak grid connections. In Proceedings of the 2010 IEEE International Conference on Industrial Technology, Vina del Mar, Chile, 14–17 March 2010; pp. 937–942.

16. Mahela, O.P.; Gupta, N.; Khosravy, M.; Patel, N. Comprehensive Overview of Low Voltage Ride Through Methods of Grid Integrated Wind Generator. *IEEE Access* **2019**, *7*, 99299–99326. [CrossRef]

17. He, G.; Chen, Q.; Kang, C.; Xia, Q.; Poolla, K. Cooperation of Wind Power and Battery Storage to Provide Frequency Regulation in Power Markets. *IEEE Trans. Power Syst.* **2017**, *32*, 3559–3568. [CrossRef]

18. Nguyen, C.L.; Lee, H.H. A Novel Dual-Battery Energy Storage System for Wind Power Applications. *IEEE Trans. Ind. Electron.* **2016**, *63*, 6136–6147. [CrossRef]

19. Kundur, P. *Power System Stability & Control*; McGraw-Hill Inc.: New York, NY, USA, 1994.

20. Rosyadi, M.; Umemura, A.; Takahashi, R.; Tamura, J.; Uchiyama, N.; Ide, K. Simplified Model of Variable Speed Wind Turbine Generator for Dynamic Simulation Analysis. *IEEJ Trans. Power Energy* **2015**, *135*, 538–549. [CrossRef]

21. Muyeen, S.M.; Tamura, J.; Murata, T. *Stability Augmentation of a Grid Connected Wind Farm*; Springer: London, UK, 2009.

22. Hazari, M.R.; Mannan, M.A.; Muyeen, S.M.; Umemura, A.; Takahashi, R.; Tamura, J. Stability Augmentation of a Grid-Connected Wind Farm by Fuzzy-Logic-Controlled DFIG-Based Wind Turbines. *Appl. Sci.* **2018**, *8*, 20. [CrossRef]

23. E. ON NETZ GmbH. *Grid Connection Regulation for High and Extra High Voltage*; E. ON NETZ GmbH: Essen, Germany, 2006.

 *electronics*

Article

Simple and Low-Cost Photovoltaic Module Emulator

Massimo Merenda [1,2,*], **Demetrio Iero** [1,2], **Riccardo Carotenuto** [1] and **Francesco G. Della Corte** [1,2]

[1]   Department of Information Engineering, Infrastructure and Sustainable Energy (DIIES),
University Mediterranea of Reggio Calabria, 89124 Reggio Calabria, Italy; demetrio.iero@unirc.it (D.I.);
r.carotenuto@unirc.it (R.C.); francesco.dellacorte@unirc.it (F.G.D.C.)

[2]   HWA srl-Spin Off dell'Università Mediterranea di Reggio Calabria, Via Reggio Campi II tr. 135,
89126 Reggio Calabria, Italy

*   Correspondence: massimo.merenda@unirc.it; Tel.: +39-0965-1693-441

Received: 30 October 2019; Accepted: 26 November 2019; Published: 1 December 2019

Abstract: The design and testing phase of photovoltaic (PV) power systems requires time-consuming and expensive field-testing activities for the proper operational evaluation of maximum power point trackers (MPPT), battery chargers, DC/AC inverters. Instead, the use of a PV source emulator that accurately reproduces the electrical characteristic of a PV panel or array is highly desirable for in-lab testing and rapid prototyping. In this paper, we present the development of a low-cost microcontroller-based PV source emulator, which allows testing the static and dynamic performance of PV systems considering different PV module types and variable operating and environmental conditions. The novelty of the simple design adopted resides in using a low-cost current generator and a single MOSFET converter to reproduce, from a fixed current source, the exact amount of current predicted by the PV model for the actual load conditions. The I–V characteristic is calculated in real-time using a single diode exponential model under variable and user-selectable operating conditions. The proposed method has the advantage of reducing noise from high-frequency switching, reducing or eliminating ripple and the demand for output filters, and it does not require expensive DC Power source, providing high accuracy results. The fast response of the system allows the testing of very fast MPPTs algorithms, thus overcoming the main limitations of state-of-art PV source emulators that are unable to respond to the quick variation of the load. Experimental results carried on a hardware prototype of the proposed PV source emulator are reported to validate the concept. As a whole result, an average error of ±1% in the reproduction of PV module I–V characteristics have been obtained and reported.

Keywords: photovoltaic emulator; photovoltaic panel; single diode model; MPPT

---

1. Introduction

The design of electronic power converters for photovoltaic (PV) applications requires a stable and repeatable PV source for experimental testing under realistic operating conditions, which can accurately reproduce the relationship between the output voltage and current of a given PV module.

Furthermore, the possibility to test variable conditions of the PV source is crucial as it is part of the validation of the final product, assessing the behavior in the broadest range of operative condition modifications as temperature, irradiation, and shadowing.

PV real installation does not satisfy at all the requirements hereafter: the I–V characteristics are linked to slowly varying operational parameters like temperature and irradiation, they require in-field test systems and apparatus, test conditions are defined by the actual meteorological conditions that should be jointly acquired [1] and which can rapidly change during the test [2].

Therefore, a PV source emulator is required to complete the on-lab assessments under variable operating conditions in a reasonable amount of time.

Typically, a PV source emulator reproduces the I–V curve of an actual PV module starting from a constant DC source, using both different conversion strategies and power sizes.

Commercial PV source emulators are available in the market, enabling the user to select different PV module or PV array emulations with variable power range [3–5]. However, commercial PV source emulators present several drawbacks, as high cost and a limitation in the rapidly changing atmospheric conditions emulation [6].

Many researchers tried to overcome the limitation of commercial products or to develop low-cost and affordable PV source emulators. Many possible approaches to performing the task of PV source emulator design were found in the literature.

The simplest model of photovoltaic generator emulator can be obtained by connecting in series a DC voltage generator and a variable resistor [7]. The open-circuit voltage V_{OC} is set by the maximum output voltage of the DC generator, while the short-circuit current I_{SC} depends both on the output voltage of the DC generator and on the value of the resistance of the variable resistor. For a given value of the series resistance R_S, if the load resistance is varied from its minimum value to its maximum value, a linear I–V characteristic with a negative slope will be obtained. The main advantage of this technique is the simplicity of implementation, however, the characteristic I–V obtained differs significantly from that of a real photovoltaic source. Moreover, this type of PV source emulator is characterized by a low efficiency (maximum 50%) due to dissipative losses on the series resistance.

A further method is based on the use of an analog amplification technique that allows to independently amplify the low current and voltage values typical of a photodiode operating in the photovoltaic mode so as to make them coincide with those of a standard photovoltaic module [8,9]. Since this type of emulator is made entirely with analog components, it has a high bandwidth, which allows using this circuit to test photovoltaic inverters with maximum power point tracker (MPPT) algorithms operating at high frequency [10–12]. On the other hand, the main disadvantage of this circuit is the high power dissipation that involves the use of heat sinks with a large surface area.

In [11] the logarithmic approximation of the ideal single diode model is used, and the power stage consists of a DC power supply feeding a linear voltage regulator.

To overcome the disadvantage of the high power consumption of PV source simulators based on analog electronics, several solutions based on the use of DC/DC static converters (choppers) have been proposed, such as [13–20]. To determine the output voltage and current values of the DC/DC converter, this is controlled by a feedback control where the output current of the chopper is compared with a reference current. The reference current can be determined in real-time by means of a mathematical model of the photovoltaic module (PV-model) [19] or extrapolated from data stored in a look-up-table (LUT). The first approach requires knowledge of the parameters, which is difficult to achieve in some situations. The most commonly used approach is to derive an analytical model to represent the I–V curves from the data available from the datasheet of the photovoltaic panel manufacturer [21]. The choice of whether to use a mathematical model of the photovoltaic module or a LUT must be made considering several factors such as the speed of response of the system, accuracy, and use of hardware resources in terms of both computational and memory. The complexity of the problem increases if there is the necessity to emulate modules with different I–V characteristics since each of them will require its own look-up-table. Another disadvantage of emulators using look-up-tables is that the I–V characteristic of the module, between two successive points stored in the look-up-table, is obtained by linear interpolation, which makes the system less accurate than the mathematical model-based approach. In contrast, emulators based on the resolution of a mathematical model of the PV module are more accurate and do not require a large amount of memory [22]. However, in order to obtain a detailed representation of the I–V characteristic, the mathematical model will contain high order equations, leading to an increase in computational time and thus to a slower system.

In [23] is proposed a solution with a field programmable analog array, characterized by great ease of reconfiguration and programming with respect to field programmable gate array (FPGA) or digital signal processing (DSP) based implementations. No digital to analog converters (DAC) or analog to digital converters (ADC) is needed while the final cost remains quite high due to the need for at least a DC/DC converter.

In [24] is proposed a solution that adopts modular hardware, configurable software, systematic modeling, and design methods, requiring a PC running Matlab/Simulink to determine the controller parameters.

Another method builds an equivalent photovoltaic source using an unlighted photovoltaic panel and a DC current power supply [10]. However, this approach requires the use of an actual PV panel that must be changed if it is necessary to emulate a different panel; also, the temperature effect is not easily simulated.

In [25] a dual-mode regulator consisting of a voltage regulator and a current regulator, connected by two diodes for power hybridization, is proposed. The system switches between voltage and current regulation, thus requiring complex and costly electronics.

In this paper, in order to overcome high realization costs, reduced accuracy and versatility, we present a low-cost, microcontroller-based PV source emulator, which allows for testing the performance of PV systems including different PV module types at user-selectable operating conditions. The I–V characteristic is calculated in real-time using a simple diode model, and it does not use any DC-DC converter, reducing the noise from high-frequency switching, and reducing or eliminating ripple and the demand of output filters. Moreover, it does not require the use of expensive DC power sources and can be used for laboratory tests and rapid prototyping by researchers and students. In addition, the proposed solution provides accurate emulations over the full span of emulated power source, differently from other state-of-art solutions that provide good results only closer to the maximum power point (MPP). The model implemented takes into account fast-changing environmental conditions that can be accurately tracked and/or emulated with the use of actual temperature, humidity and illumination sensors or by software techniques.

The paper is organized as follows. Section 2 introduces the photovoltaic cell model. Section 3 explains the working principle of the proposed PV source emulator and the experimental setup. Section 4 reports the experimental results and compares them with simulations, at different conditions. Section 5 draws the paper conclusions.

2. PV Cell Model and Characteristics

Each photovoltaic cell is characterized by some basic parameters provided by the manufacturers, referring to the standard test conditions (STC) which specifies an irradiance of 1000 W/m^2, a cell temperature of 25 °C and an air mass 1.5 (AM1.5) spectrum:

- Short-circuit current I_{SC}: the maximum current that can be supplied by a photovoltaic cell under short-circuit conditions ($V = 0$);
- Open circuit voltage V_{OC}: the maximum voltage drop at the cell terminals when the supplied current is null;
- Maximum power point MPP: the point of the I–V characteristic (V_{max}, I_{max}) where the power supplied to the load is maximum;
- Fill Factor $FF = (V_{max} \times I_{max})/(V_{OC} \times I_{SC})$: a parameter that measures the quality of the cell;
- Conversion efficiency η, defined as the ratio between the maximum power produced by the cell and the total power incident on its surface at STC.

The Short-circuit current depends linearly on the value of the irradiance, whereas the influence of the temperature can be expressed by the coefficient k_I (Temperature coefficient of short-circuit current) that indicates the percentage variation of the short circuit current as the temperature changes from the value determined under STC. Contrary to the short-circuit current, the open-circuit voltage remains relatively constant when the radiation changes but is strictly dependent on the cell temperature. The coefficient k_V (Temperature coefficient of open-circuit voltage) expresses the variation of the open-circuit voltage as the temperature changes with respect to the reference value calculated under

STC. The photovoltaic cell absorbs most of the incident solar radiation but there is a significant portion of the absorbed radiation that is not converted into electricity but generates heat, which causes an increase of the temperature of the cells that affects I_{SC} and V_{OC}, and consequently the conversion efficiency.

In the dark, the I–V characteristic of a photovoltaic cell shows an exponential shape like the I–V characteristic of a diode. As a result, a photovoltaic cell exposed to solar radiation can be assimilated from a circuital point of view to a current generator with a diode in parallel (Figure 1). Various models are available in literature [26–30] and the single diode model is one of the simplest models.

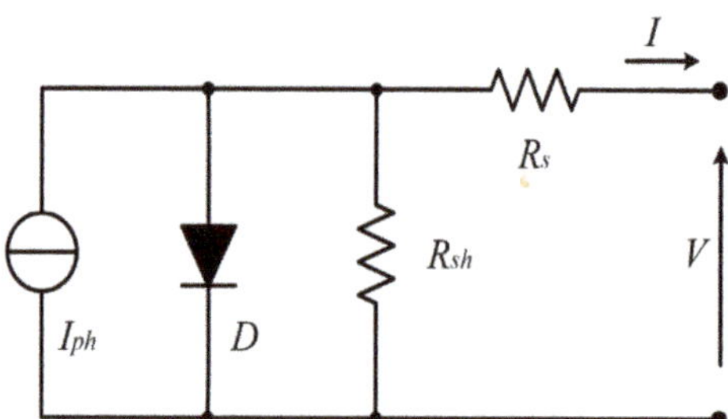

Figure 1. Photovoltaic cell: equivalent circuit of the single diode model.

The current generator delivers an I_{PH} current directly proportional to the solar radiation incident on the photovoltaic cell. Two resistors have been added to take into account the internal losses of the photovoltaic cell:

- R_S is the series resistance that models the internal losses of the cell due to the sum of the resistive contributions inside the cell and the contact resistance;
- R_{SH} is the shunt resistance that models the effects of leakage currents in the p-n junction mainly due to manufacturing defects in the photovoltaic cell.

This model can be extended to model the operation of a photovoltaic module by specifying the number of cells connected in series N_s and in parallel N_p.

Various mathematical models are available describing the electrical behavior of a photovoltaic cell, which differs according to the precision of the mathematical model to be obtained and the number of parameters available. Simplified models have been developed taking into account only the parameters that can be measured practically [31]; in this work, the mathematical model follows equations in [32–34]. The thermal voltage V_t is defined as

$$V_t = \frac{kT_{op}}{q},\tag{1}$$

where q is the charge of an electron, k is the Boltzmann constant, T_{op} is the temperature in K.

V_{OC} depends on the saturation current density of the solar cell I_S and the photo-generated current I_{ph}:

$$V_{oc} = ln\frac{I_{ph}}{I_s}V_t,\tag{2}$$

The Shockley equation that relates the current and voltage of the cell in zero-illumination condition is adjusted for a photovoltaic module [32] by specifying the number of cells connected in series N_s and in parallel N_p:

$$I_d = I_s\left(e^{\frac{V+R_sI}{nV_tN_s}} - 1\right)N_p,\tag{3}$$

where n is the ideality factor which is defined as how closely a diode follows the ideal diode equations. The reverse saturation current can be obtained by:

$$I_s = I_{rs}\left(\frac{T_{op}}{T_{ref}}\right)^3 e^{\frac{-qE_g}{nk}\left(\frac{1}{T_{op}}-\frac{1}{T_{ref}}\right)},$$ (4)

with E_g that is the extrapolated energy bandgap at 0 K. I_{rs} is the value of the saturation current at T_{op}:

$$I_{rs} = \frac{I_{SC}}{e^{\frac{V_{oc}q}{nkT_{op}}}-1}.$$ (5)

The current flowing through the shunt resistance R_{SH} is defined as:

$$I_{sh} = \frac{V+R_SI}{R_{sh}},$$ (6)

and the photo-generated current is defined as:

$$I_{ph} = G_k\left[I_{sc}+k_I\left(T_{op}-T_{ref}\right)\right],$$ (7)

with G_k being the solar irradiance and I_{sc} the short circuit current, $k_I = (I_{SC(Top)}-I_{SC(Tref)})/(T_{op}-T_{ref})$. Finally, the characteristic equation of a photovoltaic panel using the equivalent circuit of Figure 1 is deduced:

$$I = I_{ph}N_p - I_d - I_{sh}.$$ (8)

3. Description of the PV Source Emulator

3.1. System Overview

Starting from a circuit model of a photovoltaic module and from the knowledge of the parameters of that model, it is possible to create a MATLAB Simulink model useful to test the behavior of a generic photovoltaic module and to trace its I–V and P–V characteristics when the load conditions or the environmental parameters to which the module is subjected, such as temperature and radiation, vary. The Simulink model of the photovoltaic module used in this work is based on the single diode circuit model of the Solarex MSX-60 photovoltaic module model available on [35].

It is worth noting that the novelty proposed in this work resides in the accurate emulation and curve-fitting of the characteristic of a real PV module by the proposed PV source emulator, and not in the model itself.

The model parameters have been adapted to simulate the output characteristic of a 12 W photovoltaic module with the parameters reported in Table 1. The block diagram of the proposed photovoltaic source emulator is shown in Figure 2.

Table 1. Basic Parameter of the Simulated PV module at Standard Test Conditions (STC).

Parameter	Value
Typical Peak Power (P_{MPP})	12 W
Short circuit Current (I_{SC})	0.683 A
Open circuit Voltage (V_{OC})	25 V
Temperature coefficient of Open-Circuit Voltage (k_V)	$-(80 \pm 10)$ mV/°C
Temperature coefficient of Short circuit Current (k_I)	(0.065 ± 0.015) %/°C
Temperature coefficient of Power	(0.5 ± 0.05) %/°C
Nominal Operating Cell Temperature (NOCT)	47 ± 2 °C

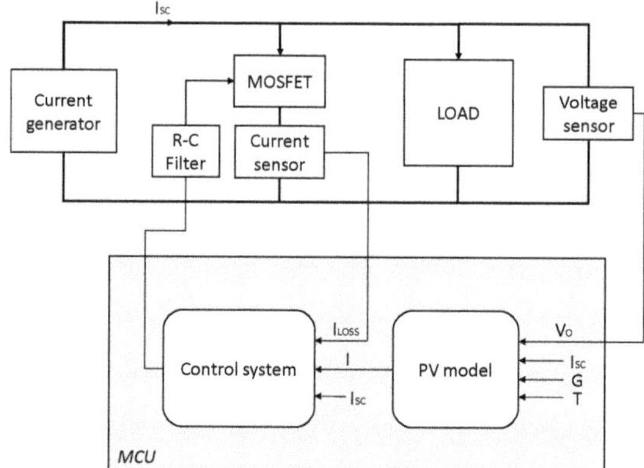

Figure 2. Block diagram of the PV source emulator.

3.2. Emulation Technique

The I–V characteristic of a photovoltaic module is a monotonous decreasing function, in fact, the current supplied by the photovoltaic module is maximum $I = I_{SC}$ when it is in a short circuit condition, and decreases as the voltage across the module increases. Referring to the I–V characteristic in Figure 3, it can be seen that, for each voltage value, the current supplied by the photovoltaic module can be determined as the difference between the short circuit current I_{SC} and a loss current I_{LOSS}:

$$\forall V \in (0, V_{oc}) \rightarrow I(V) = I_{sc} - I_{loss}(V). \tag{9}$$

The technique proposed in this paper is based on the use of:

- A current generator capable of delivering a constant current equal to I_{SC} in the voltage range from 0 to V_{OC};
- A control system capable of determining, for each voltage, the value of the current I_{LOSS} required to subtract to the load;
- A power MOSFET that drain I_{LOSS} from the branch in which the I_{SC} flows.

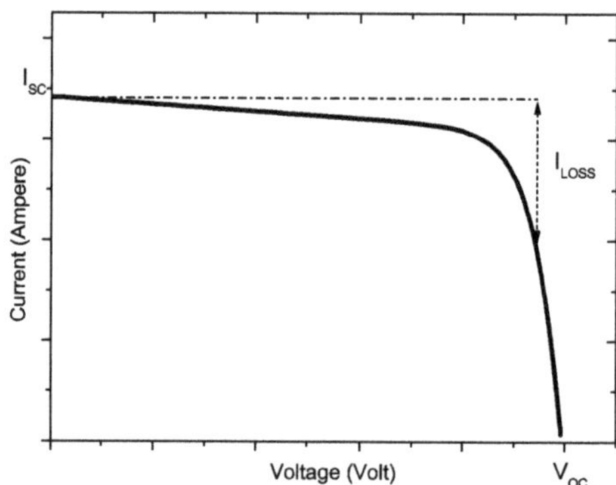

Figure 3. Graphic representation of the emulation technique; the current on the load is obtained as the difference between the I_{SC} and the I_{LOSS}.

Based on the data returned by the mathematical model of the photovoltaic module, the control system determines the value of the current I_{LOSS} and generates a control signal that is applied to the gate of the MOSFET in order to modulate the drain current so that it results in $I_{DRAIN} = I_{LOSS}$. The current circulating in the MOSFET is sensed by the control system that aims to minimize the error between the target I_{LOSS} and the one that actually flows on the MOSFET. A proportional-integral-derivative (PID) controller allows generating the suitable control signal to achieve the required current regulation.

3.3. Experimental Setup

The control logic and the mathematical model of the photovoltaic module have been implemented on an STM32F401RE microcontroller (STMicroelectronics), mounted on the STM32 Nucleo board. A specifically made shield board attached on the Nucleo board implements the necessary hardware that allows to control the MOSFET and sense current and voltage (Figure 4).

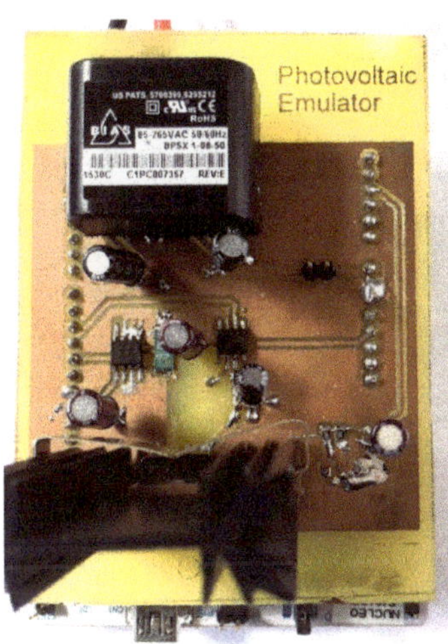

Figure 4. A picture of the realized photovoltaic source emulator prototype: the custom made shield is mounted on top of an STM32 Nucleo board by STMicroelectronics.

The microcontroller implements the PID controller that allows modulating the control signal to achieve the required current regulation. It generates a 3.3 V PWM control signal and a TC4424A (Microchip, Chandler, AZ, USA) driver produces a PWM signal between 0 V and 8 V, which is filtered by a low-pass RC filter to extract the average value and used to control an IRF820 MOSFET (International Rectifier Semiconductors) in linear mode. The average value of the PWM signal depends on the duty cycle $D = Ton/T$ of the signal, defined as the ratio between the pulse active time *Ton* and the period T of the signal. The average value of the gate signal can, therefore, be adjusted by acting on the duty cycle according to the relationship:

$$V_{out} = V_H D,$$ (10)

where V_H is the maximum voltage of the PWM signal.

The PWM frequency of the control signal is 50 kHz and the RC filter is set with a resistance R = 150 kΩ and a capacitance C = 10 μF for a cut-off frequency of 0.106 Hz to minimize the control signal high-frequency components.

The microcontroller firmware configures and initializes the microcontroller peripherals needed to interact with the PV source emulator hardware, and implements the photovoltaic emulator code. The internal 12-bit ADC is used to measure the voltage on the load through a voltage divider, and the current through a shunt resistor of 10 mΩ and a current sense amplifier INA283 (Texas Instruments, Dallas, TX, USA) with a gain of 200 V/V; this configuration allows to measure up to 1.65 A of current with very low power dissipation.

After the initialization, the firmware cyclically: (1) reads the voltage on the load and current I_{LOSS} passing through the MOSFET, (2) calculates the target $I_{LOAD,ref}$ by using the PV model equations and Equation (9), and (3) determines, by mean of a digital PID controller, the variation of duty cycle of the PWM control signal required to produce the desired $I_{LOSS,ref}$ current.

The constant current generator is chosen such that the voltage and current ratings are higher than the I_{SC} and V_{OC} of the panel to be emulated reported in Table 1. A Jolight KL824-04 power supply that delivers 700 mA with a voltage range up to 30 V and its output has been connected to a variable resistive load of a maximum of 110 Ω. The system is supplied by an AC-DC miniature switching power supply (Bias Power BPSX 1-08-50) that converts 230 V AC main voltage to 8 V and 5 V DC voltage to supply the boards.

The digital multimeter Agilent U1272A and the LeCroy WaveSurfer 343 oscilloscope were used to monitor and acquire the signals.

4. Results

4.1. Simulation Results

In order to test the Simulink model of the photovoltaic module, simulations have been carried out at different solar irradiances and operating temperatures. Figure 5 shows the I–V and P–V characteristics generated by the model for different values of irradiance and temperature.

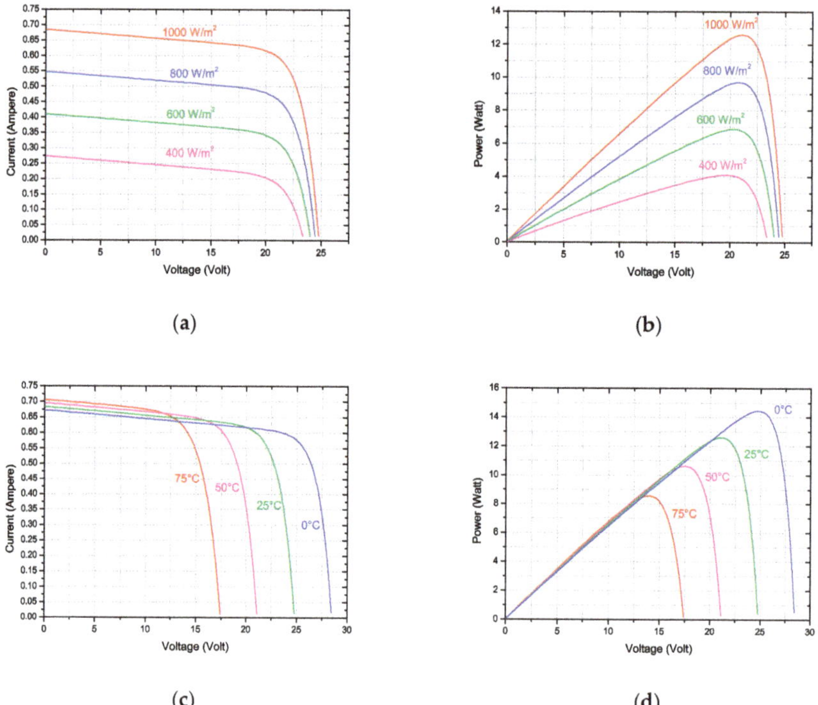

Figure 5. I–V and P–V characteristics generated by the model: (**a**) I–V characteristic at 25 °C, AM1.5, at different irradiance values; (**b**) P–V characteristic at 25 °C, AM1.5, at different irradiance values; (**c**) I–V characteristic at different temperatures, G = 1000 W/m², AM1.5; (**d**) P–V characteristic at different temperatures, G = 1000 W/m², AM1.5.

4.2. Experimental Results

The first test conducted aims to verify the ability of the photovoltaic emulator to follow the I–V characteristic of the simulated photovoltaic module in the Simulink environment under STC conditions. For the test, the photovoltaic source emulator has been programmed to $V_{OC} = 25$ V and $I_{SC} = 0.7$ A. Figure 6 shows the comparison of the I–V characteristics generated by the Simulink model and that of the proposed photovoltaic source emulator. It is possible to note that the emulator faithfully reproduces the characteristic of the simulated photovoltaic module. The right terminal part of the I–V characteristic cannot be reproduced by the emulator because the value of the power resistor used as a load in the tests could not assume sufficiently high resistance values, being limited to 110 Ω.

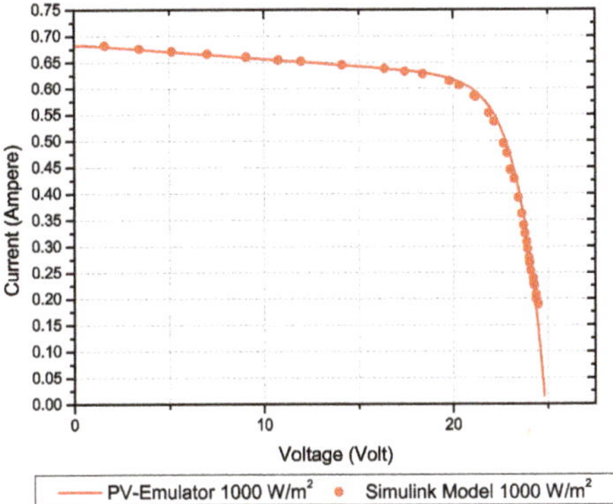

Figure 6. Comparison between the simulated I–V characteristic and the output I–V characteristic of the photovoltaic source emulator under STC conditions (1000 W/m², 25 °C).

The absolute deviation value between the simulated I–V characteristic and the one reproduced by the photovoltaic emulator has been calculated, according to:

$$Abs(Error)\% = Abs\left(I_{load_{Model}} - I_{load_{PVEmulator}}\right)100, \tag{11}$$

and the results are shown in Figure 7. The deviation for a large part of the characteristic is less than 1% and overall is below 5%. The maximum error is reported very close to V_{OC} where a slight variation of the voltage causes a sudden and large current variation. The accuracy of the system demonstrates the effectiveness of the proposed PV source emulator in reproducing the I–V characteristic.

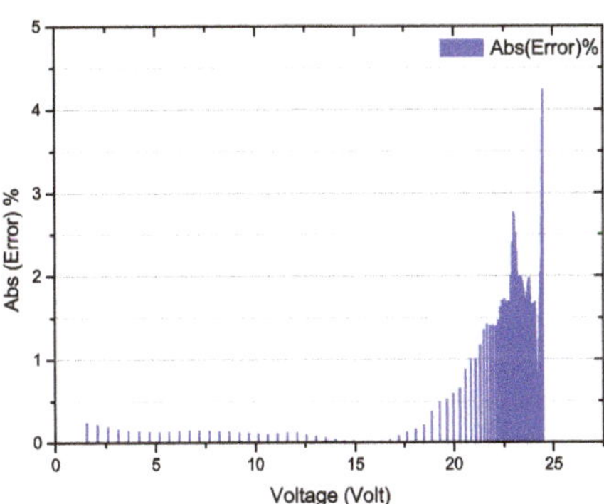

Figure 7. Maximum absolute deviation between the simulated I–V characteristic and the one reproduced by the proposed photovoltaic source emulator.

4.2.1. Results at Different Environmental Conditions

Other tests have been carried out to verify that the photovoltaic source emulator is able to follow the characteristics of I–V and P–V at different values of the environmental parameters. From the graphs shown in Figures 8 and 9, it can be seen that even when irradiation and temperature conditions vary, the photovoltaic source emulator can fit the target I–V and P–V characteristics.

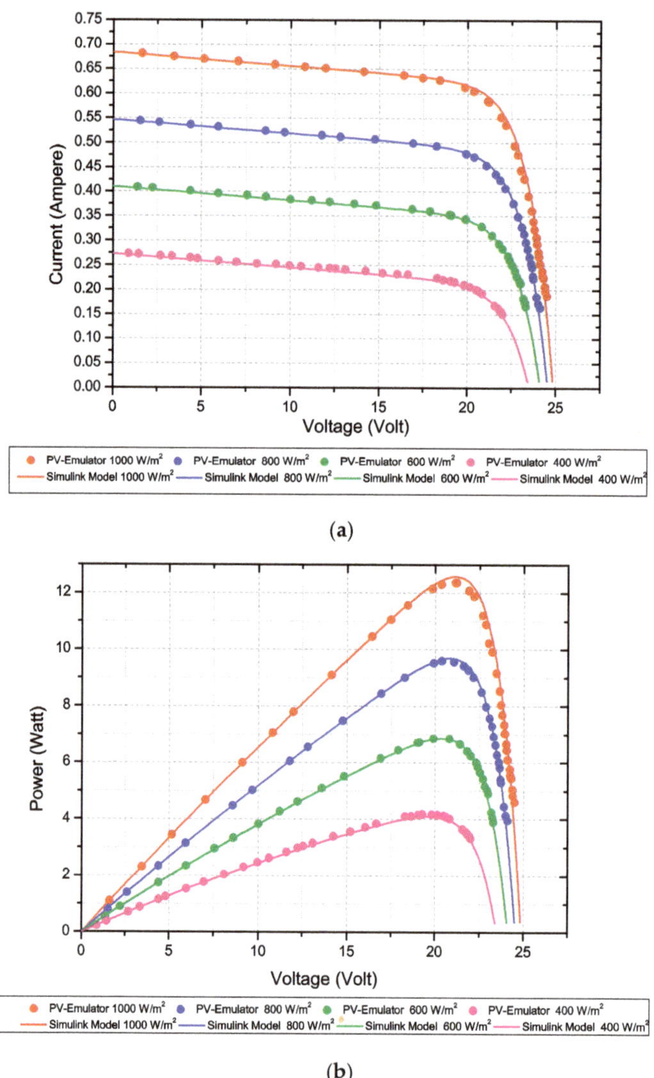

Figure 8. Comparison of simulated and emulated characteristics at different irradiation conditions: (**a**) I–V characteristic; (**b**) P–V characteristic.

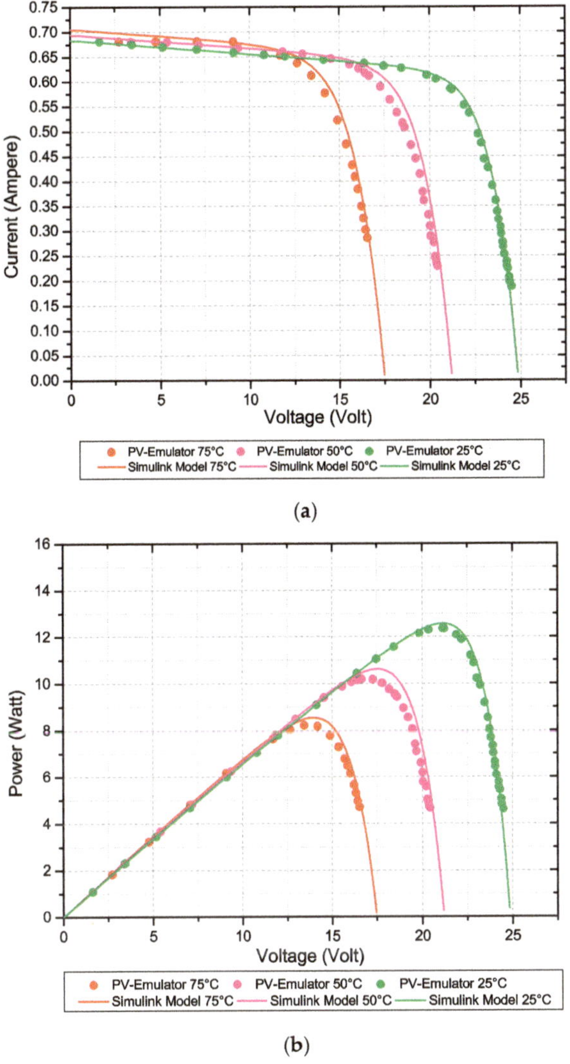

(a)

(b)

Figure 9. Comparison of simulated and emulated characteristics at different temperatures: (**a**) I–V characteristic; (**b**) P–V characteristic.

4.2.2. Dynamic Performance

In previous paragraphs, the ability of the photovoltaic source emulator to reproduce the I–V and P–V characteristics under very slowly varying load and fixed environmental conditions has been shown. On the other hand, a photovoltaic emulator must also be characterized by a dynamic point of view, evaluating the time necessary to track the sudden changes in load resistance.

To test the settling time of the control signal, the filter has been set to a cut-off frequency of 10.6 Hz and the value of the load resistance was changed using a rheostat to increase the load resistance from 5 Ω to 30 Ω. The results are reported in Figure 10 and show a response time of less than 150 ms. The system has also been tested as a power source with two different MPPTs, a commercial SolarEdge

Power Optimizer and an experimental MPPT [36], showing a correct behavior within the limits of its power range.

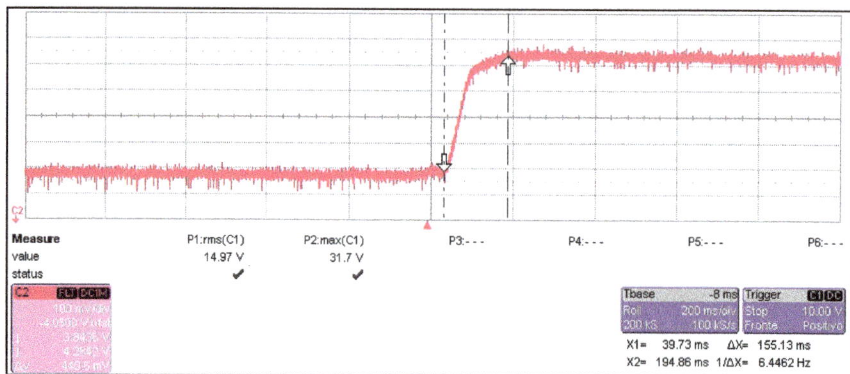

Figure 10. Transient response of the control signal following an instantaneous increase of the load resistance from 5 Ω to 30 Ω.

5. Conclusions

In this paper, we presented the design, test, and results of the development of a low-cost microcontroller-based PV source emulator, which allows testing the static and dynamic performance of PV systems considering different PV module types and variable operating and environmental conditions. The photovoltaic source emulator is based on a completely new technique, which consists in subtracting an adequate amount of current from a fixed direct current source so as to reproduce the desired I–V characteristic. Direct current sources can be found on the market at a very low price, in comparison with systems based on expensive DC voltage sources. Moreover, the proposed method has the advantage of reducing noise from high-frequency switching, reducing or eliminating ripple and the demand of output filters, and it does not require expensive DC Power source, providing high accuracy results. In fact, very good accuracy in the reproduction of PV module I–V characteristics has been obtained and reported with an average and maximum error of, respectively ±1% and ±5%. Experimental results on a hardware prototype of the proposed PV source emulator validate the concept, showing a very good adherence to the simulation.

The fast dynamic response of the system (150 ms) allows the testing of very fast MPPTs algorithms, thus overcoming the main limitations of state-of-art PV source emulator that is unable to respond to the quick variation of the load. The system has been tested as a power source with two different MPPTs showing a correct behavior within the limited power range.

A drawback of the proposed system is mostly the heat dissipation over the transistor used in linear region, which in fact reduces the efficiency of the system. However, for the purpose of this work, such drawback is considered acceptable, allowing the reduction of the Bill of Material of the board cost to less than 20 dollars and providing, at the same time, an accurate yet simple method for emulating photovoltaic sources.

Author Contributions: Conceptualization, M.M.; Data curation, M.M. and D.I.; Formal analysis, M.M.; Investigation, M.M., D.I., R.C. and F.G.D.C.; Methodology, M.M., R.C.; Software, M.M. and D.I.; Supervision, M.M.; Writing—original draft, M.M.; Writing—review & editing, M.M., D.I., R.C. and F.G.D.C.

Funding: This research received no external funding.

Acknowledgments: PAC Calabria 2014–2020 Asse Prioritario 12, Azione 10.5.12, is gratefully acknowledged by one of the authors (D.I.).

Conflicts of Interest: The authors declare no conflict of interest.

References

1. Piliougine, M.; Carretero, J.; Mora-López, L.; Sidrach-De-Cardona, M. Experimental system for current-voltage curve measurement of photovoltaic modules under outdoor conditions. *Prog. Photovolt.* **2011**, *19*, 591–602. [CrossRef]
2. Camino-Villacorta, M.; Egido-Aguilera, M.A.; Díaz, P. Test procedures for maximum power point tracking charge controllers characterization. *Prog. Photovolt. Res. Appl.* **2012**, *20*, 310–320. [CrossRef]
3. Elgar TerraSAS ETS Series. Available online: https://www.powerandtest.com (accessed on 24 October 2019).
4. Keysight E4368A Solar Array Simulator. Available online: https://www.keysight.com (accessed on 24 October 2019).
5. Magna-Power Programmable DC Power Supply. Available online: http://www.magna-power.com (accessed on 24 October 2019).
6. Sanchis, P.; López, J.; Ursúa, A.; Gubía, E.; Marroyo, L. On the testing, characterization, and evaluation of PV inverters and dynamic MPPT performance under real varying operating conditions. *Prog. Photovolt.* **2007**, *15*, 541–556. [CrossRef]
7. Mukerjee, A.; Dasgupta, N. DC power supply used as photovoltaic simulator for testing MPPT algorithms. *Renew. Energy* **2007**, *32*, 587–592. [CrossRef]
8. Nagayoshi, H. I–V curve simulation by multi-module simulator using I–V magnifier circuit. *Sol. Energy Mater. Sol. Cells* **2004**, *82*, 159–167. [CrossRef]
9. Midtgard, O.-M. A simple photovoltaic simulator for testing of power electronics. In Proceedings of the 2007 European Conference on Power Electronics and Applications, Aalborg, Denmark, 2–5 September 2007; pp. 1–10.
10. Zhou, Z.; Macaulay, J. An Emulated PV Source Based on an Unilluminated Solar Panel and DC Power Supply. *Energies* **2017**, *10*, 2075. [CrossRef]
11. Moussa, I.; Khedher, A.; Bouallegue, A. Design of a Low-Cost PV Emulator Applied for PVECS. *Electronics* **2019**, *8*, 232. [CrossRef]
12. Sanaullah, A.; Khan, H.A. Design and implementation of a low cost Solar Panel emulator. In Proceedings of the 2015 IEEE 42nd Photovoltaic Specialist Conference (PVSC), New Orleans, LA, USA, 14–19 June 2015.
13. Koutroulis, E.; Kalaitzakis, K.; Tzitzilonis, V. Development of an FPGA-based system for real-time simulation of photovoltaic modules. *Microelectron. J.* **2009**, *40*, 1094–1102. [CrossRef]
14. Rana, A.V.; Patel, H.H. Current Controlled Buck Converter based Photovoltaic Emulator. *J. Ind. Intell. Inf.* **2013**, *1*, 91–96. [CrossRef]
15. Walker, D.; Fisher, T.; Liu, J.; Schrimpf, R. Thermal modeling of single event burnout failure in semiconductor power devices. *Microelectron. Reliab.* **2001**, *41*, 571–578. [CrossRef]
16. González-Medina, R.; Patrao, I.; Garcerá, G.; Figueres, E. A low-cost photovoltaic emulator for static and dynamic evaluation of photovoltaic power converters and facilities. *Prog. Photovolt. Res. Appl.* **2014**, *22*, 227–241. [CrossRef]
17. Di Piazza, M.C.; Vitale, G. Photovoltaic field emulation including dynamic and partial shadow conditions. *Appl. Energy* **2010**, *87*, 814–823. [CrossRef]
18. Koran, A.; LaBella, T.; Lai, J.S. High Efficiency Photovoltaic Source Simulator with Fast Response Time for Solar Power Conditioning Systems Evaluation. *IEEE Trans. Power Electron.* **2014**, *29*, 1285–1297. [CrossRef]
19. Khouzam, K.; Hoffman, K. Real-time simulation of photovoltaic modules. *Sol. Energy* **1996**, *56*, 521–526. [CrossRef]
20. Chang, C.-H.; Chang, E.-C.; Cheng, H.-L. A High-Efficiency Solar Array Simulator Implemented by an LLC Resonant DC–DC Converter. *IEEE Trans. Power Electron.* **2013**, *28*, 3039–3046. [CrossRef]
21. Ortiz-Rivera, E.I.; Peng, F.Z. Analytical model for a photovoltaic module using the electrical characteristics provided by the manufacturer data sheet. In Proceedings of the IEEE 36th Power Electronics Specialists Conference, Recife, Brazil, 16 June 2005; pp. 2087–2091. [CrossRef]
22. Zhang, H.; Zhao, Y. Research on a novel digital photovoltaic array simulator. In Proceedings of the International Conference on Intelligent Computation Technology and Automation, Changsha, China, 11–12 May 2010; Volume 2, pp. 1077–1080. [CrossRef]
23. Balato, M.; Costanzo, L.; Gallo, D.; Landi, C.; Luiso, M.; Vitelli, M. Design and implementation of a dynamic FPAA based photovoltaic emulator. *Sol. Energy* **2016**, *123*, 102–115. [CrossRef]

Electronics **2019**, *8*, 1445

24. Zhang, J.; Wang, S.; Wang, Z.; Tian, L. Design and Realization of a Digital PV Simulator with a Push-Pull Forward Circuit. *J. Power Electron.* **2014**, *14*, 444–457. [CrossRef]
25. Kim, Y.; Lee, W.; Pedram, M.; Chang, N. Dual-mode power regulator for photovoltaic module emulation. *Appl. Energy* **2013**, *101*, 730–739. [CrossRef]
26. Iero, D.; Carbone, R.; Carotenuto, R.; Felini, C.; Merenda, M.; Pangallo, G.; Della Corte, F.G. SPICE modelling of a complete photovoltaic system including modules, energy storage elements and a multilevel inverter. *Sol. Energy* **2014**, *107*, 338–350. [CrossRef]
27. Protogeropoulos, C.; Brinkworth, B.J.; Marshall, R.H.; Cross, B.M.; Santos, G. Evaluation of Two Theoretical Models in Simulating the Performance of Amorphous—Silicon Solar Cells. In Proceedings of the International Conference, Lisbon, Portugal, 8–12 April 1991; Springer Science and Business Media LLC.: Berlin, Germany, 1991; pp. 412–415.
28. Chan, D.; Phang, J. Analytical methods for the extraction of solar-cell single- and double-diode model parameters from I-V characteristics. *IEEE Trans. Electron Devices* **1987**, *34*, 286–293. [CrossRef]
29. Gow, J.; Manning, C. Development of a photovoltaic array model for use in power-electronics simulation studies. *IEE Proc. Electr. Power Appl.* **1999**, *146*, 193. [CrossRef]
30. Iero, D.; Carbone, R.; Carotenuto, R.; Felini, C.; Merenda, M.; Pangallo, G.; Della Corte, F.G. One-shot SPICE simulation of photovoltaic modules, storage elements, inverter and load. In Proceedings of the 2015 AEIT International Annual Conference (AEIT), Naples, Italy, 14–16 October 2015; pp. 1–4.
31. Bellia, H.; Youcef, R.; Fatima, M. A detailed modeling of photovoltaic module using MATLAB. *NRIAG J. Astron. Geophys.* **2014**, *3*, 53–61. [CrossRef]
32. Ramos-Hernanz, J.; Campayo, J.; Larranaga, J.; Zulueta, E.; Barambones, O.; Motrico, J.; Gamiz, U.; Zamora, I. Two photovoltaic cell simulation models in Matlab/Simulink. *Int. J. Tech. Phys. Probl. Eng.* **2012**, *4*, 45–51.
33. Walker, G. Evaluating MPPT converter topologies using a matlab PV model. *J. Electr. Electron. Eng. Aust.* **2001**, *21*, 49–55.
34. Izadian, A.; Pourtaherian, A.; Motahari, S. Basic model and governing equation of solar cells used in power and control applications. In Proceedings of the IEEE Energy Conversion Congress and Exposition (ECCE), Raleigh, NC, USA, 15–20 September 2012; pp. 1483–1488.
35. Pukhrem, S. *A Photovoltaic Panel Model in Matlab/Simulink*; The MathWorks, Inc.: Natick, MA, USA, 2013; pp. 20–23. [CrossRef]
36. Merenda, M.; Iero, D.; Pangallo, G.; Falduto, P.; Adinolfi, G.; Merola, A.; Graditi, G.; Della Corte, F.G. Open-Source Hardware Platforms for Smart Converters with Cloud Connectivity. *Electronics* **2019**, *8*, 367. [CrossRef]

 *electronics*

Article

Coordinated Frequency Stabilization of Wind Turbine Generators and Energy Storage in Microgrids with High Wind Power Penetration

Moses Kang [1], Gihwan Yoon [1], Seonri Hong [1,2], Jinhyeong Park [2], Jonghoon Kim [2] and Jongbok Baek [1,*]

[1] Department of Electrical Engineering and Korea Institute of Energy Research, Deajeon 34129, Korea; moseskang@kier.re.kr (M.K.); ghyoon@kier.re.kr (G.Y.); hsr@kier.re.kr (S.H.)

[2] Department of Electrical Engineering and Chungnam National University, DaeJeon KS015, Korea; pig25t@naver.com (J.P.); whdgns0422@cnu.ac.kr (J.K.)

* Correspondence: jongbok.baek@kier.re.kr; Tel.: +82-10-3309-0378

Received: 5 November 2019; Accepted: 18 November 2019; Published: 21 November 2019

Abstract: This paper proposes a coordinated control scheme for wind turbine generators (WTGs) and energy storage in microgrids with high wind power penetration. The proposed scheme aimed to reduce the system frequency deviation caused by variations in wind power and loads. To stabilize the frequency, the WTG and energy storage system (ESS) are used for kinetic energy generation and electrical energy storage, respectively. When the WTG contributes excessively to frequency stabilization in the microgrid with a high wind power penetration, the system frequency may fluctuate considerably. Thus, it is necessary to adjust the contribution of a WTG and to share it with other sources. To achieve our objective, we proposed a coordinated control scheme between the WTG and ESS that shares their releasable and absorbable energies. The coordinated control consistently calculated the releasable and absorbable energies of the WTG and ESS and determined weight factors related to the energy ratios. Accordingly, the weight factors improved the ability of providing supporting frequency stabilization of the WTG and ESS by increasing the stored energy utilization. The performance of the scheme was investigated using MATLAB Simulink Electrical. The results show that the proposed coordinated control successfully stabilized the system frequency by calculating the appropriate contributions required from the WTG and ESS.

Keywords: frequency stabilization; coordinated control; wind turbine generator; high-fidelity battery model; releasable and absorbable energy

1. Introduction

The system frequency in a power system is indicative of the balance between the generation and consumption of active power and must be maintained within the normal range at all times. In a conventional power grid, synchronous generators that have a spinning reserve increase their mechanical power by relying on the deviation of the frequency [1]. However, in a power grid with high wind power penetration, the output power of a wind turbine generator (WTG) is critical for the maintenance of the system frequency. Hence, a WTG should contribute the system frequency stabilization because active power from the maximum power point tracking (MPPT) control results in large fluctuations in the system frequency [2]. Some countries specify requirements for the reduction of the ramping rates of the output power of a WTG to overcome fluctuations in the system frequency [3,4].

Several control schemes for smoothing the output power of a WTG by controlling the pitch angle have been reported [5–10]. However, these methods substantially decrease the captured power from the incoming wind power, and control schemes that entail releasing the kinetic energy stored in the rotating masses of WTG blade and gearbox have been proposed [11–13]. Therefore, a WTG can temporarily absorb or release electric energy to the rotating masses and contribute to the system frequency stabilization. However, in a power system with a high wind power penetration, an output with an excessive ramp rate of the WTG might adversely affect the system frequency stabilization.

Accordingly, control schemes for coordinated WTGs and energy storage have also been proposed [14–16]. Chunghun et al. [14] proposed a coordinated control of a WTG and ESS to reduce the output power fluctuation of a WTG. The WTG operated based on the wind speed variation and size of the ESS capacity when de-loaded and they successfully improved the grid reliability, especially in the case of a large wind speed variation. Ziping et al. [15] proposed maximizing the inertial response of a WTG to arrest the system frequency nadir when the large disturbances occurred in a power system. They used a small-scale battery ESS to reduce the second frequency dip that can occur while recovering the reduced rotor speed through the inertial response of the WTG. Akie et al. [16] proposed the coordination control of the WTG and ESS using load estimation via a disturbance observer in an isolated grid. In their study, frequency stabilization was supported through mitigating the fluctuation of a WTG using a pitch angle control and an active power control of the ESS that was applied to the low-frequency and high-frequency domains, respectively. These conventional papers propose coordinated controls of a WTG and ESS, and they have responded to issues such as power smoothing, primary frequency control based on large disturbance, and frequency stabilization. However, they do not deal with concerns about the high wind power penetration level.

This study proposes a coordinated control scheme for a WTG and ESS in a microgrid with a high wind power penetration to improve the frequency stabilization. This was achieved through the continuous calculation of the releasable and absorbable energy of a WTG and ESS, along with the determined weight factors related to the energy ratios. Thus, the weight factors improved the ability for supporting frequency stabilization of a WTG and ESS by increasing the stored energy utilization. Contrary to Chunghun et al. [14] and Akie et al. [16], in this study, an ESS had a high utilization to maintain the system reliability in an isolated microgrid with a high wind power penetration. The performance of the scheme was investigated using MATLAB R2018b Simulink Electrical.

The rest of the paper is structured as follows. In Section 2, in addition to the typical variable speed WTG model, the high-fidelity battery model is briefly described. In Section 3, the proposed coordinated control of the WTG and ESS is explained, including the aim of the proposed scheme, the control strategy for frequency stabilization, and its advantages. In Section 4, we describe three cases that were conducted to demonstrate the superior operation for system frequency regulation of the coordinated control of WTG and ESS under varying wind speed conditions and set values of the initial state of charge (SOC). In Section 5, in-depth conclusions are provided.

2. The WTG and Battery Models

The WTG and battery models system supports the frequency stabilization in isolated microgrids by controlling the active powers based on the system frequency variation. A WTG is a permanent magnet synchronous generator (PMSG) and it can control the active power related to its rotor speed. In this paper, the high-fidelity battery model was adjusted to verify the correct performance of the proposed scheme.

2.1. WTG Model

The PMSG configuration is presented in Figure 1a. The mechanical output power of the turbine, P_m, is given by:

$$P_m = \frac{1}{2}\rho\pi R^2 v_w^3 c_p(\lambda, \beta),\tag{1}$$

where ρ, R, v_W, C_P, λ, and β are the air density, blade length, wind speed, power coefficient, tip-speed ratio of the rotor blade tip speed to wind speed, and blade pitch angle, respectively. According to Siegfried [17], c_P used in this paper can be represented by:

$$C_p(\lambda, \beta) = c_1\left(\frac{c_2}{\lambda_i} - c_3\beta - c_4\right) \times e^{-\frac{c_5}{\lambda_i}} + c_6\lambda, \tag{2}$$

where

$$\frac{1}{\lambda_i} = \frac{1}{\lambda + 0.08\beta} - \frac{0.035}{\beta^3 + 1} \tag{3}$$

and the coefficients c_1 to c_6 are $c_1 = 0.5176$, $c_2 = 116$, $c_3 = 0.4$, $c_4 = 5$, $c_5 = 21$, and $c_6 = 0.0068$. In this study, λ_{opt} and $C_{P,max}$ were set to 8.1 and 0.48, respectively.

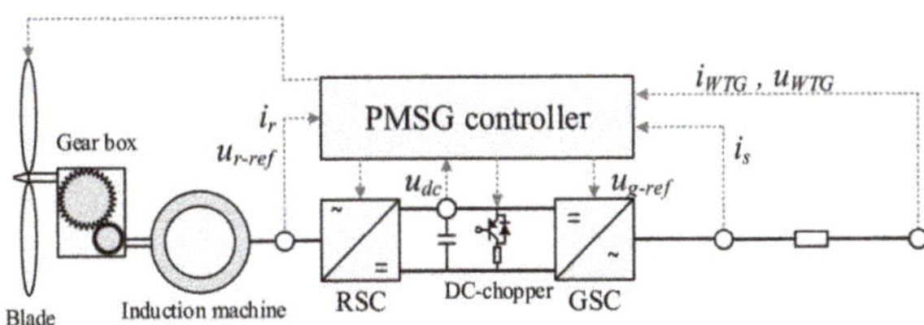

(a)

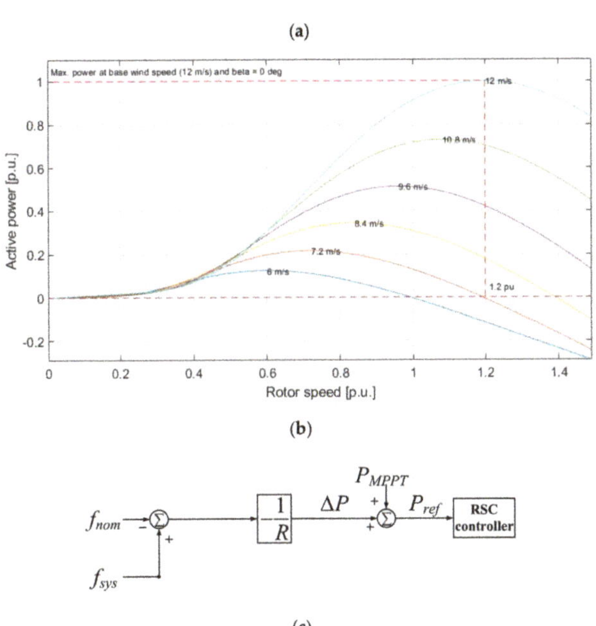

(b)

(c)

Figure 1. The PMSG model. (**a**) Typical configuration of a PMSG; V_r, I_r: voltage and current in the rotor circuit; V_t, I_t: voltage and current at the terminal; $V_{r,ref}$, $V_{t,ref}$: reference RSC and GSC voltage; V_{DC}: DC-link voltage; V_g, I_g: voltage and current at the point of common coupling (PCC); ω_r: rotor speed; β: pitch angle. (**b**) Operational characteristics of a PMSG. (**c**) Operational characteristics of a frequency stabilization scheme.

The maximum power point tracking (MPPT) output, P_{MPPT}, is represented as:

$$P_{MPPT} = k_g \omega_r^3,\tag{4}$$

where ω_r is the rotor speed and k_g is a constant that is set to 0.512 in this paper. The rotor-side converter (RSC) in the PMSG controls the active and reactive power injected into a grid., and the grid-side converter (GSC) controls the DC-link and terminal voltages.

Figure 1b shows the mechanical power curves at different wind speeds as indicated by the thin solid lines. The maximum power limit was set to 1.1 p.u. and the operating range of the rotor speed of the PMSG was between 0.6 p.u. to 1.2 p.u. [17], which is represented in Figure 1b by the red dashed lines.

A controllable WTG (of type III and type IV) can aid in frequency stabilization by using the kinetic energy stored in its rotor. The WTG is connected to the power system through power electronic devices, which means the frequency from the generator side is decoupled from the system frequency, unlike a synchronous generator, which is directly connected to the power system. Hence, for it to regulate the system frequency, the reference power based on the system frequency must be adjusted. Figure 1c illustrates the frequency stabilization control scheme, which is based on the frequency deviation. In Figure 1c, the total active power reference, P_{ref}, consists of the active power reference for the MPPT control, P_{MPPT}, and additional power reference based on the droop loop, ΔP, which can be expressed as:

$$\Delta P = -\frac{1}{R}(f_{sys} - f_{nom}),\tag{5}$$

where f_{sys}, f_{nom}, and $1/R$ are the system frequency, nominal frequency, and loop gain for droop, respectively. When the system frequency is larger than the nominal frequency, WTG reduces the output power through the droop loop. Thus, the system frequency automatically reduces and the rotor speed of a WTG increases. When the rotor speed reaches the maximum, WTG reduces the output power by increasing the pitch angle, consequentially rejecting the wind energy.

2.2. Battery Model

Battery models integrated in the wind farm are commonly used for simplified models that only have fixed parameters regardless of the SOC variation. However, in this paper, a high-fidelity battery model with parameters of a battery equivalent circuit based on the SOC was used to accurately examine the performance change of the battery due to the proposed coordinated control. Figure 2 shows the equivalent circuit models that can be used to represent the electrical property of Li-ion batteries and Figure 3 shows the experimental data of an open circuit voltage and line current. These data were extracted from a lithium-ion battery when the battery was discharged by inversely pulsating in 5% increments of SOC.

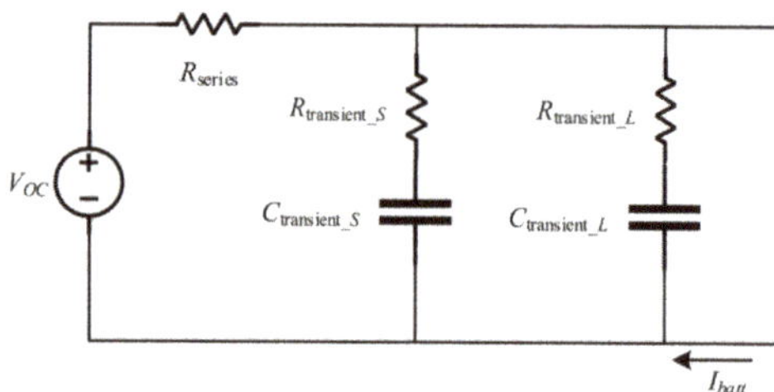

Figure 2. General equivalent circuit model (two R-C branches).

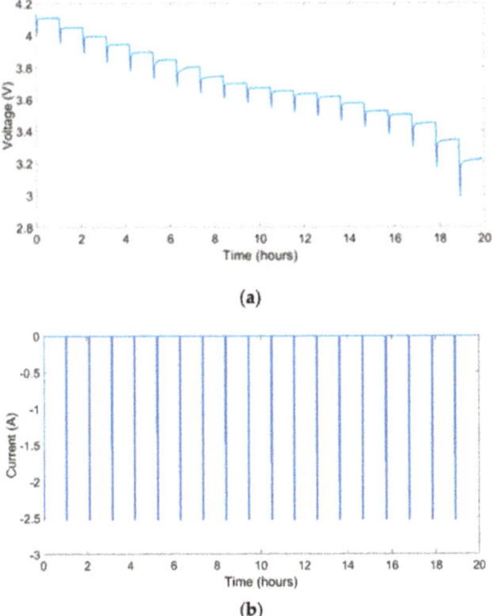

(a)

(b)

Figure 3. Pulse discharge in 5% increments of the SOC. (**a**) Experimental data of open circuit voltage. (**b**) Experimental data of discharged current.

The pulse type tests cause the circuit dynamics of battery, which can provide necessary data about the performance of the battery cell at different points of the SOC. Figure 4 shows one pulse of the discharge test. To estimate R_{series}, the voltage drops and rated current were used, which can be represented as:

$$R_{series} = V_{drop} \times I_{rated},$$ (6)

where V_{drop} and I_{rated} are the voltage drop and rated current, respectively. Using the experimental data, in this paper, the R-C ladder parameters were estimated using MATLAB Curve Fitting. The resistance and capacitance of the two ladders was estimated using the equation for curve fitting determined by:

$$U_{OCV} = ae^{-t/b} + ce^{-t/d},$$ (7)

where a, b, c, and d are denoted as $R_{transient\_S}$, $\tau_{transient\_S}$, $R_{transient\_L}$, and $\tau_{transient\_L}$, respectively. The $C_{transient\_S}$ and $C_{transient\_L}$ were calculated using:

$$C_{transient\_S} = R_{transient\_S} \times \tau_{transient\_S}$$ (8)

$$C_{transient\_L} = R_{transient\_L} \times \tau_{transient\_L}$$ (9)

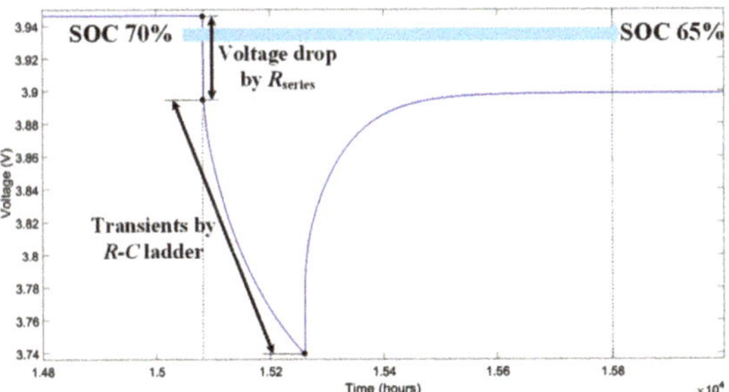

Figure 4. One pulse from the discharge test.

3. Coordinated Frequency Stabilization between the WTG and Energy Storage

The aim of the proposed coordinated frequency stabilization scheme was to reduce the system frequency deviation in an isolated microgrid with a high wind power penetration. This was achieved by calculating the releasable and absorbable energy of a WTG and ESS and determining the related weighting factors from the coordinated control. The weighting factors were calculated using the releasable and absorbable energy of the WTG and ESS to the power system. Figure 5 shows the flowchart for the coordinated frequency stabilization between the WTG and ESS. The algorithm presents the host controller for the coordination between the WTG and ESS that regulates the system frequency to the nominal frequency and increases the system reliability.

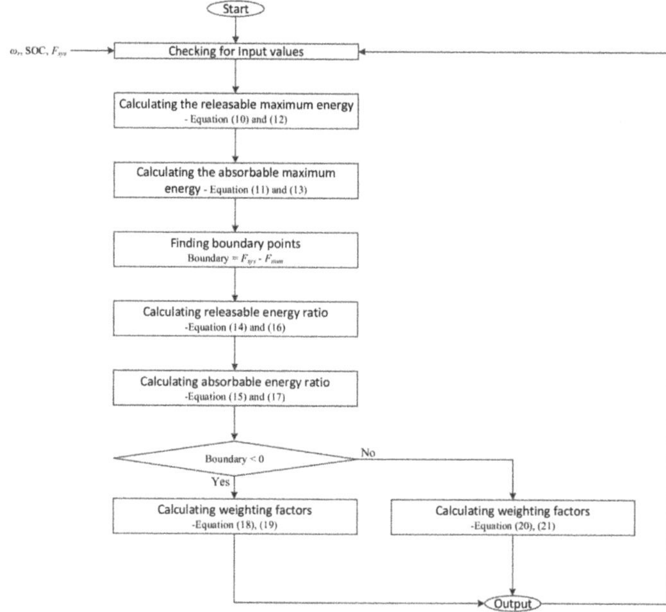

Figure 5. Flowchart for the coordinated frequency stabilization between a WTG and ESS.

The resource and system values, such as w_r, SOC, and F_{sys}, were required when implementing the proposed algorithm. It performed the following series of steps:

1) The algorithm calculated the releasable and absorbable maximum energy of the WTG and ESS. The releasable and absorbable maximum energy of the WTG were represented as

$$E_{WTG\_rele\_max} = \frac{1}{2}J\left(\omega_{max}^2 - \omega_{min}^2\right), \tag{10}$$

$$E_{WTG\_abso\_max} = \frac{1}{2}J\left(\omega_{min}^2 - \omega_{max}^2\right), \tag{11}$$

where J, w_{max}, and w_{min} are the inertia constant, maximum rotor speed, and minimum rotor speed of a WTG, respectively. The releasable and absorbable maximum energy of an ESS were represented using:

$$E_{ESS\_rele\_max} = C(SOC_{max} - SOC_{min}), \tag{12}$$

$$E_{ESS\_abso\_max} = C(SOC_{min} - SOC_{max}), \tag{13}$$

where C, SOC_{max}, and SOC_{min} are the capacity, maximum SOC, and minimum SOC of a battery, respectively.

2) To adjust the determined energy ratios for the WTG and ESS, the algorithm calculated the frequency deviation, $\Delta f = f_{sys} - f_{nom}$.

3) The releasable and absorbable energy ratios of the WTG and ESS were calculated to increase the utilized energy and improve the frequency stabilization. The releasable and absorbable energy ratio of the WTG were represented as:

$$E_{WTG\_rele\_ratio} = \frac{1}{2}J\left(\omega_r^2 - \omega_{min}^2\right)/E_{WTG\_rele\_max}, \tag{14}$$

$$E_{WTG\_abso\_ratio} = \frac{1}{2}J\left(\omega_r^2 - \omega_{max}^2\right)/E_{WTG\_abso\_max}. \tag{15}$$

The releasable and absorbable energy ratio of ESS were represented as:

$$E_{ESS\_rele\_ratio} = C(SOC - SOC_{min})/E_{ESS\_rele\_max}, \tag{16}$$

$$E_{ESS\_abso\_ratio} = C(SOC - SOC_{max})/E_{ESS\_abso\_max}. \tag{17}$$

4) The energy ratio consisted of two values from Step 1 to Step 3, i.e., the ratio for the releasable energy and the ratio for the absorbable energy. The weighting factors of the WTG and ESS were calculated for two cases, i.e., one without a power system and the other with excessive electricity, which could be confirmed through the system frequency in Step 2.

The weighting factors of the releasable energy were represented as:

$$\alpha_{WTG} = p \times E_{WTG\_rele\_ratio}/(E_{WTG\_rele\_ratio} + E_{ESS\ rele\ ratio}), \tag{18}$$

$$\alpha_{ESS} = q \times E_{ESS\_rele\_ratio}/(E_{WTG\_rele\_ratio} + E_{ESS\_rele\_ratio}). \tag{19}$$

The weighting factors of absorbable energy were represented as:

$$\alpha_{WTG} = p \times E_{WTG\_abso\_ratio}/(E_{WTG\_abso\_ratio} + E_{ESS\_abso\_ratio}), \tag{20}$$

$$\alpha_{ESS} = q \times E_{ESS\_abso\_ratio}/(E_{WTG\_abso\_ratio} + E_{ESS\_abso\_ratio}), \tag{21}$$

where p and q are the coefficients related to the penetration levels of WTG and ESS, respectively. The penetration levels of the WTG and ESS needed to be considered when determining the proper weight of the control for the frequency stabilization of the power system.

The weighting factors calculated using the coordinated control were multiplied and converted to the conventional frequency stabilization scheme as shown in Figure 6. Figure 6a shows the operational characteristics of a WTG; to support the frequency stabilization, the additional reference power, ΔP, was calculated and added to the reference of the MPPT control. When the frequency fluctuated, ΔP was calculated using the multiplication of the inverse of the system frequency and weighting factor according to the releasable and absorbable energy. The reference active power of ESS was calculated using the droop loop multiplied by a weighting factor. Thus, the WTG and ESS could contribute to the frequency stabilization and also adequately utilize the stored energy by sharing their releasable and absorbable energies.

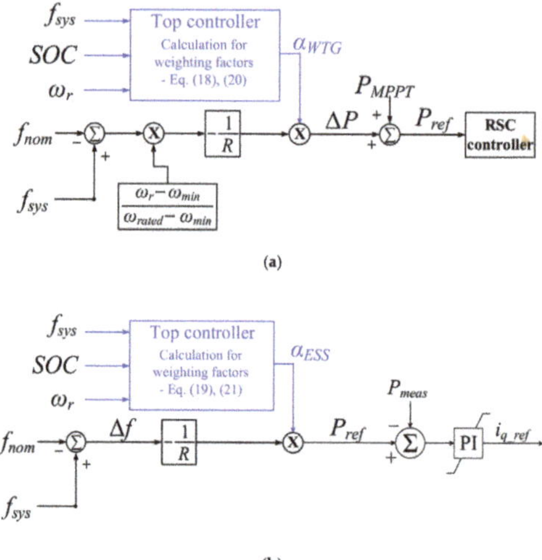

Figure 6. Operational characteristics of the proposed coordination control scheme. (**a**) Active power control of a WTG. (**b**) Active power control of an ESS.

4. Case Studies

The isolated microgrid was modeled to investigate the performance of the coordinated frequency stabilization scheme. It was simulated using MATLAB Simulink Electrical simulator. The model system consisted of the static load, asynchronous motor, ESS, diesel generator, and PMSG, and they were connected in parallel, as shown in Figure 7. Furthermore, the initial values of the microgrid are shown in Table 1. In the Appendix A, the parameters that affected the dynamics of the model system are explained, namely the parameters of the asynchronous motor and diesel generator. In this study, the system frequency was computed using zero crossing detection [18]. The WTG and ESS had a droop loop to allow them to aid in the stabilization of the system frequency (see Figure 1c). In this model, the wind power penetration level, which is defined as the capacity of wind power divided by the peak load, was calculated to be 40.0%.

The performance of the frequency stabilization control of a WTG is affected by the system frequency, which depends on the balance between supply power and demand power. Thus, we investigated the performance of the frequency stabilization schemes under various wind speeds. In addition, the proposed scheme had a positive performance, even in cases where the wind speed was kept very low. Furthermore, in the case of load variations, the performance of the proposed scheme was verified.

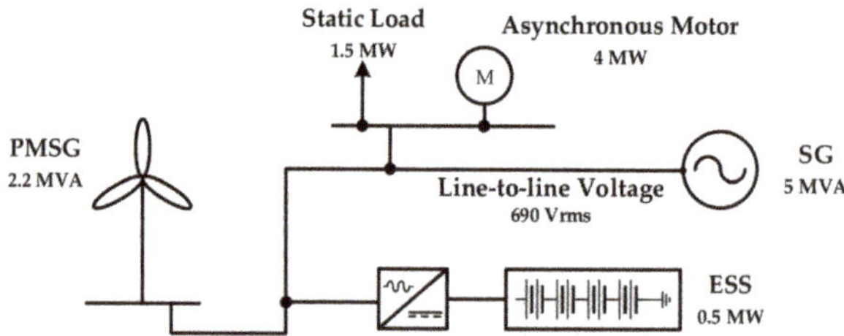

Figure 7. A single-line-diagram of the isolated microgrid model.

Table 1. Initial values of the microgrid consistence.

BESS	Value	WTG	Value
Capacity of a single cell	10 Ah	Rated generator power	2.2 MVA
Number of cells in series	500 EA	Rated turbine power	2.0 MW
Number of stacks in parallel	80 EA	Rated rotor speed	1.2 p.u.
Rated active power	0.5 MW	Minimum rotor speed	0.6 p.u.
Diesel	**Value**	Rated wind speed	12.0 m·s^{-1}
Rated generator power	5 MVA	Cut-in speed	4.5 m·s^{-1}
Line-to-line voltage	690 Vrms	Generator inertia	35000 kg·m^2

The performance of the coordinated control scheme was compared to a case where the WTG and ESS participated in frequency stabilization. In addition, it was also compared to a case where the WTG did not support the frequency stabilization but ESS supported it. In the conventional case, the droop gains of the WTG and ESS were set to −200 and −20, respectively. Furthermore, the coefficients p and q in the coordinated control were set to 1.5 and 3, respectively.

Table 2 shows the organization of the case studies to performance of the frequency stabilization control. Cases 1–3 analyze the performance of the proposed scheme according to the retention status of the releasable and absorbable energy of the WTGs and ESS. The main cause of the system frequency fluctuations in this case was the wind speed variation. Case 4 analyzes the performance of the proposed scheme due to rapid load changes. The following subsections describe the comparative analysis results for the frequency stabilization for the four cases.

Table 2. Organization of the case studies.

Cases	Wind Speeds	Initial SOCs	Loads
Case 1	Variable	50%	Constant
Case 2	Variable	25%	Constant
Case 3	Variable and low	50%	Constant
Case 4	Constant	50%	Variable

4.1. Case 1: Variable Wind Speed, 50% of Initial SOC, and Constant Loads

In this case, the initial SOC of the ESS was set at 50% and both the WTG and ESS supported the system frequency stabilization. Figure 8 shows the wind speed variation profile, and based on the fluctuation of the output power of a WTG from wind speed variation, the system frequency fluctuated. Figure 9 shows the simulation results, including the effects of the wind speed variation. The frequency deviation for the proposed scheme was less than that of the conventional scheme. The peak-to-peak of the system frequency for the MPPT, conventional scheme, and proposed scheme were 0.3930 Hz, 0.3966 Hz, and 0.3404 Hz, respectively. This was because the coordinated control arbitrated the releasable and absorbable energy of WTG and ESS; thus, in the results for the proposed scheme, the active power of the ESS was released more in line with the conventional scheme (see Figure 9c). The active power of the WTG rapidly reduced as the wind speed significantly decreased between 80.0 s and 82.0 s. The system frequency decreased to 59.71 Hz for the conventional scheme, and the decrease was less than that of the proposed scheme by 0.063 Hz. This was because the output power of a WTG in the conventional scheme was abruptly reduced and the output of an ESS could not easily and rapidly compensate for the decrease whereas the stored kinetic energy of the WTG was significantly released. However, the proposed scheme determined the weighting factors by considering their releasable and absorbable energy. The sudden decrease in wind power output due to wind speed decrease resulted in a decrease in releasable energy. Thus, α_{WTG} decreased, whereas α_{ESS} increased; accordingly, ESS could compensate for the reduced output power of the WTG by releasing its electrical energy.

The reduced amount of SOC for the conventional and proposed schemes were 0.2984% and 0.4802%, respectively (see Figure 9e). Because the coordinated control increased the additional active power using weighting factors related to SOC, the weighting factors of the WTG and ESS were calculated for the releasable and absorbable energy. In this case, the ESS had enough releasable and absorbable energy, and thus, α_{ESS} was determined such that the ESS had a greater contribution compared to the WTG, as shown in Figure 8f. Note that the high value of coefficient p could adversely affect the frequency stabilization when the penetration level of the WTG was high.

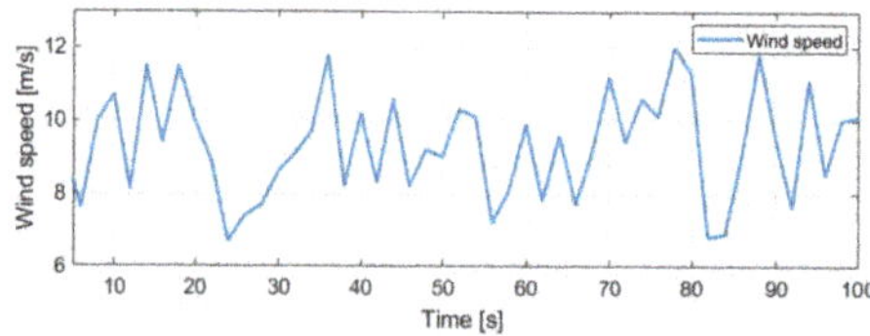

Figure 8. Wind speed (m/s) (variation profile for case 1 and case 2.

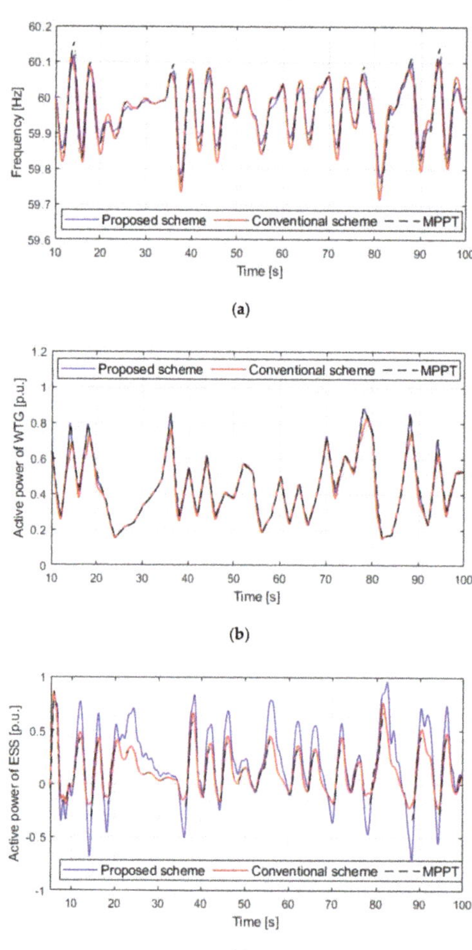

(a)

(b)

(c)

Figure 9. *Cont.*

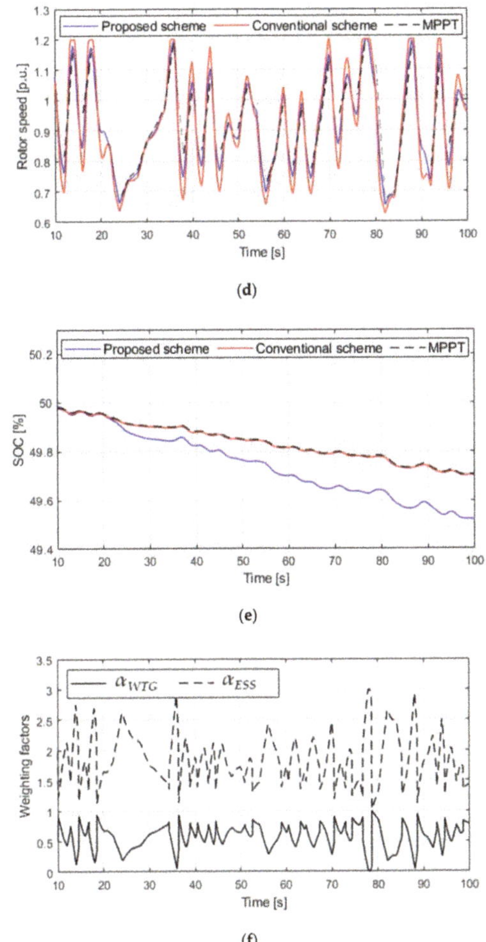

(d)

(e)

(f)

Figure 9. Simulation results for case 1: (**a**) system frequencies, (**b**) active powers of the WTG, (**c**) active powers of the ESS, (**d**) rotor speeds of the WTG, (**e**) SOCs of the ESS, and (**f**) weighting factors of the coordinated control.

4.2. Case 2: Variable Wind Speed, 25% of Initial SOC, and Constant Loads

The results for case 2 are presented in Figure 10. In this case, the initial SOC was set at 25% and was less than that of case 1. The peak-to-peak of the system frequency for the proposed scheme was 0.3556 Hz, which was less than in the MPPT by 0.0365 Hz, and less than in the conventional scheme by 0.0405 Hz. When the frequency deviation was smaller than zero, WTG and ESS increased their output powers. However, the results of the peak-to-peak of the system frequency for the proposed scheme was greater than in case 1. This was because the releasable energy of not only the ESS, but also the WTG, was smaller than in case 1. However, in situations where the energy was absorbed, the active power of the ESS was larger than that of case 1 (see Figure 10c), since the initial SOC was set at a low value. Thus, the reduced amount of SOC for the proposed scheme was 0.3411% (see Figure 10e), lower than that of the proposed scheme in case 1 by 0.1391%.

Figure 10f shows the weighting factors of the coordinated control, and because of the low SOC, α_{WTG} was determined for a high value in comparison to the results for case 1 when the frequency deviation was lower than zero. Thus, the WTG released more kinetic energy stored in the rotor; accordingly, w_r slightly decreased (see Figure 10d). In addition, the determined α_{ESS} was lower in comparison to the results in case 1. Hence, the released energy of the ESS decreased and the absorbed energy increased in comparison to the results in case 1.

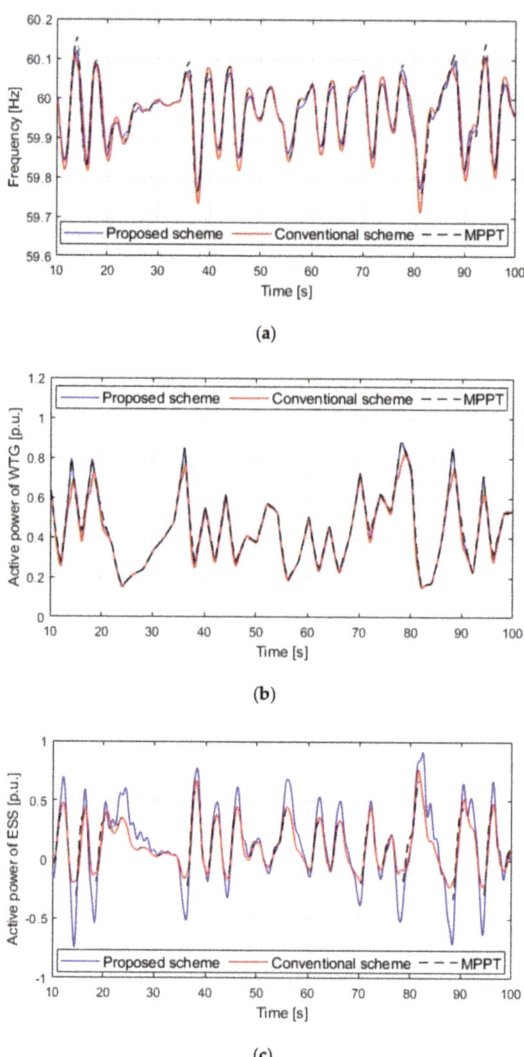

(a)

(b)

(c)

Figure 10. *Cont.*

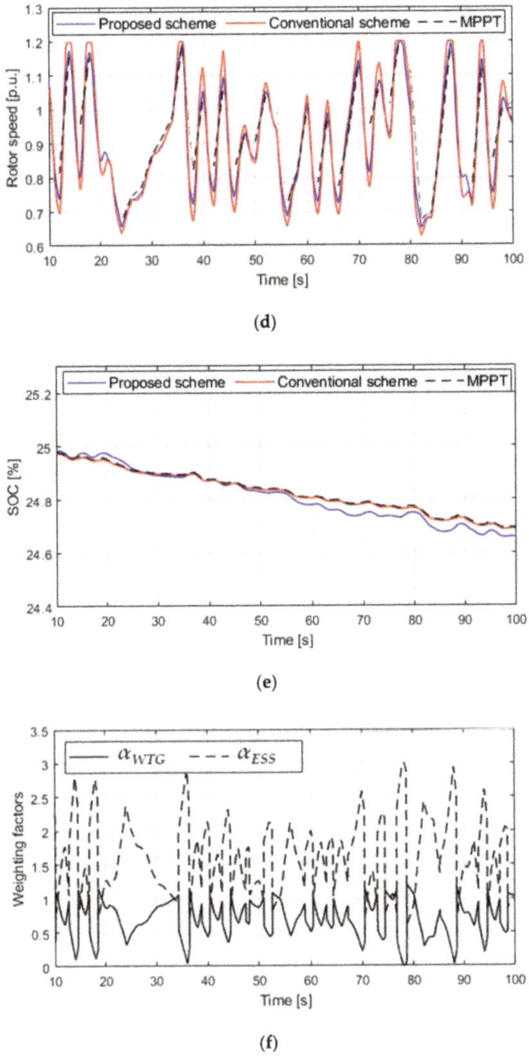

(d)

(e)

(f)

Figure 10. Simulation results for case 2: (**a**) system frequencies, (**b**) active powers of the WTG, (**c**) active powers of the ESS, (**d**) rotor speeds of the WTG, (**e**) SOCs of the ESS, and (**f**) weighting factors of the coordinated control.

4.3. Case 3: Variable Wind Speed and Low Wind Speed, 50% of Initial SOC, Constant Loads

The performance of the coordinated control was also affected as the wind speed remained low, which meant that WTG had low stored kinetic energy to release when the system frequency was lower than the nominal frequency. Figure 11 shows the wind speed variation profile for case 3, and in this profile, the wind speed was maintained at a low speed after 54 s.

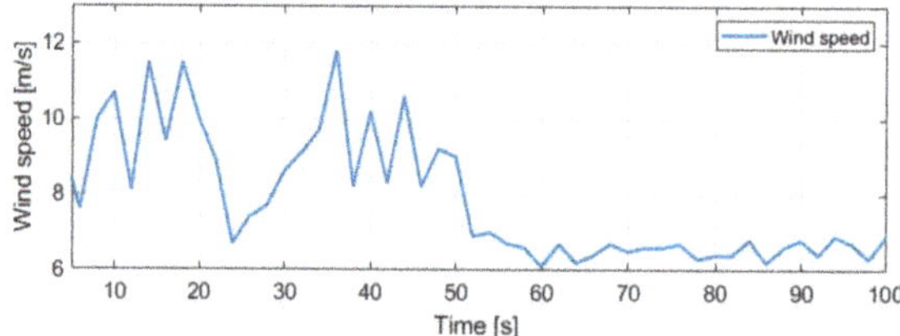

Figure 11. Wind speed (m/s) variation profile for Case 3.

Figure 12 shows the results for case 3. The peak-to-peak of the system frequency before 54 s for the MPPT, conventional scheme, and proposed scheme were 0.3930 Hz, 0.3770 Hz, and 0.3325 Hz, respectively. In addition, the peak-to-peak of the system frequency after 54 s for the MPPT, conventional scheme, and proposed scheme were 0.0282 Hz, 0.0285 Hz, and 0.0184 Hz, respectively. This was because the output power fluctuations of a WTG were reduced with the low wind speed, consequently resulting in temporary fluctuations in the system frequency from 50 s onwards, as shown in Figure 12a. The amount of SOC for the proposed scheme was reduced by 0.6613%, which was larger than the results of the proposed scheme in case 1 by 0.1811%. In this case, the static load resulted in a deviation of the system frequency deviation after 54 s. However, if a larger disturbance were to occur in a power system, the system frequency would decrease because the WTG had no releasable kinetic energy. The proposed scheme could calculate the weighting factor of the ESS due to the shortage of the releasable energy that could be emitted from the wind turbine, thus preventing the large decrease in frequency in such a situation.

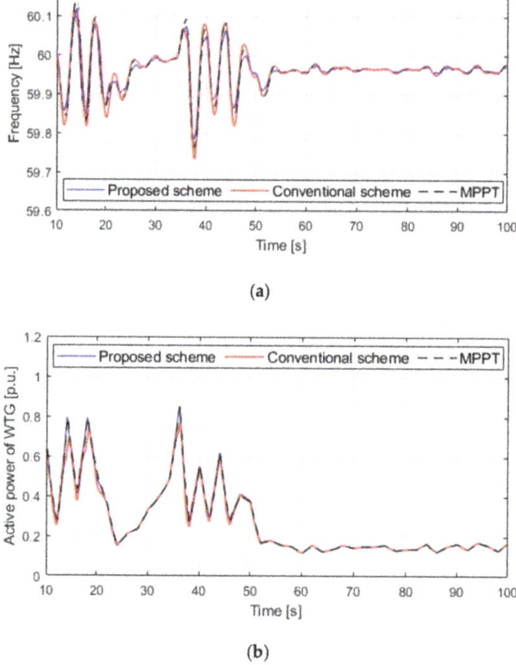

(a)

(b)

Figure 12. *Cont.*

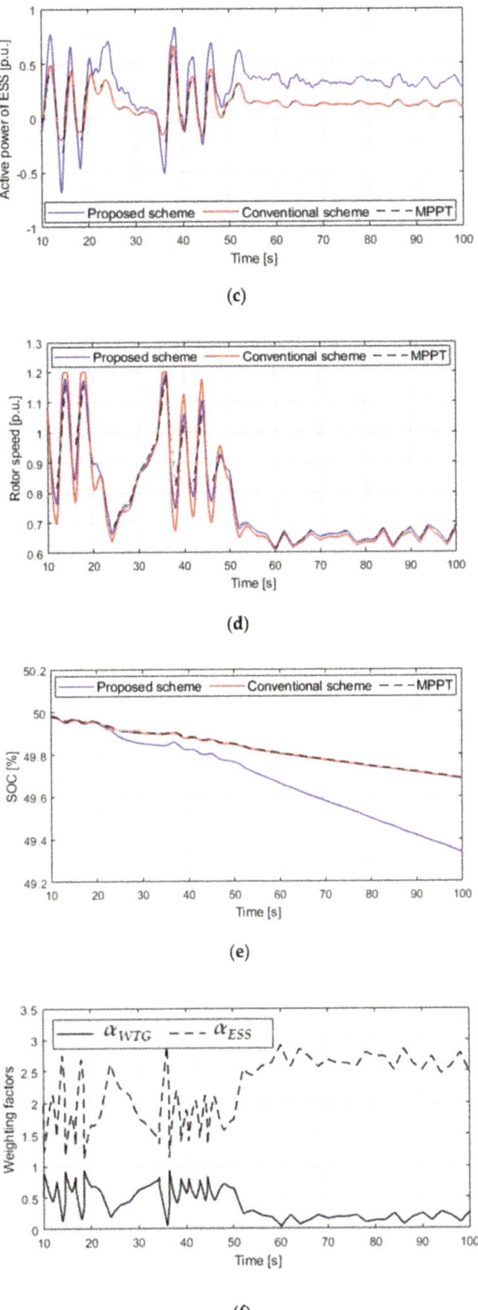

(c)

(d)

(e)

(f)

Figure 12. Simulation results for case 3: (**a**) system frequencies, (**b**) active powers of the WTG, (**c**) active powers of the ESS, (**d**) rotor speeds of the WTG, (**e**) SOCs of the ESS, and (**f**) weighting factors of the coordinated control.

4.4. Case 4: Load Variation, Constant Wind Speed, and 50% of Initial SOC

The system frequency deviation was also based on the load variation. Figure 13 shows the load variation profile for case 4. The load fluctuated between 2 MW and 5.5 MW, and the peak load occurred at 48 s for 5.5 MW. The load variation resulted in significant system frequency fluctuations. In this case, the coefficients p and q in the coordinated control were set at 0.8 and 3.4, respectively.

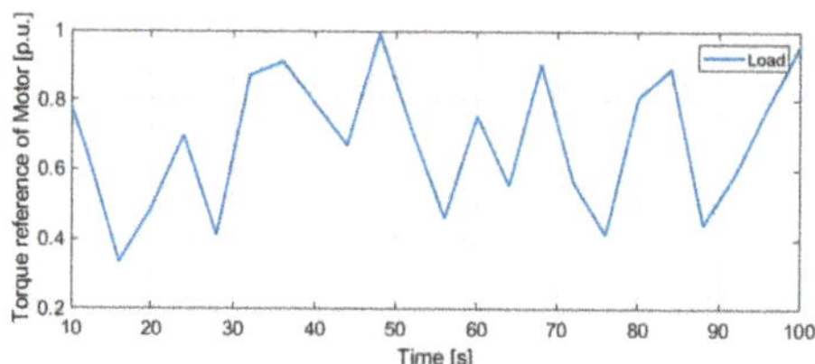

Figure 13. Load variation (p.u.) profile for Case 4.

Figure 14 shows the results for case 4. In this case, the initial SOC was set at 50% and the wind speed was constant at 10 m/s. The peak-to-peak of the system frequency for the proposed scheme was 0.2841 Hz, less than that of the MPPT by 0.030 Hz, and smaller than in the conventional scheme by 0.087 Hz. This resulted in a higher peak-to-peak system frequency for the conventional scheme in comparison to the MPPT because the WTG excessively contributed to the frequency stabilization, as shown in Figure 14b,d. In particular, the system frequency drastically decreased in the process of recovering the kinetic energy since the WTG significantly increased the output power to compensate for the load variation at 48.0 s. The proposed scheme prevented the excessive contribution of the WTG in the frequency stabilization because the weighting factor was accurately calculated based on the kinetic energy of the wind turbine; this is important since significant increases in wind turbine output due to frequency drops can have a negative impact on isolated microgrids with a high wind power penetration.

Figure 14e shows the SOC of the ESS. The reduced amount of SOC for the proposed scheme and conventional scheme were 0.4225% and 0.2731%, respectively. In the proposed scheme, the WTG had a lower contribution to frequency stabilization than in the conventional scheme. Thus, the contribution of the ESS was increased to perform the frequency stabilization, as shown in Figure 14f. As a result, the frequency stabilization control was performed stably by preventing an excessive contribution toward the frequency stabilization by the WTG.

In all four cases, the results show that the releasable and absorbable energy of a WTG and the ESS was shared well in the proposed scheme and this enabled it to successfully control the frequency stabilization. Hence, the accurate determination of the contribution to the frequency stabilization of a WTG prevented its excessive contribution, and thus, in cases of high wind power penetration levels, the system frequency deviation successfully decreased in the proposed scheme.

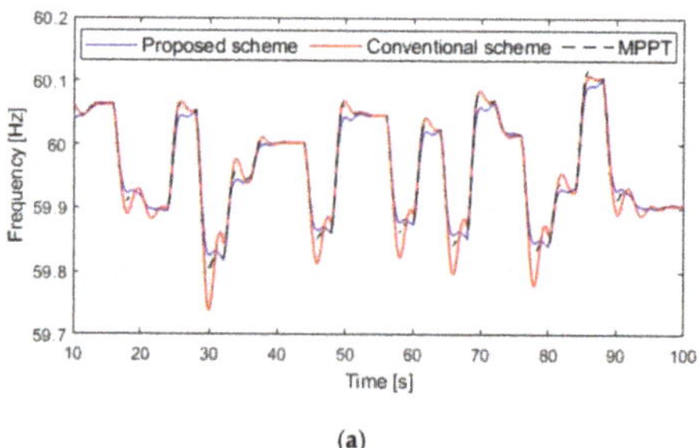

(a)

Figure 14. *Cont.*

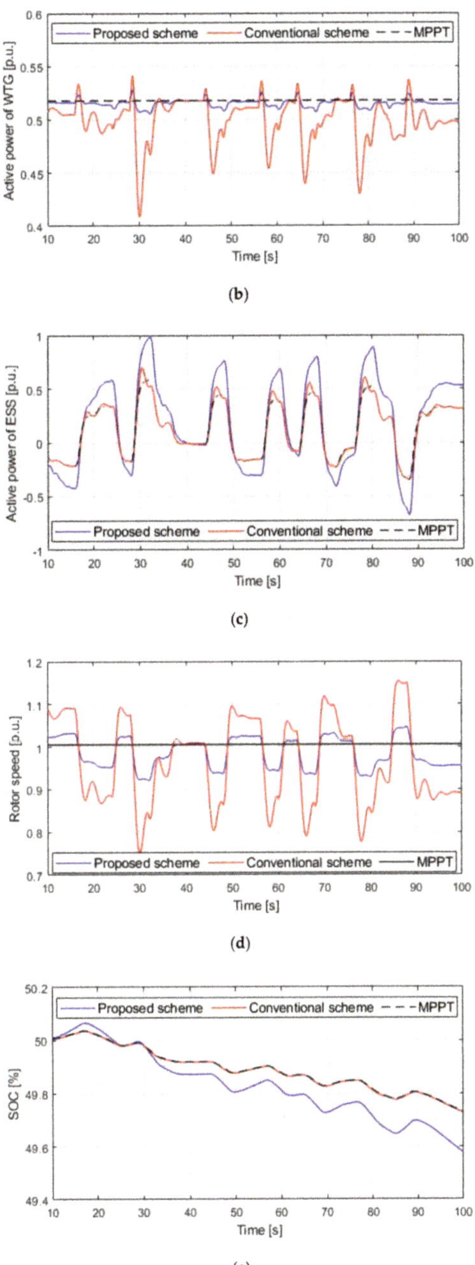

(b)

(c)

(d)

(e)

Figure 14. *Cont.*

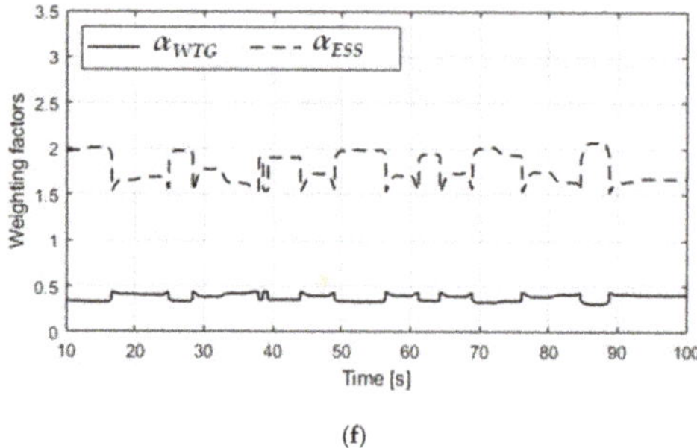

(f)

Figure 14. Simulation results for case 4: (**a**) system frequencies, (**b**) active powers of the WTG, (**c**) active powers of the ESS, (**d**) rotor speeds of the WTG, (**e**) SOCs of the ESS, and (**f**) weighting factors of the coordinated control.

5. Conclusions

The coordinated frequency stabilization of a WTG and ESS for increasing utilized energy to improve the frequency stabilization were investigated. The ability of a WTG and ESS to support the frequency stabilization support was dependent on how the stored kinetic energy of the WTG and ESS was used. The proposed scheme consistently calculated the releasable and absorbable energy of the WTG and ESS and determined weighting factors related to the energy ratios. In each frequency support control loop of the WTG and ESS, the determined weighting factors produced an additional active power reference. Therefore, the active power references for supporting frequency stabilization were based on the ability for the utilized energy of WTG and ESS.

The simulation results showed that the proposed coordinated control scheme successfully improved the system frequency and prevented the excessive contribution of a WTG in power systems with a high wind power penetration. Furthermore, the proposed scheme ensured minimized resource losses of the participating frequency support as they had low releasable or absorbable energy capacities.

The advantages of the proposed coordinated scheme are that it can ensure adequate utilized stored energy. Therefore, the proposed scheme may provide potential solutions to ancillary services, especially in isolated microgrids, by increasing the reserve power in an electric power grid.

Author Contributions: M.K., G.Y., and J.B. mainly proposed the coordinated scheme. M.K., G.Y., S.H., J.K. and J.B. performed the simulation tests and analyzed the results. J.P. and J.K. contributed the design for the high-fidelity battery model and revised the original scheme. All authors contributed toward writing the paper.

Funding: This research was funded by Korea Institute of Energy Technology Evaluation and Planning (KETEP) and Ministry of Trade, Industry & Energy (MOTIE), No. 20182410105280.

Acknowledgments: This work was supported by the KETEP and the MOTIE of the Republic of Korea. (no. 20182410105280).

Conflicts of Interest: The authors declare no conflict of interest.

Appendix A Appendix

The detailed parameters of an asynchronous were used as introduced in Ziping et al. [15]. The capacity of the asynchronous motor was 4 MW. The stator resistance and inductance of the motor were set to 0.02 p.u. and 0.04 p.u., respectively. The rotor resistance and inductance of the motor were set to 0.02 p.u. and 0.04 p.u., respectively. The mutual inductance and inertia constant were set to 1.36 p.u. and 0.1 p.u., respectively.

Figure A1 shows the typical governor and diesel engine block diagram. In this Figure, Tr_1, Tr_2, Tr_3, K, Ta_1, Ta_2, Ta_3, and T_d were set to 0.01, 0.02, 0.2, 40, 0.25, 0.009, 0.0384, and 0.024, respectively. The inertia time constant of a diesel generator was set to 5 s.

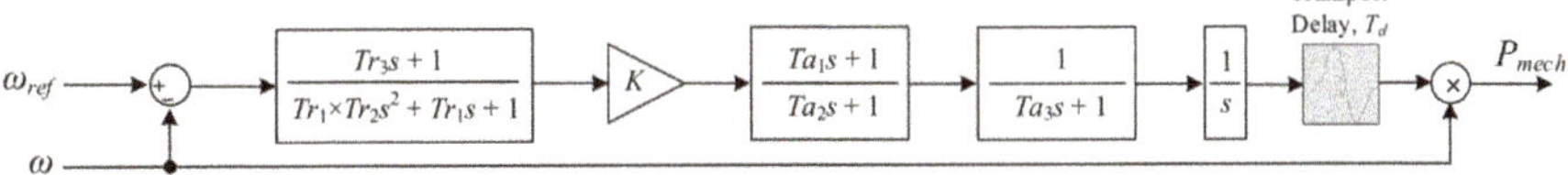

Figure A1. A governor and diesel engine block diagram.

References

1. Machowski, J.; Bialek, J.W.; Bumby, J.R. *Frequency Stability and Control*; John Wiley & Sons: Hoboken, NJ, USA, 2008.
2. Eto, J.H. *Use of Frequency Response Metrics to Assess the Planning and Operating Requirements for Reliable Integration of Variable Renewable Generation*; Lawrence Berkeley National Laboratory: Berkeley, CA, USA, 2011.
3. Fagan, E.; Grimes, S.; McArdle, J.; Smith, P.; Stronge, M. Grid Code Provisions for Wind Generators in Ireland. In Proceedings of the IEEE PES General Meeting, San Francisco, CA, USA, 12–16 June 2005; pp. 1241–1247.
4. EirGrid. *Controllable Wind Farm Power Station Grid Code Provisions*; Grid Code Version 3.1; EirGrid: Dublin, Ireland, 2008.
5. Muljadi, E.; Butterfield, C.P. Pitch-Controlled variable speed wind turbine generation. *IEEE Trans. Ind. Appl.* **2001**, *37*, 240–246. [CrossRef]
6. Senjyu, T.; Sakamoto, R.; Urasaki, N.; Funabashi, T.; Fujita, H.; Sekine, H. Output power leveling of wind turbine generator for all operating regions by pitch angle control. *IEEE Trans. Energy Convers.* **2006**, *21*, 467–475. [CrossRef]
7. Haixin, W.; Junyou, Y.; Zhe, C.; Weichun, G.; Shiyan, H.; Yiming, M.; Yunlu, L.; Guanfeng, Z.; Lijian, Y. Gain scheduled torque compensation of PMSG-based wind turbine for frequency regulation in an isolated grid. *Energies* **2018**, *11*, 1623–1642.
8. Lin, J.; Sun, Y.Z.; Song, Y.H.; Gao, W.Z.; Sørensen, P. Wind power fluctuation smoothing controller based on risk assessment of grid frequency deviation in an isolated system. *IEEE Trans. Sustain. Energy* **2013**, *4*, 379–392. [CrossRef]
9. Margaris, I.D.; Papathanassiou, S.Λ.; Hatziargyriou, N.D.; Hansen, A.D.; Sørensen, P. Frequency control in autonomous power systems with high wind power penetration. *IEEE Trans. Sustain. Energy* **2012**, *3*, 189–199. [CrossRef]
10. Tingting, C.; Sutong, L.; Gangui, Y.; Hongbo, L. Analysis of doubly fed induction generators participating in continuous frequency regulation with different wind speeds considering regulation power constraints. *Energies* **2019**, *12*, 635–655.
11. Morren, J.; Pierik, J.; de Haan, S.W.H. Intertial response of variable speed wind turbines. *Elect. Power Syst. Res.* **2006**, *76*, 980–987. [CrossRef]
12. Ullah, N.R.; Thiringer, T.; Karlsson, D. Temporary primary frequency control support by variable speed wind turbine—Potential and applications. *IEEE Trans. Power Syst.* **2008**, *23*, 601–612. [CrossRef]
13. Kim, Y.; Kang, M.; Muljadi, E.; Park, J.-W.; Kang, Y.C. Power smoothing of a variable-speed wind turbine generator in association with the rotor-speed-dependent gain. *IEEE. Trans. Sustain. Energy* **2017**, *8*, 990–999. [CrossRef]
14. Chunghun, K.; Muljadi, E.; Chungchoo, C. Coordinated control of wind turbine and energy storage system for reducing wind power fluctuation. *Energies* **2018**, *11*, 52–70.
15. Ziping, W.; Wenzhong, G.; Huaguang, Z.; Shijie, Y.; Xiao, W. Coordinated control strategy of battery energy storage system and PMSG-WTG to enhance system frequency regulation capability. *IEEE Trans. Sustain. Energy* **2017**, *8*, 1330–1343.
16. Akie, U.; Tomonobu, S.; Atsushi, Y.; Toshihisa, F. Frequency Control by Coordination Control of WTG and Battery Using Load Estimation. In Proceedings of the 2009 International Conference on Power Electronics and Drive Systems, Taipei, Taiwan, 2–5 November 2010.
17. Siegfried, H. *Grid Integration of Wind Energy Conversion System*; John Wiley & Sons: Hoboken, NJ, USA, 1998.
18. Vainio, O.; Ovaska, S.J. Digital filtering for robust 50/60 Hz zero-crossing detectors. *IEEE Trans. Instrum. Meas.* **1996**, *45*, 426–430. [CrossRef]

Article

Development and Calibration of an Open Source, Low-Cost Power Smart Meter Prototype for PV Household-Prosumers

F. Sanchez-Sutil [1], A. Cano-Ortega [1], J.C. Hernandez [1,*] and C. Rus-Casas [2]

[1]   Department of Electrical Engineering, University of Jaen, 23071 Jaen, Spain
[2]   Department of Electronic and Automatic Engineering, University of Jaen, 23071 Jaen, Spain
*   Correspondence: jcasa@ujaen.es; Tel.: +34-953-212463

Received: 1 July 2019; Accepted: 6 August 2019; Published: 7 August 2019

Abstract: Smart meter roll-out in photovoltaic (PV) household-prosumers provides easy access to granular meter measurements, which enables advanced energy services. The design of these services is based on the training and validation of models. However, this requires temporal high-resolution data for generation/load profiles collected in real-world household facilities. For this purpose, this research developed and successfully calibrated a new prototype for an accurate low-cost On-time Single-Phase Power Smart Meter (OSPPSM), which corresponded to these profiles. This OSPPSM is based on the Arduino open-source electronic platform. Not only can it locally store information, but can also wirelessly send these data to cloud storage in real-time. This paper describes the hardware and software design and its implementation. The experimental results are presented and discussed. The OSPPSM demonstrated that it was capable of in situ real-time processing. Moreover, the OSPPSM was able to meet all of the calibration standard tests in terms of accuracy class 1 (measurement error $\leq$1%) included in the International Electrotechnical Commission (IEC) standards for smart meters. In addition, the evaluation of the uncertainty of electrical variables is provided within the context of the law of propagation of uncertainty. The approximate cost of the prototype was 60 € from eBay stores.

Keywords: advanced metering infrastructure; data acquisition; IEC standards; low-cost; open source; power measurement; smart meter; uncertainty evaluation

1. Introduction

Smart meter roll-out in households with PV distributed generation, hereafter known as PV household-prosumers, provides easy access to on-time detailed meter measurements, which enable advanced energy services. The range of services provided include the application of demand response measures [1–6], smart home/building automation [7,8], and the provision of balancing services such as frequency control services (frequency containment reserve [9–12] and frequency restoration reserve [13]). The design of these services is based on the training and validation of models. However, this requires temporal high-resolution data for generation/load profiles collected in real-world household facilities. The optimal sizing of storage and generation facilities for these PV household-prosumers [3,14–17] also depends on the availability of reliable PV household profile data. The criteria for this sizing are based on technical, economical, and hybrid indicators [18]. Furthermore, the monitoring of PV household generation/load profiles has experienced an exponential growth in recent years [19–31]. PV household-prosumers generally include a battery energy storage system (BESS) [18,32].

BESSs, which manage household appliances [17,33,34] and/or renewable generation (e.g., PV [17,33–37]), often experience significant power fluctuations (range of 0.01–5 Hz [35,36,38–40]). These involve high charge/discharge powers. These rapid fluctuations should be taken into account in the design and assessment of advanced energy services and/or their sizing. For example, when longer time frames are envisaged in profile data (e.g., 1 or 5 min vs. 1 s), the sizing of storage and generation facilities for PV household-prosumers can lead to the miscalculation of costs [41]. In this context, the results may overestimate self-consumption and self-sufficiency ratios of these PV prosumers [3,18] and underestimate battery aging [42].

Regarding the availability of household consumption profile data, there are currently a small number of open-source datasets with varying levels of detail and scale. The projects in References [19,20] provided active power data at a sampling rate of 6 s and 8 s, respectively. Certain datasets, such as those in References [21,22], sampled data at 10 kHz, but only for a few weeks. Others, such as References [23–26], recorded data for at least a year, but at sampling intervals of 1 min or more. In addition, Reference [27] aggregated current and voltage at 6 s. Some publicly available datasets focused on capturing many individual appliance signatures [28,29], whereas other datasets offered aggregate and submetering measurements, sampled at 1 Hz, such as References [27,30,31].

The absence of reliable household consumption profile data, namely, data of temporal high-resolution (timeframe for data availability about 4 Hz or 0.25 s) was evidently a problem. To overcome this obstacle, this research study developed and calibrated a prototype of an accurate low-cost OSPPSM. In short, major contributions of the prototype (see Section 2) are the following: (i) it is able to monitor and store the fundamental AC electrical variables (v, i, PF (cos φ)) and the derived variables (active and reactive power p, q) within an non-intrusive load monitoring (NILM) scheme, (ii) it is able to monitor at temporal high-resolution (4 Hz, 0.25 s), and to store data both locally and in the cloud. In this way, PV household-prosumer profiles are available to accurately assess different advanced energy services provided by PV household-prosumers.

The rest of the paper is structured as follows. Section 2 overviews the state of the art regarding prototypes of power smart meters for households. Section 3 reviews the theoretical framework for electrical measurements. Section 4 describes the prototype from both a hardware and software perspective. Section 5 outlines the standard procedure for calibrating the prototype and describes the uncertainty assessment. Section 6 presents the results and discusses them. Lastly, the conclusions derived from this study as well as plans for future research are given in Section 7.

2. Research on Power Smart Meter Prototypes for Households

This section assesses the state of the art regarding prototypes of power smart meters for households. Depending on the support platform, these prototypes can be classified as follows: (i) Arduino open-source platform [3,7,14,43–57], (ii) field-programmable gate array (FPGA) technology [4,41], (iii) Lopy [5], and (iv) others [6,8,51,52].

The following research studies on the Arduino platform are presented in ascending order, depending on the number of functions offered by each prototype. Reference [43] developed a general–purpose voltage and current monitoring system designed for mobile devices via Bluetooth communication. It used an Arduino Nano board [53] to monitor instantaneous values. The research in Reference [47] developed a prototype for a power factor (PF) compensation monitoring system, capable of compensating at the industry or household level. This involved the measurement of voltage, current, and active power by means of a D1R1 Arduino microcontroller, based on an ESP-8266EX platform. The study in Reference [48] remotely controlled an energy meter by disconnecting and reconnecting the service of a particular consumer, based on an ATMega328P microcontroller. Reference [49] designed a meter that monitored energy and also sent data to the Internet by means of a wireless transmission system. The meter used global system mobile (GSM) and ZigBee wireless communication protocols. The study in Reference [44] developed a low-cost system for monitoring and remotely controlling greenhouses. The system used fuzzy logic to adapt to environmental conditions by means of an Arduino Mega board [54].

On the other hand, focusing on applications for PV systems on the Arduino platform, the research in Reference [45] developed a household monitoring system with a data logger, based on an Arduino platform. Hall-effect sensors with an analogical to digital converter (ADC) (LTS-15NP) were planned for currents. Reference [3] arranged a set of sensors based on an Arduino platform for monitoring and controlling household appliances with PV and BESS, which provided significant insights into their respective benefits. The monitoring system in Reference [46] had an Arduino Mega 2560 microcontroller board [58] and different sensors, with a clock speed of 16 MHz. It offered considerable flexibility to acquire data, and interface with the computer. The current, from the INA 219 DC [55] sensor, had a resolution of 0.1 mA and 1% accuracy.

Concerning FPGA technology, the smart meter developed in Reference [4] permitted a reconfigurable architecture. It allowed users to select the proper processing modules, depending on their application. The meter had voltage and current signals at a high sampling rate under a non-intrusive load-monitoring (NILM) scheme.

With regard to Lopy technology, the research in Reference [5] presented the proof-of-principle of a user-friendly monitoring system for household power consumption that made consumers aware of its consumption and impact. It was designed with a Lopy 4 module based on ESP32, and Wi-Fi connection to upload data to the Internet.

As for other platforms, the study in Reference [51] designed a monitoring system for stand-alone PV systems that provided pulse width modulation (PWM) signals for the battery charge controller. Reference [8] developed the prototype of a low-cost power smart meter, based on an ADE7913 chip. This meter was able to adapt its behavior to the grid with a high level of accuracy. Reference [6] designed an open-source, low-cost single-phase energy smart meter and power quality (PQ) analyzer that could be easily set up and used by inexperienced users at home. This instrument was able to retrieve a large amount of information related to energy consumption and PQ variables, which complied with national and international standards.

Regarding the calibration of meter prototypes, only Reference [52] addressed this issue with a PQ meter, according to the IEC standard 61000-4-7 [56]. Reference [50] devised a method for calibrating a ZMPT101B voltage sensor, using polynomial regression.

Concerning the evaluation of uncertainty, Reference [57] compared three methods to assess uncertainty in impedance monitoring. Moreover, Reference [58] applied a method to evaluate the uncertainty of data aggregation for root mean square (R.M.S) voltage [59].

This literature review [3–8,43–51] reflects that an accurate low-cost prototype for monitoring PV household-prosumers at temporal high-resolution has still not been developed and validated. Major shortcomings found in the prototypes proposed thus far include the following.

- Timeframes for monitoring were not adjusted to a temporal high-resolution [59], i.e., in the range 0.25 s [35,36,38–40], namely, 300 milliseconds [44], 0.5 s [47], 3 s [58], 5 s [49], 30 s [45], 1 min [3,44], 10 min [58], and 0.5 h [39].
- With the exception of References [6,8,42,47], data were not uploaded to the Internet. Even though Bluetooth communication provided an alternative communication channel, it was rarely planned [4,43]. Other researchers used GSM technology [48,49].
- If the prototype design included a data upload to the Internet, its cadence was low, e.g., 1 s [47], 3 s [6], 5 s [5], 10 s [4], and 5 min [49].
- Redundant storage means were rarely considered, e.g., local storage and upload to the Internet in Reference [4] even though the storage capacity was limited to two days.
- Even when the NILM scheme was included [4,6,8,47], it was inaccurately designed.
- Open source software was only used in References [6,43].
- None of the prototypes were calibrated according to standard calibration tests, and the uncertainty evaluation was only partially carried out in References [45,51].
- The analysis window for the stationary analysis was not frequently discussed, except for Reference [8] with 200 ms.

- Most of the meter prototypes measured DC variables, with the exception of the AC prototypes in References [4–6,8,43,47,57].
- Although various references [4–7,44,46–49] state that a low-cost prototype was developed, its actual cost was rarely evaluated, except for Reference [8] with a partial cost of 13 €, [43] with a cost of 25 €, and [45] with a cost of 60 €.

This study developed and calibrated a prototype of an accurate open-source, low-cost, OSPPSM in order to acquire the PV household-prosumer profiles at temporal high-resolution. This prototype has an Arduino low-cost, open-source platform and is able to monitor and store the fundamental AC electrical variables and the derived variables within an NILM scheme. This prototype has none of the previously mentioned shortcomings. The OSPPSM is, thus, able to monitor at temporal high-resolution (4 Hz, 0.25 s), and stores data both locally and in the cloud. The stationary analysis has 10-cycle analysis window with a sampling rate of 1 kHz. In this way, PV household-prosumer profiles are available to accurately assess different advanced energy services provided by PV household-prosumers. These services account for the effects of both the fast short-term fluctuations of input profiles (<4 Hz) and their hourly/daily/weekly/monthly variability.

3. Theoretical Background for Electrical Measurement

In electrical systems, time-invariable voltages or currents are almost impossible. Therefore, the window size of the analysis is assumed to be stationary. The windowing results in a given time localization and the spectrum this obtained is called a local spectrum. This window moves along the entire length of the signal in order to calculate the localized spectra. The window that measures electrical variables (e.g., voltage, current, harmonics, etc.) is a 10-cycle time interval for a 50-Hz power system (Class-A performance [59]).

To accurately measure an electrical signal, the sampling frequency should be at least twice the highest frequency of the signal. Since this study used a 1 kHz sampling frequency, the number of samples n_s for the 10-cycle analysis window was, thus, 200.

The R.M.S. value of an electrical variable (e.g., v, i) in a specified analysis window is given by the aggregation using the square root of the arithmetic mean of the squares of the n_s instantaneous values taken [4,5,59]. The instantaneous value of the active power is given by the product of the instantaneous values of voltage and current [60]. The average active power for a set of samples n_s is given by the equation below [60].

$$p^{avg} = \frac{\sum_{n=1}^{n_s} v^{ins,n} \cdot i^{ins,n}}{n_s} \tag{1}$$

The variable power factor *PF* can be expressed as the ratio of the average active power to the product of the R.M.S. values of voltage and current, respectively [4,5,47]. Lastly, the R.M.S. reactive power q is given by the equation below.

$$q^{r.m.s.} = v^{r.m.s.} \cdot i^{r.m.s.} \cdot \sin(\arccos\varphi) \tag{2}$$

4. Design of the On-Time Single-Phase Power Smart Meter (OSPPSM)

4.1. Hardware Design

This study designed and developed an OSPPSM for households with both local and cloud storage, based on AUR3 [61] and AD1R1 [62] Arduino boards. The OSPPSM (Figure 1) is modularly integrated, which means that, in the case of malfunction, parts can be replaced without affecting the general operation of this electrical measuring instrument (MI).

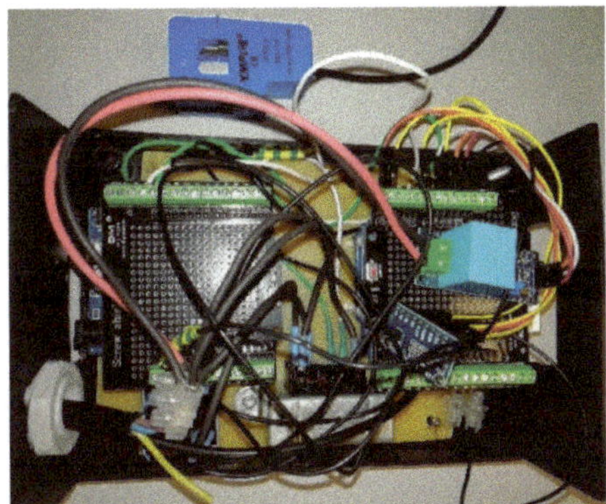

Figure 1. OSPPSM prototype.

In the first phase, the analog voltage and current sensors [63] on the AUR3 Arduino board of this electrical MI capture and process fundamental electrical variables, such as the voltage v, current i, and *PF (cos φ)*. This is followed by the calculation of derived variables, active p, and reactive q power. In a second phase, the AD1R1 Arduino board uploads the data to Firebase [64] using a Wi-Fi connection.

This OSPPSM receives two signals from the consumer unit to which it is connected: voltage and current. Each signal comes from the sensors, and is read through three analog inputs, including one for voltage and two for current.

Although it was possible for the the MKR WiFI 1010 Arduino board [65] to have more than one analog input and Wi-Fi connection for access to the Firebase, this option was disregarded because of its high price. Moreover, other low-cost Arduino Wi-Fi boards (e.g., wemos D1 R1 [62], wemos d1 mini [62], and NodeMCU [66], etc.) usually have only one analog input, and, therefore, could not be used. Even a single board could perform all of the tasks (i.e., acquisition of electrical signals, processing, and upload data to the cloud). This had the drawback of requiring high board features. This research, thus, decided on a dual board configuration in order to obtain the following: (i) significant cost reduction, (ii) improved device performance in terms of time of computation, thanks to dual processing, which provided shorter interval times for managing recordings. Figure 2 shows a hardware block diagram of the OSPPSM.

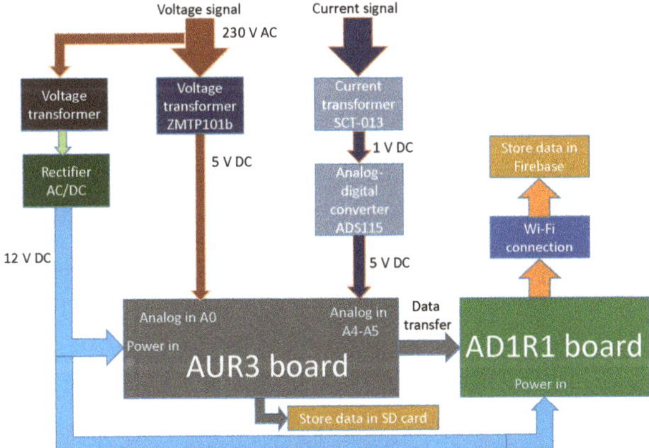

Figure 2. Hardware block diagram of the OSPPSM.

Figure 3 shows the wiring diagram of the OSPPSM. The Arduino boards are fed through one of the two 12 V AC outputs, which is rectified to DC to match the supply voltage of the AUR3 and AD1R1 boards (range 7–12 V DC). For the voltage signal, the ZMTP101b [67] voltage transformer is used to transform the 230 V AC signal to 5 V DC accepted by the AUR3 analog input. The A0 input is reserved for the voltage.

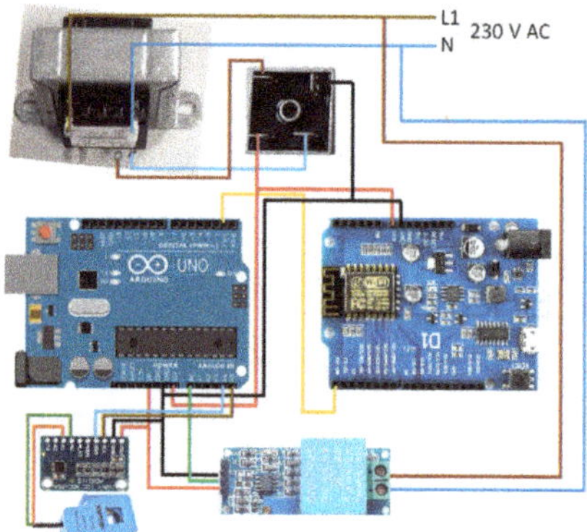

Figure 3. Wiring diagram of the OSPPSM.

The STC-013 current sensor [68] has a 1-V DC output. This is largely due to the ADS1115 analog digital converter [69], which adapts to the 5-V DC voltage for the analog inputs of the AUR3 board. The A4 and A5 inputs are reserved for current.

4.1.1. Microcontroller

The microcontroller is a small computer inside of a single integrated circuit, and contains one or more CPUs, RAM memory, and programmable input/output peripherals. They are widely used in industrial and residential equipment, because they are able to control signals and devices.

The rapid evolution of electronic devices had led to the availability of low-cost powerful hardware tools, which provide a viable solution for measuring and monitoring applications. In this context, the AUR3 board performs data digitization, processing, and transmission. It has a high 16-MHz clock speed, which obtains measurements in shorter time intervals (0.25 s). This is one of the aims of our OSPPSM.

The AUR3 board is based on the ATmega328P microcontroller on a platform for open-source electronic prototypes. Its technical specifications are given in Reference [61].

4.1.2. Wireless Communication

The wireless communication module is based on the AD1R1 board and serves as an interface between the microcontroller and the cloud data storage (i.e., Firebase). This board uses the ESP8266 platform as the core of operations, which allows WEP (Wired Equivalent Privacy) or WPA/WPA2 (Wi-Fi Protected Access) authentication for secure Wi-Fi communication. Furthermore, it operates with 802.11 b/g/n wireless systems, which are supported by most routers and modems on the market. These features signify that the average data upload time to the cloud is 0.15 s, which is shorter than the 0.25-second time interval of the planned power. The technical specifications of AD1R1 are given in Reference [62].

4.1.3. Current Sensor

Both invasive and non-invasive sensors are on the market. Invasive sensors require modification of the electrical installation, whereas non-invasive sensors measure current without any modification.

Techniques used to measure electrical currents include Hall Effect sensors and current transformers, which all transform the electrical current signal into a proportional voltage signal.

The STC-013 non-invasive current sensor [68] from YHDC is planned for the OSPPSM. It has a core to be installed at the service cable of the consumer unit of the monitored household. Options range from 5 to 100 A. The 30-A option is tuned for households, since it reaches a 6600-W power, which is higher than the average of most households. The 100-A limit allows 23,000-W household power.

An ADC, model ADS1115 [69] (Texas Instruments brand) matches the STC-013 output voltage to the 5-V DC level of the AUR3 board.

4.1.4. Voltage Sensor

There are various options for measuring and adapting voltage to the analog input signal of the AUR3 board: (i) a 230/12 V transformer, AC/DC rectifier, and voltage divider for adapting to a 5 V DC, (ii) a 230/24 V transformer, AC/DC rectifier, and DC meter FZ0430 that gives a maximum of 5 V DC, (iii) a 230/24 V transformer, AC/DC rectifier, and INA219 DC that gives a maximum of 5 V DC, and (iv) a ZMPT101b voltage transformer that directly provides a maximum of 5 V DC. Because most of these possibilities require several components, option (iv) was selected because it adapted best to the AUR3 board level. The technical specifications of the ZMPT101b voltage transformer are given in References [67].

4.1.5. Datalogger Shield

One problem that can arise is that, depending on network usage, the Internet connection for uploading data to the cloud is not always guaranteed, or may be excessively slow. For this reason, the OSPPSM was equipped with a datalogger shield with an 8 GB SD memory card. The storage capacity adopted has an autonomy of two years as well as five floating point data types for variables v, i, PF, p, and q, every 0.25 s. In addition, the datalogger shield includes a real-time clock that records the date and time of the measurements.

4.2. Software Design

Figure 4 shows a process timeline for the OSPPSM. In a first parallel process, the software in the AUR3 main microcontroller (Section 4.2.1) determines the fundamental and derived electrical variables, sends them through the serial port, and, lastly, stores them on the data logger. In a second parallel process, the software in the AD1R1 board (Section 4.1.2) uploads data to the Internet through the Wi-Fi connection.

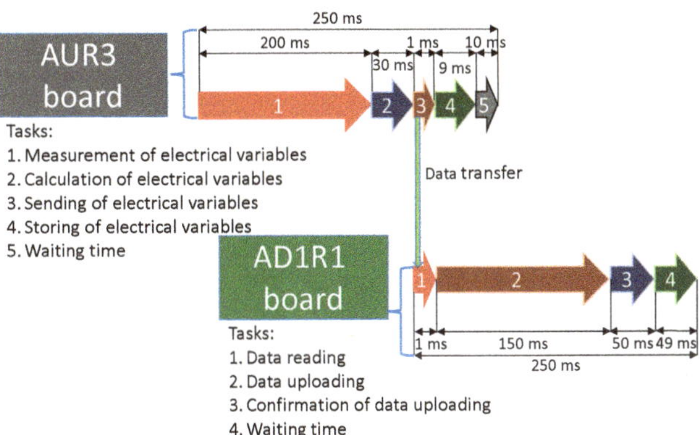

Figure 4. Process timeline for the OSPPSM.

4.2.1. Measurement and Computation of the Electric Variable

The microcontroller in the AUR3 board determines the fundamental and derived electrical variables, as reflected in the flowchart in Figure 5. The first step involves the initialization of the system. This includes putting the serial port in the data-sending mode by resetting analog inputs, the initialization of the SD memory card system, and the initiation of the clock in real time. However, these processes are only performed when the meter is connected. The second step is the measurement of fundamental electrical variables through analog inputs A0, A4, and A5. The fundamental and derived electrical variables are computed as specified in Section 3. Electrical variables are then sent through the serial port to the AD1R1 board, and, lastly, the fundamental and derived electrical variables are stored in the SD memory card as a backup copy. The measurement and storage of these variables are continuously performed while the meter is connected.

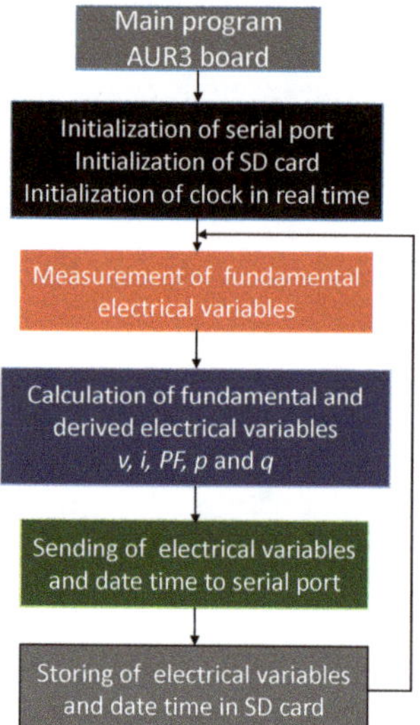

Figure 5. Flowchart for the measurement and computation of the electric variable: AUR3 board.

The maximum times for each task are shown in Figure 4: (i) 200 ms for the 10-cycle analysis window, (ii) 30 ms for calculating fundamental and derived variables, (iii) 1 ms for sending data to the AD1R1 board, (iv) 9 ms for storing all of the variables in a backup SD memory, and (v) a waiting time of 10 ms.

The software is implemented in the Arduino open-source platform [70], which is able to acquire data each millisecond, which signifies a 1-kHz sampling frequency. As a result, 200 measurement samples are acquired in the 10-cycle analysis window.

4.2.2. Cloud Data Uploading

The IoT has various options for storing data records in the cloud, such as ThingSpeak [50,71–73] and MQTT [74–77]. In their free version, both platforms store data records every 15 s. However, for shorter time intervals, it is necessary to purchase a standard commercial license. Even in that case, the shortest possible rate limit is 1 s.

Our study required data record storage and availability every 0.25 s. The only platform that provides this rate in its free version is Google's Firebase [64] platform with compatible data record storage times of every 0.1 s.

Cloud data uploading follows the flowchart in Figure 6. The upload program is located in the AR1D1 board [62], and performs the following tasks: (i) initialization of the system, which includes the preparation of the serial port in the data reading mode, the initialization of the Wi-Fi system to connect the household wireless network, and the initialization of the Firebase system, (ii) data reading from the AUR3 board, (iii) data uploading to the cloud using Firebase, and (iv) confirmation of data uploading. Tasks (ii-iv) are continuously performed while the meter is connected.

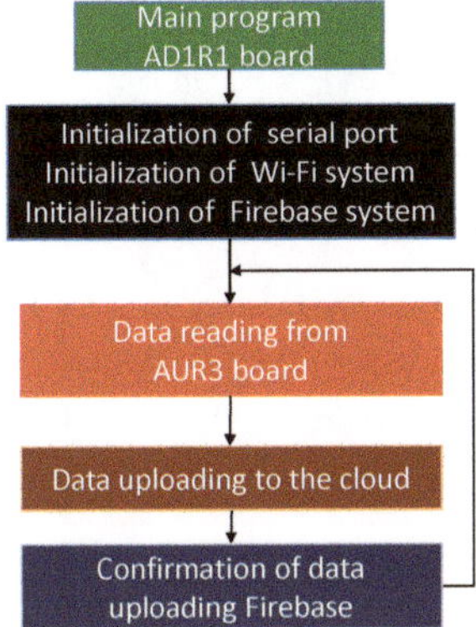

Figure 6. Flowchart for cloud data uploading: AD1R1 board.

The maximum times for each task are shown in Figure 4: (i) 1 ms for data reading from the serial port, (ii) 150 ms for data uploading to the cloud, (iii) 50 ms for confirming the data uploading by the Firebase server, and (iv) a wait of 49 ms.

5. Standard Guidance on Calibration and Uncertainty Evaluation for Power Smart Meters

This section provides the theoretical background for the characterization of errors, which includes the standard tests that should be used to calibrate power smart meters and the evaluation procedure of the uncertainty in measurements.

5.1. Characterization of Errors

The error of an MI (e.g., a power smart meter) is obtained by subtracting the true value from the indicated value [78]. In particular, the intrinsic error [79] of an nth measurement of the variable x_i is the error of the MI as compared to the reference measurement standard (RMS) when used under reference conditions.

$$E_{x_j^n} = 100 \times \frac{\left(x_{j,ref}^{n,MI} - x_{j,ref}^{n,RMS}\right)}{x_{j,ref}^{n,RMS}} \tag{3}$$

The mean absolute percentage error (MAPE) and the mean relative error (MRE) for a set of measurements ns of the variable $x_j = [x_j^1, x_j^2, \dots, x_j^{n_s}]$ are given by Reference [51].

$$MAPE_{x_j} = \frac{100}{n_s} \times \sum_{n=1}^{n_s} \left| \frac{\left(x_{j,ref}^{n,MI} - x_{j,ref}^{n,RMS}\right)}{x_{j,ref}^{n,RMS}} \right| \tag{4}$$

$$MRE_{x_j} = \frac{100}{n_s} \times \sum_{n=1}^{n_s} \frac{\left(x_{j,ref}^{n,MI} - x_{j,ref}^{n,RMS}\right)}{x_{j,ref}^{n,RMS}} \tag{5}$$

The characterization of the intrinsic error distribution $E_{x_j} = [E_{x_j^1}, E_{x_j^2}, \ldots, E_{x_j^{n_s}}]$ can be performed by moments, which are a set of descriptive constants of the distribution. Thus, the first moment mean (μ) of the distribution E_{x_j} is given by the equation below.

$$\mu_{E_{x_j}} = \frac{\sum_{n=1}^{n_s} E_{x_j^n}}{n_s} \tag{6}$$

The second moment is the variance (σ) of the distribution.

$$\sigma^2_{E_{x_j}} = \frac{\sum_{n=1}^{n_s} \left(E_{x_j^n} - \mu_{E_{x_j}}\right)^2}{n_s - 1} \tag{7}$$

Alternatively, the standard deviation can be obtained as follows.

$$\sigma_{E_{x_j}} = \sqrt{\sigma^2_{E_{x_j}}} \tag{8}$$

5.2. Standard Calibration Test

This section provides a brief description of the common standard tests that should be used to calibrate a power smart meter. It is necessary to apply standard calibration tests for voltmeters [80], ammeters [80], watt meters [81], varmeters [81], and PF meters [82]. Additionally, functional tests for measuring, recording, and displaying PQ parameters in instruments for the distribution grid are relevant [83].

The accuracy of an electrical MI characterizes the degree of proximity between the indicated value and the true value. In particular, the accuracy class categorizes the potential errors within specified limits [78]. In the calibration of an electrical MI, the indicated values [79] of the MI are compared with those of a RMS (indicative of the highest metrological quality) under different working points at reference conditions [78]. As a result of these tests, a maximum intrinsic error is obtained, which defines the accuracy class of the calibrated electrical MI [78]. In what follows, the common standard tests are described.

The intrinsic value test determines the intrinsic error of an electrical MI for the fundamental and derived electrical variables under reference conditions [78]. The measuring range of the fundamental variables (v, i, *PF* (*cos φ*)) includes the interval of 0% to 120% of the rated value at different points on the scale [83]. For ammeters (current measuring [80]) and voltmeters (voltage measuring [80]), both variable magnitudes must be modified. For PF meters (PF measuring), the PF should be modified [82]. For watt meters (active power measuring) and varmeters (reactive power measuring [81]), once the rated voltage and reference PF are fixed, the current should be changed [83].

The current magnitude distortion test [83] superposes a 20% harmonic third-magnitude wave on the sinusoidal fundamental electrical variables, while adapting the fundamental component to maintain the resulting R.M.S. value. In the case of watt meters, varmeters, and PF meters, a waveform with 20% of the third harmonic to the rated voltage and current is superposed, and the fundamental component is adapted to maintain the resulting R.M.S. value [80–82]. The intrinsic error is then determined.

The alternating current frequency variation test applies a frequency variation from 40 to 60 Hz to fundamental electrical variables [83]. Particularly for PF meters, the frequency change applies to voltage and current, with different PFs (PF: 0.5 lagging, 1, 0.5 leading). For watt meters and varmeters, the PF is set at the reference value [81].

The alternating current/voltage component variation test performs two assessments on PF meters [82]. The first one focuses on the voltage variation, and fixes the current at 50% of its rated value with three voltage magnitudes: (i) rated value, (ii) lower limit of the nominal range of use (NRU), and (iii) higher limit of the NRU. In addition, it is necessary to include different PFs (0 lagging, 1, 0.5 lagging, and 0.5 leading). The second assessment focuses on the current variation, and fixes the voltage at a rated value, with three current magnitudes: (i) rated value, (ii) lower limit of the NRU, and (iii) higher limit of the NRU. Once again, the different PFs should be analyzed [83]. In the case of watt meters and varmeters [81], the current is set to 80% of the upper limit of the NRU and PF at reference conditions with three voltage magnitudes: (i) rated value, (ii) lower limit of the NRU, and (iii) higher limit of the NRU. The intrinsic error is then evaluated [78].

The PF variation test focuses only on the variable active power (wattmeter) and reactive power (varmeter) [81]. This includes two assessments at reference frequency. The first is at the nominal current, where the voltage measuring range is from the lower to upper limit of the NRU, with different PFs (1, 0.5 lagging, 0.5 leading). The second is at nominal voltage, where the current fulfills the relevant previous variation range [83]. The intrinsic error [78] is then evaluated.

The continuous overload test applies an overload of 120% for two hours on fundamental electrical variables. For watt meters, varmeters, and PF meters, the 120% overload applies to the current, once the rated voltage and reference PF have been fixed [83]. The intrinsic error [78] applies to the evaluated quantities.

5.3. Uncertainty in Measurements

The objective of a measurement is to determine the true value of a measurand [84]. However, the result of a measurement is an estimate of this true value. It should, thus, be accompanied by a statement of the uncertainty of that estimate [85]. Consequently, the uncertainty reflects the lack of exact knowledge of the value of the measurand [84].

5.3.1. Uncertainty of Fundamental Variables and Standard Uncertainty

The uncertainty evaluation associated with the measurements of a fundamental variable is characterized by the standard uncertainty [84]. This can be obtained from a Type A or Type B evaluation. These are essentially two ways of evaluating uncertainty components [84], which are based on different procedures and probabilistic distributions.

The standard uncertainty type A for a series of observations n_s of the variable $x_j = [x_j^1, x_j^2, \ldots, x_j^{n_s}]$ is characterized by a statistical analysis, in particular, the estimated variance [84].

$$\sigma_{x_j}^s = \sigma_{x_j}^s(\mu_{x_j}) = \sqrt{\frac{\sum\limits_{n=1}^{n_s} \left(x_j^n - \mu_{x_j}\right)^2}{n_s - 1}} \tag{9}$$

where:

$$\mu_{x_j} = \frac{\sum\limits_{n=1}^{n_s} x_j^n}{n_s} \tag{10}$$

In addition to previous data, the evaluation of standard uncertainty type B requires knowledge of the MI, manufacturer specifications, and calibration and uncertainty data, defined by the manufacturer [84].

5.3.2. Uncertainty of Derived Variables and Combined Uncertainty

The combined uncertainty of a derived variable y of two or more fundamental variables $(x_j, x_m, \ldots, x_w)$ is characterized by a numerical value expressed in the form of standard deviation obtained by applying the usual method of the combination of variances [84].

$$\sigma_y^c = \sqrt{\sum_{z=1}^{n_v}\left[\frac{\partial f}{\partial x_z}\right]^2\left[\sigma_{x_z}^s(\mu_{x_z})\right]^2 + 2\sum_{z=1}^{n_v-1}\sum_{j=z+1}^{n_v}\frac{\partial f}{\partial x_z}\frac{\partial f}{\partial x_j}\left(\sigma_{x_z}^s(\mu_{x_z})\cdot\sigma_{x_z}^s(\mu_{x_j})\cdot\rho(x_z,x_j)\right)} \tag{11}$$

where:

$$\rho(\mu_{x_z},\mu_{x_j}) = \frac{\sum\limits_{n=1}^{n_s}(x_z^n-\mu_{x_z})\times\left(x_j^n-\mu_{x_j}\right)}{n_s-1}/\sigma_{x_z}^s\cdot\sigma_{x_j}^s \tag{12}$$

Regarding the OSPPSM, the combined uncertainty of the derived variables, active p, and reactive q power, can be obtained using the formulas below [84].

$$\sigma_p^c = \sqrt{\begin{array}{l}[\mu_i\mu_{PF}]^2\cdot(\sigma_v^s)^2 + [\mu_v\mu_{PF})]^2\cdot\left(\sigma_i^s\right)^2 - [\mu_v\mu_i\sin(arc\cos(\mu_{PF}))]^2\cdot\left(\sigma_\varphi^s\right)^2\\ +2\left\{\begin{array}{l}[\mu_i\mu_{PF}]\cdot[\mu_v\mu_{PF}]\cdot\sigma_v^s\cdot\sigma_i^s\cdot\rho(\mu_v,\mu_i)\\ +[\mu_i\mu_{PF}]\cdot[-\mu_v\mu_i\sin(arc\cos(\mu_{PF}))]\cdot\sigma_v^s\cdot\sigma_\varphi^s\cdot\rho\left(\mu_v,\mu_\varphi\right)\\ +[\mu_v\mu_{PF}]\cdot[-\mu_v\mu_i\sin(arc\cos(\mu_{PF}))]\cdot\sigma_i^s\cdot\sigma_\varphi^s\cdot\rho\left(\mu_i,\mu_\varphi\right)\end{array}\right\}\end{array}} \tag{13}$$

$$\sigma_q^c = \sqrt{\begin{array}{l}[\mu_i\sin(arc\cos(\mu_{PF}))]^2\cdot(\sigma_v^s)^2 + [\mu_v\sin(arc\cos(\mu_{PF}))]^2\cdot\left(\sigma_i^s\right)^2 + [\mu_v\mu_i\mu_{PF})]^2\cdot\left(\sigma_\varphi^s\right)^2\\ +2\left\{\begin{array}{l}[\mu_i\sin(arc\cos(\mu_{PF}))]\cdot[\mu_v\sin(arc\cos(\mu_{PF}))]\times\sigma_v^s\cdot\sigma_i^s\cdot\rho(\mu_v,\mu_i)\\ +[\mu_i\sin(arc\cos(\mu_{PF}))]\cdot[\mu_v\mu_i\mu_{PF}]\cdot\sigma_v^s\cdot\sigma_\varphi^s\cdot\rho\left(\mu_v,\mu_\varphi\right)\\ +[\mu_v\sin(arc\cos(\mu_{PF}))]\cdot[\mu_v\mu_i\mu_{PF}]\cdot\sigma_i^s\cdot\sigma_\varphi^s\cdot\rho\left(\mu_i,\mu_\varphi\right)\end{array}\right\}\end{array}} \tag{14}$$

5.3.3. Confidence Level of the Uncertainty Evaluation

The uncertainty value when a measurement is taken depends on the confidence level and the sample number n_s of the measurement. Nonetheless, given a certain confidence level, the sample number can be determined. The definition of the confidence interval [84] for the mean μ_{x_j} of a measurement set n_s of variable x_j is given by the formula below.

$$\left(\mu_{x_j} - Z_{\frac{\alpha}{2}}\cdot\frac{\sigma_{x_j}^s}{\sqrt{n_s}}, \mu_{x_j} + Z_{\frac{\alpha}{2}}\cdot\frac{\sigma_{x_j}^s}{\sqrt{n_s}}\right) \tag{15}$$

and the probability that the mean will be in this confidence interval is shown below.

$$P\left(\mu_{x_j} - Z_{\frac{\alpha}{2}}\cdot\frac{\sigma_{x_j}^s}{\sqrt{n_s}} < \mu_{x_i} < \mu_{x_j} + Z_{\frac{\alpha}{2}}\cdot\frac{\sigma_{x_j}^s}{\sqrt{n_s}}\right) = 1 - \alpha \tag{16}$$

When evaluating the uncertainty of the different levels in Reference [84], the 99% level is usually set. This confidence level $(1-\alpha)$ is defined as the probability that the measurement mean will be within this range, which is shown in Equation (16).

Assuming an N(0,1) normal Gaussian distribution for the measurements and a set confidence level α^{set} of 1%, the corresponding value of $Z_{\frac{\alpha^{set}}{2}}$ is 2.58. Therefore, once the value $Z_{\frac{\alpha^{set}}{2}}$ is known, the minimum number of samples needed to attain a confidence level α^{set} is shown below.

$$n_{s\text{-min}} \geq \left(\frac{Z_{\frac{\alpha^{set}}{2}}\cdot\sigma_{x_j}^s}{\alpha^{set}}\right)^2 \tag{17}$$

6. Results

The OSPPSM was tested in the electrical engineering laboratory at the University of Jaen (Spain), Figure 7. This meter was connected to Wi-Fi and Firebase during all of the tests to store the data.

In order to create the testing conditions, a grid emulator and a programmable electronic load were used. This equipment had larger regulation possibilities that significantly exceeded the requirements to perform the calibration standard tests. Moreover, an electrical RMS was used as a reference for the calibration. The following subsections describe the test equipment.

Figure 7. Test equipment and OSPPSM in the laboratory at Jaen university.

The tests allowed us to determine the accuracy class reached by the OSPPSM, based on the maximum error observed. Additional tests were also conducted to evaluate the uncertainty that fully defined the OSPPSM.

6.1. Test Equipment

6.1.1. Electrical Reference Measurement Standard (RMS)

The electrical RMS to perform the calibration of the OSPPSM was the Class-A power quality (PQ) analyzer (Fluke^{TM} 1760TR Fluke Corporation, Everett, WA, USA), which is a high-end model of the brand, that allows the recording of a multitude of electrical variables compatible with calibration standard tests. Table 1 shows the main characteristics of Fluke 1760TR.

Table 1. Main features of electrical RMS and model Fluke 1760TR.

Parameter	Range
Continuous recording	Voltage, current, active, reactive, and apparent power, power factor, energy, harmonics, etc.
Measuring intervals	10, 20, 200, 500 ms, or 3 s

Parameter	Range
Sampling rate	10–24 kHz
Resolution	16 ppm
Uncertainty for frequency	<20 ppm
Uncertainty for voltage	0.1% at 230 V
Intrinsic uncertainty for harmonics	Class I [56]
Accuracy class	Class I

6.1.2. Grid Emulator

The Cinergia GE+15 unit emulates a low voltage grid. This allowed us to modify parameters such as frequency, phase angle, magnitude, and harmonic context in the main voltage. This capability was required for calibration standard tests. Table 2 shows the main characteristics of the used grid emulator.

Table 2. Main features of grid emulator, model Cinergia GE+15.

Parameter	Range
Voltage	0 to 277 V phase-neutral 0 to 480 V phase-phase
Current	66 A max
Phase angle	0° to 360° resolution 0.01°
Power	15 kW
Frequency	10 to 100 Hz
Harmonics	Up to 50th 15 harmonics independent/phase
Accuracy	±0.1% voltage, ±0.2% current

6.1.3. Programmable Electronic Load

Calibration standard tests require variable loads. Therefore, a programmable electronic load, model Cinergia EL+15, made it possible to adapt the required load for each specific standardized test. Table 3 shows the characteristics of the programmable electronic load used in this study.

Table 3. Main features of electronic load, model Cinergia EL+15.

Parameter	Range
Voltage	0 to 277 V phase-neutral 0 to 480 V phase-phase
Current	66 A max
Phase angle	−90° to 90° resolution 0.01°
Power	15 kW
Frequency	10 to 100 Hz
Harmonics	Up to 50th 15 harmonics independent/phase
Accuracy	±0.1% voltage, ±0.2% current

6.2. Calibration Standard Test

6.2.1. Intrinsic Value Test

Figure 8a, Figure 9a, Figure 10a,b, Figure 11a, and Figure 12a portray the correlation between the values in the OSPPSM for the variables, v_{OSPPSM}, i_{OSPPSM}, PF_{OSPPSM}, p_{OSPPSM}, and q_{OSPPSM}. They also show the reference values in the RMS for v_{RMS}, i_{RMS}, PF_{RMS}, p_{RMS}, and q_{RMS} under different intrinsic value tests on a 0.25 s basis. The intrinsic errors obtained for the variables are reflected in Figure 8b, Figure 9b, Figure 10c, Figure 11b, and Figure 12b. As can be observed, the intrinsic error did not exceed 1% for each electrical variable measured (v, i, PF, p, and q), which signified that the OSPPSM is in the accuracy class 1.

The results of the Kolmogorov-Smirnov test (95% confidence level) show that the intrinsic error follows a beta distribution for voltage measurement (p-value = 0.104), and a uniform distribution for both current measurement (p-value = 0.312) and PF measurement (p-value = 0.311). This uniformity in the intrinsic error distribution highlights the accuracy of the OSPPSM, with a maximum intrinsic error value of 0.9783% as a voltmeter and 0.91% as a PF meter, wattmeter, and varmeter.

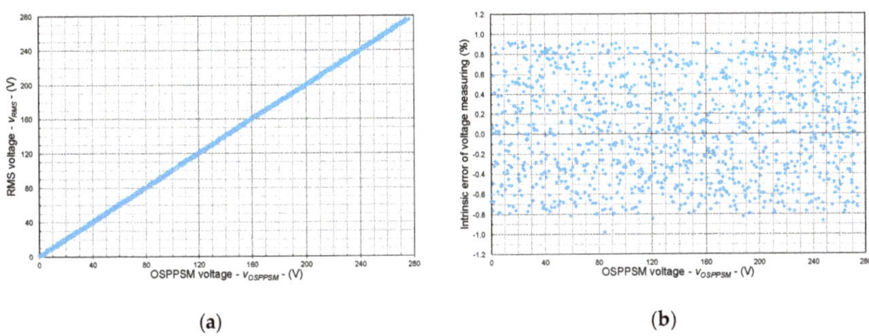

(**a**)   (**b**)

Figure 8. Voltmeter test: (**a**) OSPPSM vs. RMS voltage. (**b**) Intrinsic error of voltage measuring.

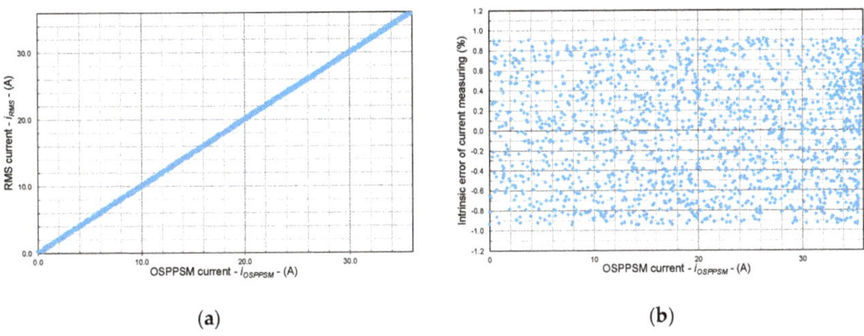

(**a**)   (**b**)

Figure 9. Ammeter test: (**a**) OSPPSM vs. RMS current. (**b**) Intrinsic error of current measuring.

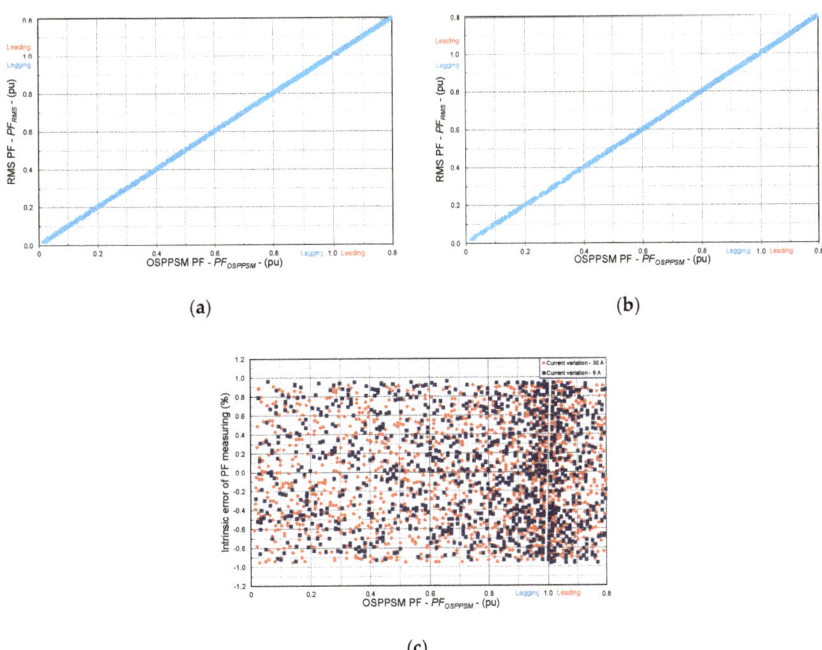

(**a**)   (**b**)

(**c**)

Figure 10. PF meter test: (**a**) 30 A, OSPPSM vs. RMS PF. (**b**) 8 A, OSPPSM vs. RMS PF. (**c**) Intrinsic error of PF measuring, 8 A and 30 A.

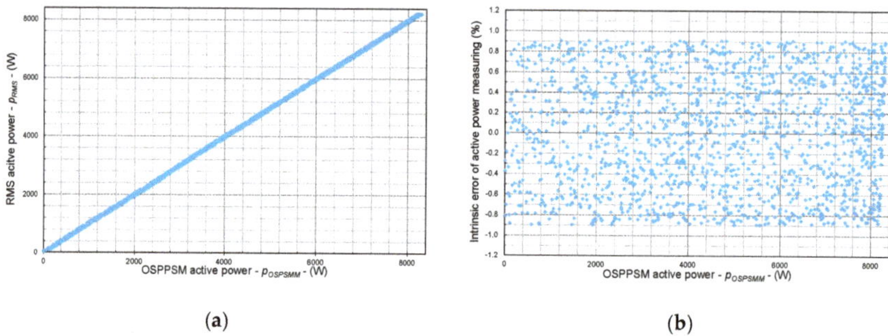

(**a**) (**b**)

Figure 11. Wattmeter test: (**a**) OSPPSM vs. RMS active power. (**b**) Intrinsic error of active power measuring.

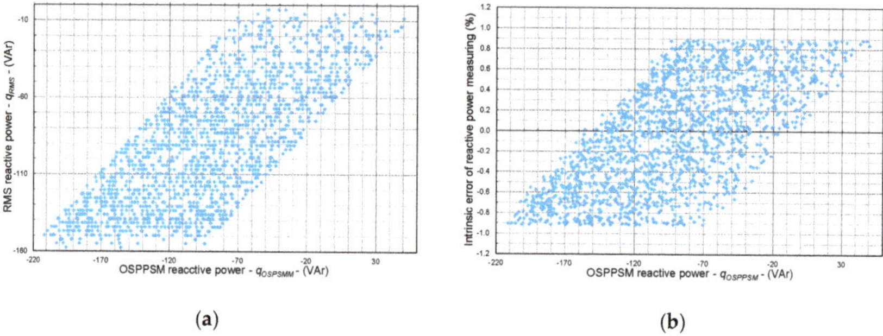

(**a**) (**b**)

Figure 12. Varmeter test: (**a**) OSPPSM vs. RMS reactive power. (**b**) Intrinsic error of reactive power measuring.

Table 4 displays the MAPE and MRE statistical error indicators as well as the standard deviation. The MAPE was less than 0.48%, and MRE was below 0.07%. During all of the intrinsic value tests, the standard deviation did not exceed 0.55%.

Table 4. Intrinsic value test: maximum intrinsic error (%), MAPE (%), MRE (%), and standard deviation (%).

Test	Maximum Intrinsic Error	MAPE	MRE	Standard Deviation
Voltmeter	0.9783	0.4303	0.0643	0.4957
Ammeter	0.9400	0.4712	−0.0080	0.5428
PF meter, 8 A	0.9500	0.4218	−0.0133	0.5123
PF meter, 30 A	0.9100	0.4675	0.0133	0.5366
Wattmeter	0.9100	0.4675	0.0133	0.5366
Varmeter	0.9100	0.4540	−0.0280	0.5240

6.2.2. Current Magnitude Distortion Test

Figures 13 and 14 display the results of the voltage variation, which show the relationship between the voltage v_{OSPPSM} and the reference voltage v_{RMS}. They also show the intrinsic error for variables v_{OSPPSM}, p_{OSPPSM}, and q_{OSPPSM}. Figures 15 and 16 focus on the current variation, which reflects the relationship between the current value i_{OSPPSM} and the reference current i_{RMS}. The intrinsic error for the variables i, p_{OSPPSM}, and q_{OSPPSM} is also shown.

The error graphs underline the fact that the intrinsic error was less than 1%, which signifies that the OSPPSM belongs to accuracy class 1. Furthermore, the maximum intrinsic error for the voltage variation and current variation was lower than 0.99%. Once again, the intrinsic error had a uniform distribution.

Tables 5 and 6 show the MAPE and MRE statistical error indicators as well as the standard deviation for the voltage and current variation, respectively. The voltage variation test did not exceed 0.49% for the MAPE, 0.49% for the MRE, and 0.56% for the standard deviation. Meanwhile, the current variation test indicated that the MAPE and MRE values and standard deviation were similar to those of the voltage variation test.

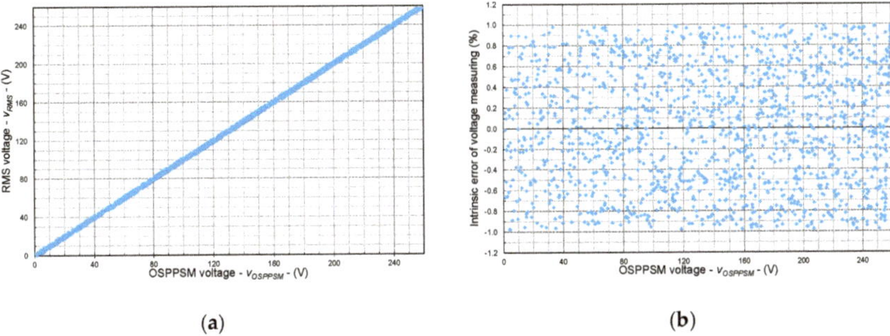

Figure 13. Voltmeter test: (a) OSPPSM vs. RMS voltage. (b) Intrinsic error of voltage measuring.

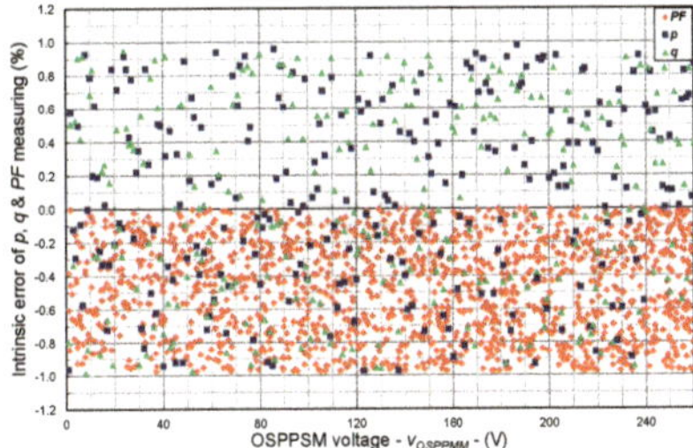

Figure 14. PF meter, wattmeter, and varmeter test: intrinsic error of different variables measuring the voltage variation.

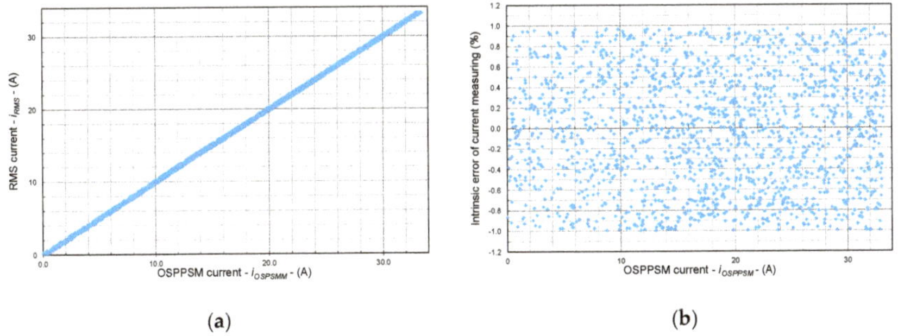

Figure 15. Ammeter test: (a) OSPPSM vs. RMS current. (b) Intrinsic error of current measuring.

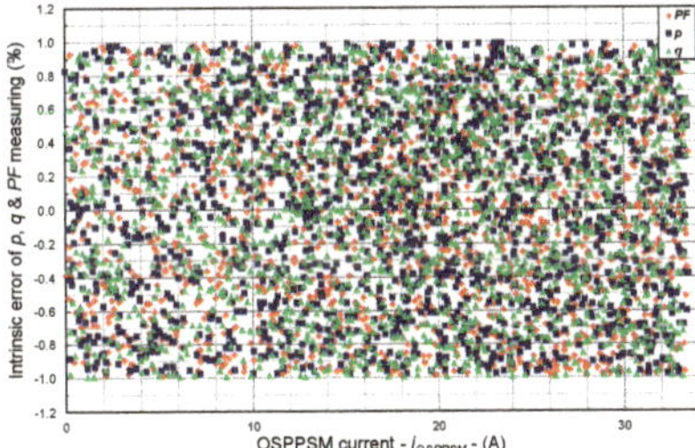

Figure 16. PF meter, wattmeter, and varmeter test: intrinsic error of different variable measuring for current variation.

Table 5. Current magnitude distortion test: maximum intrinsic error (%), MAPE (%), MRE (%), and standard deviation (%): voltage variation.

Test	Maximum Intrinsic Error	MAPE	MRE	Standard Deviation
Voltmeter	0.9900	0.4869	−0.0048	0.5593
PF meter	0.9700	0.4820	−0.4820	0.2860
Wattmeter	0.9900	0.4208	0.0225	0.5239
Varmeter	0.9900	0.3738	0.0076	0.5251

Table 6. Current magnitude distortion test: maximum intrinsic error (%), MAPE (%), MRE (%), and standard deviation: current variation.

Test	Maximum Intrinsic Error	MAPE	MRE	Standard Deviation
Ammeter	0.9900	0.4820	−0.0121	0.5594
PF meter	0.9800	0.4906	−0.0300	0.5721
Wattmeter	0.9900	0.4949	0.0288	0.5698
Varmeter	0.9900	0.2069	0.0038	0.5663

6.2.3. Alternating Current Frequency Variation Test

Figures 17a and 18a illustrate the values of the variables v_{OSPPSM} and i_{OSPPSM} in relation to the reference values v_{RMS} and i_{RMS}. The results of the intrinsic error for variables v, i, PF, p, and q are depicted in Figure 17b, Figure 18b, Figure 19, and Figure 20. As a result, the intrinsic error for none of the variables studied exceeded 1%, which confirmed the inclusion of the OSPPSM in accuracy class 1. The distribution of errors was again uniform with a maximum value of 0.98%.

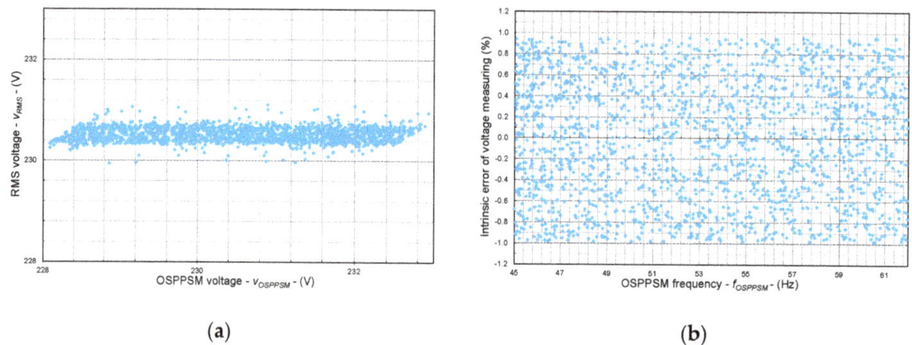

Figure 17. Voltmeter test: (**a**) OSPPSM vs. RMS voltage. (**b**) Intrinsic error of voltage measuring.

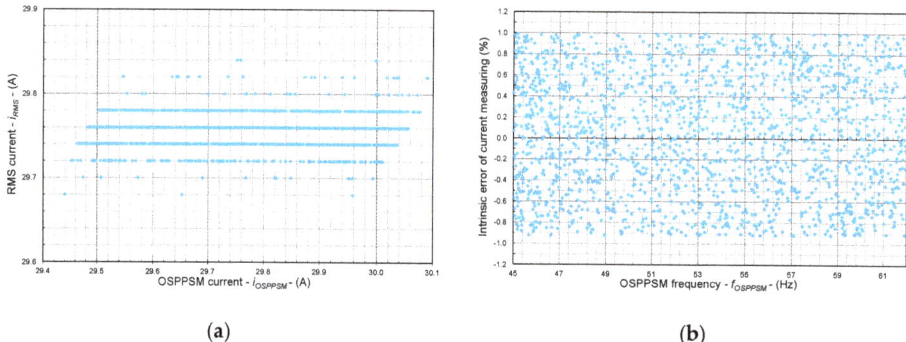

Figure 18. Ammeter test: (**a**) OSPPSM vs. RMS current. (**b**) Intrinsic error of current measuring.

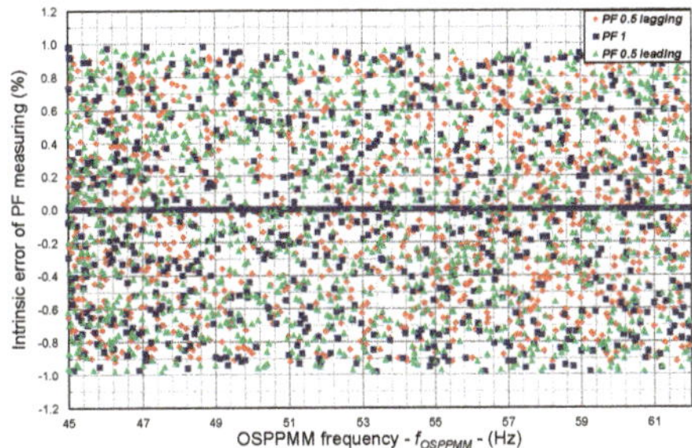

Figure 19. PF meter test: Intrinsic error of PF measuring (*PF*: 0.5 lagging, 1, 0.5 leading).

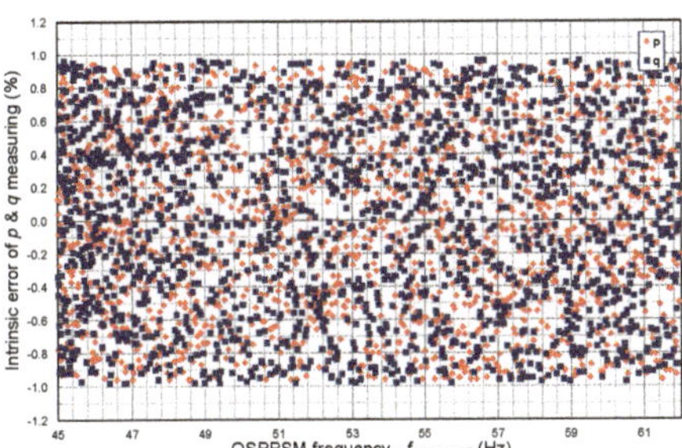

Figure 20. Wattmeter and varmeter test: Intrinsic error of different variable measuring.

Table 7 summarizes the statistical error indicators. As can be observed, the maximum MAPE value for the different electrical variables did not exceed 0.63%. The MRE did not reach values higher than 0.05%, and the standard deviation did not exceed 0.8%.

Table 7. Alternating current frequency variation test: maximum intrinsic error (%), MAPE (%), MRE (%), and standard deviation (%).

Test	Maximum Intrinsic Error	MAPE	MRE	Standard Deviation
Voltmeter	0.9800	0.4875	−0.0090	0.5619
Ammeter	0.9900	0.4778	0.0446	0.5511
PF meter, *PF*: 0.5 lagging	0.9200	0.2944	0.0025	0.4227
PF meter, *PF*: 1	0.9800	0.3296	−0.0110	0.4653
PF meter, *PF*: 0.5 leading	0.9500	0.6211	0.0119	0.7645
Wattmeter	0.9600	0.4712	−0.0206	0.5466
Varmeter	0.9800	0.4904	0.0089	0.5650

6.2.4. Alternating Current/Voltage Component Variation Test

Tables 8 and 9 show the results for the alternating current/voltage component variation test, which changed the voltage and current when OSPPSM acted as a PF meter. The maximum intrinsic error was lower than 0.98% for both assessments. This maintained the OSPPSM in accuracy class 1. The tables also include the statistical error indicators. In this regard, the maximum MAPE value was below 0.34% for both assessments, whereas the maximum MRE value did not exceed 0.26% for the voltage variation, and 0.24% for current variation. The standard deviation was lower than 0.48%.

Table 10 gives the results obtained when the OSPPSM acted as a wattmeter and varmeter. The fact that the maximum intrinsic error was 0.97% signifies that OSPPSM continued in accuracy class 1. The maximum MAPE and MRE values were 0.512% and −0.034%, respectively. The standard deviation was 0.582%.

Table 8. Alternating current/voltage component variation test (PF meter, voltage variation): maximum intrinsic error (%), MAPE (%), MRE (%), and standard deviation (%).

Test	Maximum Intrinsic Error			MAPE			MRE			Standard Deviation		
	$v = 230$ V $i = 15$ A	$v = 0$ V $i = 15$ A	$v = 253$ V $i = 15$ A	$v = 230$ V $i = 15$ A	$v = 0$ V $i = 15$ A	$v = 253$ V $i = 15$ A	$v = 230$ V $i = 15$ A	$v = 0$ V $i = 15$ A	$v = 253$ V $i = 15$ A	$v = 230$ V $i = 15$ A	$v = 0$ V $i = 15$ A	$v = 253$ V $i = 15$ A
PF meter, PF: 0 lagging	0.920	0.920	0.940	0.339	0.286	0.287	−0.002	0.008	−0.012	0.462	0.412	0.417
PF meter, PF: 1	0.920	0.920	0.940	0.222	0.259	0.275	−0.222	−0.259	−0.275	0.279	0.298	0.311
PF meter, PF: 0.5 lagging	0.980	0.950	0.970	0.332	0.294	0.316	−0.046	−0.044	−0.003	0.463	0.432	0.462
PF meter, PF: 0.5 leading	0.960	0.930	0.930	0.279	0.293	0.301	−0.018	−0.008	0.005	0.403	0.430	0.423

Table 9. Alternating current/voltage component variation test (PF meter, current variation): maximum intrinsic error (%), MAPE (%), MRE (%), and standard deviation (%).

Test	Maximum Intrinsic Error			MAPE			MRE			Standard Deviation		
	$v = 230$ V $i = 30$ A	$v = 230$ V $i = 0$ A	$v = 230$ V $i = 36$ A	$v = 230$ V $i = 30$ A	$v = 230$ V $i = 0$ A	$v = 230$ V $i = 36$ A	$v = 230$ V $i = 30$ A	$v = 230$ V $i = 0$ A	$v = 230$ V $i = 36$ A	$v = 230$ V $i = 30$ A	$v = 230$ V $i = 0$ A	$v = 230$ V $i = 36$ A
PF meter, PF: 0 lagging	0.920	0.950	0.950	0.308	0.308	0.328	0.007	−0.012	−0.027	0.439	0.445	0.464
PF meter, PF: 1	0.930	0.900	0.900	0.225	0.234	0.215	−0.225	−0.234	−0.215	0.289	0.312	0.290
PF meter, PF: 0.5 lagging	0.960	0.910	0.960	0.308	0.340	0.305	0.057	−0.030	0.075	0.439	0.455	0.440
PF meter, PF: 0.5 leading	0.970	0.970	0.930	0.313	0.317	0.372	0.039	−0.046	−0.041	0.447	0.438	0.473

Table 10. Alternating current/voltage component variation test (wattmeter and varmeter, voltage, current, and PF variation): maximum intrinsic error (%), MAPE (%), MRE (%), and standard deviation (%).

Test	Maximum Intrinsic Error			MAPE			MRE			Standard Deviation		
	$v = 230$ V $i = 24$ A $PF = 1$	$v = 0$ V $i = 24$ A $PF = 1$	$v = 256$ V $i = 24$ A $PF = 1$	$v = 230$ V $i = 24$ A $PF = 1$	$v = 0$ V $i = 24$ A $PF = 1$	$v = 256$ V $i = 24$ A $PF = 1$	$v = 230$ V $i = 24$ A $PF = 1$	$v = 0$ V $i = 24$ A $PF = 1$	$v = 256$ V $i = 24$ A $PF = 1$	$v = 230$ V $i = 24$ A $PF = 1$	$v = 0$ V $i = 24$ A $PF = 1$	$v = 256$ V $i = 24$ A $PF = 1$
Wattmeter	0.920	0.970	0.970	0.506	0.512	0.512	0.010	−0.034	−0.034	0.572	0.582	0.582
Varimeter	0.950	0.940	0.940	0.483	0.453	0.453	0.003	0.028	0.028	0.553	0.523	0.523

Table 11. PF variation test (wattmeter and varmeter, voltage variation): maximum intrinsic error (%), MAPE (%), MRE (%), and standard deviation (%).

Test	Maximum Intrinsic Error			MAPE			MRE			Standard Deviation		
	$v = 0$ V $i = 30$ A	$v = 230$ V $i = 30$ A	$v = 256$ V $i = 30$ A	$v = 0$ V $i = 30$ A	$v = 230$ V $i = 30$ A	$v = 256$ V $i = 30$ A	$v = 0$ V $i = 30$ A	$v = 230$ V $i = 30$ A	$v = 256$ V $i = 30$ A	$v = 0$ V $i = 30$ A	$v = 230$ V $i = 30$ A	$v = 256$ V $i = 30$ A
Wattmeter, PF:1	0.940	0.950	0.910	0.458	0.492	0.518	−0.011	−0.025	−0.004	0.529	0.557	0.578
Varmeter, PF:1	0.920	0.940	0.900	0.465	0.483	0.467	0.022	−0.022	0.029	0.529	0.546	0.528
Wattmeter, PF:0.5 lagging)	0.940	0.930	0.970	0.473	0.488	0.526	−0.046	0.035	0.028	0.546	0.557	0.579
Varmeter, PF:0.5 lagging	0.960	0.980	0.940	0.488	0.499	0.506	0.119	0.008	0.058	0.547	0.581	0.557
Wattmeter, PF:0.5 leading)	0.970	0.990	0.930	0.485	0.497	0.461	0.051	0.011	0.037	0.555	0.575	0.538
Varmeter, PF:0.5 leading	0.950	0.950	0.940	0.421	0.493	0.474	−0.044	0.009	0.009	0.494	0.557	0.544

Table 12. PF variation test (wattmeter and varmeter, current variation): maximum intrinsic error (%), MAPE (%), MRE (%), and standard deviation (%).

Test	Maximum Intrinsic Error			MAPE			MRE			Standard Deviation		
	$v = 230$ V $i = 0$ A	$v = 230$ V $i = 30$ A	$v = 230$ V $i = 36$ A	$v = 230$ V $i = 0$ A	$v = 230$ V $i = 30$ A	$v = 230$ V $i = 36$ A	$v = 230$ V $i = 0$ A	$v = 230$ V $i = 30$ A	$v = 230$ V $i = 36$ A	$v = 230$ V $i = 0$ A	$v = 230$ V $i = 30$ A	$v = 230$ V $i = 36$ A
Wattmeter, PF:1	0.910	0.980	0.940	0.472	0.483	0.454	0.100	−0.050	−0.048	0.507	0.552	0.520
Varmeter, PF:1	0.910	0.960	0.900	0.523	0.419	0.467	−0.020	0.028	0.064	0.581	0.502	0.531
Wattmeter, PF:0.5 lagging)	0.950	0.950	0.930	0.413	0.490	0.480	0.032	0.031	−0.035	0.493	0.555	0.542
Varmeter, PF:0.5 lagging	0.950	0.950	0.970	0.544	0.463	0.480	0.004	0.057	0.125	0.597	0.538	0.538
Wattmeter, PF:0.5 leading)	0.950	0.910	0.950	0.460	0.494	0.501	−0.014	−0.027	−0.028	0.541	0.556	0.567
Varmeter, PF:0.5 leading	0.930	0.950	0.950	0.450	0.466	0.492	0.026	−0.005	0.016	0.518	0.546	0.564

6.2.5. PF Variation Test

Table 11 shows the first assessment of the PF variation test. This test changed the voltage when the OSPPSM acted as a wattmeter and varmeter. Since the intrinsic error attained a 0.97% maximum value, the OSPPSM remained in accuracy class 1.

Table 12 shows the second assessment of the PF variation test. This test changed the current when the OSPPSM acted as a wattmeter and varmeter. Since the intrinsic error reached a 0.98% maximum value, the OSPPSM remained in accuracy class 1.

6.2.6. Continuous Overload Test

Table 13 gives the results of the continuous overload test. The maximum intrinsic error was lower than 0.99%. Consequently, the OSPPSM continued in accuracy class 1.

Table 13. Continuous overload test: maximum intrinsic error (%), MAPE (%), MRE (%), and standard deviation (%).

Test	Maximum Intrinsic Error	MAPE	MRE	Standard Deviation
Voltmeter	0.9800	0.4734	0.0352	0.5483
Ammeter	0.9900	0.4789	−0.0099	0.5582
PF meter	0.8600	0.0041	0.0003	0.0484
Wattmeter	0.9900	0.5047	−0.0066	0.5813
Varmeter	0.9800	0.4991	−0.0113	0.5725

6.3. Uncertainty Evaluation

Table 14 shows 10 independent sets of simultaneous observations for the three fundamental variables v, i, and PF ($\cos \varphi$). Because the variables were simultaneously measured, they were correlated. Evidently, these correlations should be taken into account in the uncertainty evaluation of derived variables p and q.

Table 14. Fundamental variables: input quantities from 10 sets of simultaneous observation.

Measure k/Input Quantities	v (V)	i (A)	PF (p.u.)
1	236.86	1.02	0.670
2	237.10	1.01	0.670
3	236.48	1.04	0.680
4	236.64	1.03	0.670
5	236.86	1.03	0.670
6	236.25	1.05	0.670
7	237.34	0.99	0.670
8	237.38	1.00	0.670
9	236.95	1.03	0.670
10	236.73	1.05	0.670

Table 15 summarizes the standard uncertainty results for the fundamental variables. Accordingly, Table 16 shows the absolute precision of these variables in relation to the recommendations and requirements of the JCGM guide 100:2008 [84].

Table 15. Fundamental variables and standard uncertainty.

Fundamental Variables	Mean	Standard Uncertainty	Standard Uncertainty (%)	Correlation Coefficients
v (V)	236.859	0.1130	0.00047	$\rho(\mu_v, \mu_i) = -0.900$
i (A)	1.0248	0.0064	0.0062	$\rho(\mu_i, \mu_{PF}) = 0.262$
PF(p.u.)	0.6710	0.0010	0.0015	$\rho(\mu_v, \mu_{PF}) = -0.373$

Table 16. Fundamental variables and absolute accuracy.

Fundamental Variables	Absolute Accuracy
v (V)	$AA_v = \text{InputReadingVoltage} \pm 0.1130$
i (A)	$AA_i = \text{InputReadingCurrent} \pm 0.0064$
PF (p.u.)	$AA_{PF} = \text{InputReadingPF} \pm 0.001$

The relationship between the derived and the fundamental variables is described in Section 3. After the application of Equations (13) and (14), Table 17 gives the standards uncertainties for the derived variables, according to Reference [28].

Table 17. Derived variables and calculated values for derived quantities p and q.

Relationship between Variables	Estimate Value of Derived Variables	Combined Uncertainty	Combined Uncertainty (%)
$p = v \cdot i \cdot \cos \varphi$ (W)	162.880	1.031	0.0063
$q = v \cdot i \cdot \sin \varphi$ (VAr)	179.977	1.010	0.0056
Correlation coefficient	$\rho(\mu_p, \mu_q) = 0.901$		

The results in Table 18 reflect the absolute accuracy of the derived variables, according to the JCGM guide 100:2008 [84].

Table 18. Derived variables and absolute accuracy.

Derived Variables	Absolute Accuracy
p (W)	$AA_p = \text{CalculatedActivePower} \pm 1.031$
q (VAr)	$AA_q = \text{CalculatedReactivePower} \pm 1.010$

As can be observed in the results of the uncertainty evaluation, the maximum uncertainty percentage was achieved for variable p, with a 0.0063% value, whereas the rest of the variables had lower values. In order to achieve a 99% confidence level ($\alpha^{set} = 1\%$) when evaluating uncertainty, the minimum number of samples required for the active power measurement, according to Equation (17), is:

$$n_{s\text{-min}} \geq \left(\frac{2.58 \times 1.031}{1}\right)^2 = 7.08 \tag{18}$$

and for the reactive power measurement:

$$n_{s\text{-min}} \geq \left(\frac{2.58 \times 1.011}{1}\right)^2 = 6.81 \tag{19}$$

The 10 samples taken were sufficient to fulfill the set confidence level of 99%.

7. Conclusions and Discussion

This research study developed and successfully calibrated a new prototype of an accurate low-cost OSPPSM for collecting generation/load profiles at temporal high-resolution in real-world PV household-prosumer facilities. This OSPPSM was based on the Arduino open-source electronic platform. Input data were gathered with a set of sensors based on Arduino components. The NILM approach used in the OSPPSM makes it ideal for measuring electrical variables without modifying the monitored household.

This prototype has a number of advantages. More specifically, it determines fundamental and derived electrical variables in conformity with the IEC 61000-4-30 standard. The stationary analysis has a 10-cycle analysis window with a sampling rate of 1 kHz. Thanks to its dual board, the computational time is extremely fast. In fact, the OSPPSM is able to perform real-time monitoring with temporal high-resolution data, every 0.25 s (4 Hz). Evidently, this real-time calculation capacity and the big-data support in the cloud have promising applications, especially all that concerns the provision of advanced energy services for PV household-prosumers. The design of these services is based on the training and validation of models. However, this requires temporal high-resolution data for generation/load profiles that can be collected by using this prototype in real-world household facilities. It should be highlighted that these services account for the effects of both the fast short-term fluctuations of input profiles (<4 Hz) and their hourly/daily/weekly/monthly variability.

Another promising application of the prototype is the verification of household status by using a computer or mobile device that is able to perform the appropriate actions.

This prototype was calibrated as accuracy class 1, according to IEC standards for power smart meters. This was confirmed by the results of the calibration standard tests, in which there was a maximum error of 0.99% and a maximum uncertainty of 0.0063%. Moreover, the Kolmogorov-Smirnov tests performed on the OSPPSM, working as an ammeter and PF meter, showed uniform distributions of the intrinsic error distributions. This highlights the accuracy of the OSPPSM.

Furthermore, the prototype design features easy-to-obtain hardware and open-source software. This means that it is freely accessible to researchers and users in general for their own design and use in a non-intrusive way. With no loss of accuracy and uncertainty, the cost of this prototype is considerably lower than commercially available power-energy loggers (about 1400 €). It also has equivalent measurement functionalities without cloud data uploading.

The results obtained in this research justify the continuation and further development of the OSPPSM prototype. Further research will focus on building additional OSPPSMs and testing them in real-world scenarios by deploying a network of OSPPSMs in different PV household-prosumers. For the benefit of the research community, an open web interface will be designed to visualize the main electrical variables at temporal high-resolution of the monitored PV household-prosumers.

Author Contributions: All the authors contributed substantially to this paper. F.S.-S., A.C.-O., and C.R.-C. performed the simulations and experimental work, and also wrote the paper. J.C.H. provided the conceptual approach, commented on all the stages of the simulation and experimental work, and revised the manuscript.

Funding: This research was funded by the Agencia Estatal de Investigación (AEI) and the Fondo Europeo de Desarrollo Regional (FEDER) aimed at the Challenges of Society (Grant No. ENE 2017-83860-R "Nuevos servicios de red para microredes renovables inteligentes. Contribución a la generación distribuida residencial").

Acknowledgments: This research was funded by the Agencia Estatal de Investigación (AEI) and the Fondo Europeo de Desarrollo Regional (FEDER) aimed at the Challenges of Society (Grant No. ENE 2017-83860-R "Nuevos servicios de red para microredes renovables inteligentes. Contribución a la generación distribuida residencial").

Conflicts of Interest: The authors declare no conflict of interest.

Abbreviation

Nomenclature
AA_{x_i}: absolute precision of variable x_j
AC: alternating current
ADC: analogic to digital converter
BESSs: battery energy storage systems
DC: direct current
E: intrinsic error
F: frequency
FPGA: field programmable gate array
GSM: global system mobile
i: current
IoT: Internet of Things
MAPE: mean absolute percentage error
MI: measuring instrument

MRE: mean relative error
n: index for the set of samples
n_s: number of samples
n_v: number of fundamental variables
NILM: non-intrusive load monitoring
NRU: nominal range of use
OSPPSM: on-time single-phase power smart meter
P: active power
PF:power factor
PF: power factor ($=\cos\varphi$)
PQ: power quality
PV: photovoltaic
PWM: pulse width modulation
q: reactive power
R.M.S: root mean square
RMS: reference measurement standard
s: apparent power
v: voltage
x: fundamental electrical variable
y: derived variable
WEP: wired equivalent privacy
z: index for the set of variables
Greek symbols
μ: mean
μ_{x_j}: mean of variable x_j
$\rho(\mu_{x_z}, \mu_{x_j})$: correlation coefficient of variables x_z, x_j
σ: standard deviation
σ^2: variance
$\sigma^s_{x_j}$: standard uncertainty type A for the variable x_j
σ^c_y:combined uncertainty of variable y
φ:phase angle of current
$1-\alpha$: confidence level
Subscripts
din: declared input
i: current
*j, m, w: j*th *m*th, *w*th variable
max: maximum
min: minimum
OSPPSM: on-time single-phase power smart meter
p: active power
PF: power factor
q: reactive power
ref: reference
v: voltage
x_j: variable x_j
Superscripts
avg: average
k: *k*th specified analysis window
ins: instantaneous
n: index for the set of samples
set: set
r.m.s: root mean square

References

1. Elma, O.; Tascıkaraglu, A.; Ince, A.T.; Selamogulları, U.S. Implementation of a dynamic energy management system using real time pricing and local renewable energy generation forecasts. *Energy* **2017**, *134*, 206–220. [CrossRef]
2. Hosseinnia, H.; Tousi, B. Optimal operation of DG-based micro grid (MG) by considering demand response program (DRP). *Electr. Pow. Syst. Res.* **2019**, *167*, 252–260. [CrossRef]
3. Oprea, S.V.; Bara, A.; Ileana Uță, A.; Pirjan, A.; Căruțașu, G. Analyses of distributed generation and storage effect on the electricity consumption curve in the smart grid context. *Sustainability* **2018**, *10*, 2264. [CrossRef]
4. Morales-Velazquez, L.; Romero-Troncoso, R.J.; Herrera-Ruiz, G.; Morinigo-Sotelo, D.; Osornio-Rios, R.A. Smart sensor network for power quality monitoring in electrical installations. *Measurement* **2017**, *103*, 133–142. [CrossRef]
5. Angrisani, L.; Bonavolonta, F.; Liccardo, A.; Schiano Lo Moriello, R.; Serino, F. Smart power meters in augmented reality environment for electricity consumption awareness. *Energies* **2018**, *11*, 2303. [CrossRef]
6. Viciana, E.; Alcayde, A.; Montoya, F.; Baños, R.; Arrabal-Campos, F.; Zapata-Sierra, A.; Manzano-Agugliaro, F. OpenZmeter: An efficient low-cost energy smart meter and power quality analyser. *Sustainability* **2018**, *10*, 4038. [CrossRef]
7. Robles Algarín, C.; Sevilla Hernández, D.; Restrepo Leal, D. A low-cost maximum power point tracking system based on neural network inverse model controller. *Electronics* **2018**, *7*, 4. [CrossRef]
8. Abate, F.; Carratu, M.; Liguori, C.; Paciello, V. A low cost smart power meter for IoT. *Measurement* **2019**, *136*, 59–66. [CrossRef]
9. Schlund, J.; German, R. A control algorithm for a heterogeneous virtual battery storage providing FCR power. In Proceedings of the IEEE International Conference on Smart Grid and Smart Cities, Singapore, Singapore, 23–26 July 2017; pp. 61–66. [CrossRef]
10. Megel, O.; Mathieu, J.; Andersson, G. Scheduling distributed energy storage units to provide multiple services. In Proceedings of the IEEE Power Systems Computation Conference, Wroclaw, Poland, 18–22 August 2014; pp. 1–7. [CrossRef]
11. Steber, D.; Bazan, P.; German, R. SWARM—Strategies for providing frequency containment reserve power with a distributed battery storage system. In Proceedings of the IEEE International Energy Conference, Leuven, Belgium, 4–8 April 2016; pp. 1–7. [CrossRef]
12. Hernandez, J.C.; Sanchez-Sutil, F.; Vidal, P.G.; Rus-Casas, C. Primary frequency control and dynamic grid support for vehicle-to-grid in transmission systems. *Int. J. Electr. Power Energy Syst.* **2018**, *100*, 152–166. [CrossRef]
13. Litjens, G.B.M.A.; Worrell, E.; Van Sark, W.G.J.H.M. Economic benefits of combining selfconsumption enhancement with frequency restoration reserves provision by photovoltaic-battery systems. *Appl. Energy* **2018**, *223*, 172–187. [CrossRef]
14. Braun, M.; Büdenbender, K.; Magnor, D.; Jossen, A. Photovoltaic self-consumption in Germany: Using lithium-ion storage to increase self-consumed photovoltaic energy. In Proceedings of the 24th European Photovoltaic Solar Energy Conference, Hamburg, Germany, 21–25 September 2009; pp. 1–7.
15. Bruch, M.; Müller, M. Calculation of the cost-effectiveness of a PV battery system. *Energy Procedia* **2014**, *46*, 262–270. [CrossRef]
16. Schreiber, M.; Hochloff, P. Capacity-dependent tariffs and residential energy management for PV storage systems. In Proceedings of the IEEE Power and Energy Society General Meeting, Vancouver, BC, Canada, 21–25 July 2013; pp. 1–5. [CrossRef]
17. Linssen, J.; Stenzel, P.; Fleer, J. Techno-economic analysis of photovoltaic battery systems and the influence of different consumer load profiles. *Appl. Energy* **2017**, *185*, 2019–2025. [CrossRef]
18. Luthander, R.; Widen, J.; Nilsson, D.; Palm, J. Photovoltaic self-consumption in buildings: A review. *Appl. Energy* **2015**, *142*, 80–94. [CrossRef]
19. Fridgen, G.; Kahlen, M.; Ketter, W.; Riegera, A.; Thimmel, M. One rate does not fit all: An empirical analysis of electricity tariffs for residential microgrids. *Appl. Energy* **2018**, *210*, 800–814. [CrossRef]
20. Murray, D.; Stankovic, L.; Stankovic, V. An electrical load measurements dataset of United Kingdom households from a two-year longitudinal study. *Sci. Data* **2017**, *4*, 1–12. [CrossRef] [PubMed]

21. Kolter, J.Z.; Johnson, M.J. REDD: A public data set for energy disaggregation research. In Proceedings of the KDD Workshop on Data Mining Applications in Sustainability, San Diego, CA, USA, 21 August 2011; pp. 1–6.

22. Anderson, K.; Ocneanu, A.F.; Benitez, D.; Carlson, D.; Rowe, A.; Berges, M. BLUED: A Fully labeled public dataset for event-based non-intrusive load monitoring research. In Proceedings of the 2nd KDD Workshop on Data Mining Applications in Sustainability, Beijing, China, 12–16 August 2012; pp. 1–5.

23. *Energy Saving Trust, Department of Energy and Climate Change (DECC)*; Final Report; Department for Environment, Food & Rural Affairs (DEFRA) Household Electricity Survey: London, UK, 2012.

24. Makonin, S.; Popowich, F.; Bartram, L.; Gill, B.; Bajic, I.V. A public dataset for load disaggregation and eco-feedback research. In Proceedings of the IEEE Electrical Power and Energy Conference, Halifax, NS, Canada, 21–23 August 2013; pp. 1–6. [CrossRef]

25. Makonin, S.; Ellert, B.; Bajic, I.V.; Popowich, F. Electricity, water, and natural gas consumption of a residential house in Canada from 2012 to 2014. *Sci. Data* **2016**, *3*, 160037. [CrossRef] [PubMed]

26. Hebrail, G.E.R.; Barard, A.E.R. Individual Household Electric Power Consumption Data Set (IhepcDS). Available online: https://archive.ics.uci.edu/ml/datasets/Individual+household+electric+power+consumption (accessed on 22 June 2019).

27. Kelly, J.; Knottenbelt, W. The UK-DALE dataset, domestic appliance-level electricity demand and whole-house demand from five UK homes. *Sci. Data* **2015**, *2*, 150007. [CrossRef] [PubMed]

28. Ridi, A.; Gisler, C.; Hennebert, J. ACS-F2—A new database of appliance consumption signatures. In Proceedings of the 6th International Conference on Soft Computing and Pattern Recognition, Tunis, Tunisia, 11–14 August 2014; pp. 145–150. [CrossRef]

29. Reinhardt, A.; Baumann, P.; Burgstahler, D.; Hollick, M.; Chonov, H.; Werner, M.; Steinmetz, R. On the accuracy of appliance identification based on distributed load metering data. In Proceedings of the Sustainable Internet and ICT for Sustainability, Pisa, Italy, 4–5 October 2012; pp. 1–6.

30. Beckel, C.; Kleiminger, W.; Cicchetti, R.; Staake, T.; Santini, S. The ECO data set and the performance of non-intrusive load monitoring algorithms. In Proceedings of the 1st ACM Conference on Embedded Systems for Energy-Efficient Buildings, Memphis, TN, USA, 3–6 November 2014; pp. 80–89. [CrossRef]

31. Barker, S.; Mishra, A.; Irwin, D.; Cecchet, E.; Shenoy, P. Smart: An open data set and tools for enabling research in sustainable homes. In Proceedings of the 2nd KDD Workshop on Data Mining Applications in Sustainability, Beijing, China, 12–16 August 2012; pp. 1–6.

32. Bendato, I.; Bonfiglio, A.; Brignone, M.; Delfino, F.; Pampararo, F.; Procopio, R.; Rossi, M. Design criteria for the optimal sizing of integrated photovoltaic-storage systems. *Energy* **2018**, *149*, 505–515. [CrossRef]

33. Wolisz, H.; Schütz, T.; Blanke, T.; Hagenkamp, M.; Kohrn, M.; Wesseling, M.; Müller, D. Cost optimal sizing of smart buildings' energy system components considering changing end-consumer electricity markets. *Energy* **2017**, *137*, 715–728. [CrossRef]

34. Dargahi, A.; Ploix, S.; Soroudi, A.; Wurtz, F. Optimal household energy management using V2H flexibilities. *COMPEL Int. J. Comp. Math. Electr. Electron. Eng.* **2014**, *33*, 777–793. [CrossRef]

35. Lim, Y.S.; Tang, J.H. Experimental study on flicker emissions by photovoltaic systems on highly cloudy region: A case study in Malaysia. *Renew. Energy* **2014**, *64*, 61–70. [CrossRef]

36. Saleh, M.; Meek, L.; Masoum, M.A.; Abshar, M. Battery-less short-term smoothing of photovoltaic generation using sky camera. *IEEE Trans. Ind. Inform.* **2018**, *14*, 403–414. [CrossRef]

37. Marcos, J.; Marroyo, L.; Lorenzo, E.; Alvira, D.; Izco, E. Power output fluctuations in large scale PV plants: One year observations with one second resolution and a derived analytic model. *Prog. Photovolt. Res. Appl.* **2011**, *19*, 218–227. [CrossRef]

38. Widen, J.; Wackelgard, E.; Lund, P.D. Options for improving the load matching capability of distributed photovoltaics: Methodology and application to high latitude data. *Sol. Energy* **2009**, *83*, 1953–1966. [CrossRef]

39. Dougal, R.A.; Liu, S.; White, R.E. Power and life extension of battery–ultracapacitor hybrids. *IEEE Trans. Compon. Packag. Technol.* **2002**, *25*, 120–131. [CrossRef]

40. Wright, A.; Firth, S. The nature of domestic electricity-loads and effects of time averaging on statistics and on-site generation calculations. *Appl. Energy* **2007**, *84*, 389–403. [CrossRef]

41. Omar, N.; Monem, M.A.; Firouz, Y.; Salminen, J.; Smekens, J.; Hegazy, O.; Gaulous, H.; Mulder, G.; Van den Bossche, P.; Coosemans, T.; et al. Lithium iron phosphate based battery. Assessment of the aging parameters and development of cycle life model. *Appl. Energy* **2014**, *113*, 1575–1585. [CrossRef]

42. Ruddell, A.J.; Dutton, A.G.; Wenzl, H.; Ropeter, C.; Sauer, D.U.; Merten, J.; Orfanogiannis, C.; Twidell, J.W.; Vezin, P. Analysis of battery current microcycles in autonomous renewable energy systems. *J. Power Sources* **2002**, *112*, 531–546. [CrossRef]

43. Jabbar Mnati, M.; Van den Bossche, A.; Farhood Chisab, R. A smart voltage and current monitoring system for three phase inverters using an android smartphone application. *Sensors* **2017**, *17*, 872. [CrossRef]

44. Robles Algarín, C.; Callejas Cabarcas, J.; Polo Llanos, A. Low-cost fuzzy logic control for greenhouse environments with web monitoring. *Electronics* **2017**, *6*, 71. [CrossRef]

45. Fuentes, M.; Vivar, M.; Burgos, J.M.; Aguilera, J.; Vacas, J.A. Design of an accurate, low-cost autonomous data logger for PV system monitoring using Arduino™ that complies with IEC standards. *Sol. Energy Mater. Sol. Cells* **2014**, *130*, 529–543. [CrossRef]

46. Amiry, H.; Benhmida, M.; Bendaoud, R.; Hajjaj, C.; Bounouar, S.; Yadir, S.; Raïs, K.; Sidki, M. Design and implementation of a photovoltaic I-V curve tracer: Solar modules characterization under real operating conditions. *Energy Convers. Manag.* **2018**, *69*, 206–216. [CrossRef]

47. Cano Ortega, A.; Sanchez Sutil, F.J.; Hernandez, J.C. Power factor compensation using teaching learning based optimization and monitoring system by cloud data logger. *Sensors* **2019**, *19*, 2172. [CrossRef] [PubMed]

48. Visalatchi, S.; Sandeep, K.K. Smart energy metering and power theft control using arduino & GSM. In Proceedings of the 2nd International Conference for Convergence in Technology, Mumbai, India, 7–9 April 2017; pp. 1–6. [CrossRef]

49. Arif, A.; Al-Hussain, M.; Al-Mutairi, N.; Al-Ammar, E.; Khan, Y.; Malik, N. Experimental study and design of smart energy meter for the smart grid. In Proceedings of the International Renewable and Sustainable Energy Conference, Ouarzazate, Morocco, 7–9 March 2013; pp. 1–6. [CrossRef]

50. Abubakar, I.; Khalid, S.N.; Mustafa, M.W.; Shareef, H.; Mustapha, M. Calibration of ZMPT101b voltage sensor module using polynomial regression for accurate load monitoring. *ARPN J. Eng. Appl. Sci.* **2013**, *12*, 1076–1084.

51. Jimenez-Castillo, G.; Muñoz-Rodriguez, F.J.; Rus-Casas, C.; Hernandez, J.C.; Tina, G.M. Monitoring PWM signals in stand-alone photovoltaic systems. *Measurement* **2019**, *134*, 412–425. [CrossRef]

52. Tarasiuk, T.; Szweda, M.; Tarasiuk, M. Estimator–analyzer of power quality: Part II—Hardware and research results. *Measurement* **2011**, *44*, 248–258. [CrossRef]

53. Arduino Nano. Available online: https://store.arduino.cc/arduino-nano (accessed on 22 June 2019).

54. Arduino Mega. Available online: https://store.arduino.cc/mega-2560-r3 (accessed on 15 June 2019).

55. Adafruit Industries Ltd. Available online: https://www.adafruit.com/product/904 (accessed on 22 June 2019).

56. IEC. *IEC Standard 61000-4-7. Electromagnetic Compatibility (EMC): Testing and Measurement Techniques—General Guide on Harmonics and Interharmonics Measurements and Instrumentation, for Power Supply Systems and Equipment Connected Thereto*; International Electrotechnical Commission: Geneva, Swizerland, 2008.

57. Ramos, P.M.; Janeiro, F.M.; Girao, P.S. Uncertainty evaluation of multivariate quantities: A case study on electrical impedance. *Measurement* **2016**, *78*, 397–411. [CrossRef]

58. Apetrei, D.; Silvas, I.; Albu, M.; Postolache, P.; Neurohr, R. Voltage estimation in power distribution networks. a case study on data aggregation and measurement uncertainty. In Proceedings of the International Workshop on Advanced Methods for Uncertainty Estimation in Measurement, Bucharest, Romania, 6–7 July 2009; pp. 1–6. [CrossRef]

59. IEC. *IEC Standard 61000-4-30. Electromagnetic Compatibility (EMC): Testing and Measurement Techniques—Power Quality Measurement Methods*; International Electrotechnical Commission: Geneva, Swizerland, 2015.

60. Webster, J.G. *Electrical, Measurement, Signal Processing, and Displays*; CRC Press LLC: Boca Raton, FL, USA, 2004.

61. Arduino Uno Rev3. Available online: https://store.arduino.cc/arduino-uno-rev3 (accessed on 22 June 2019).

62. WEMOS Electronics. Available online: https://wiki.wemos.cc/products:d1:d1 (accessed on 22 June 2019).

63. Dechang Electronics Co. Ltd. Available online: http://en.yhdc.com (accessed on 22 June 2019).

64. Firebase. Available online: https://firebase.google.com (accessed on 22 June 2019).

65. Arduino MKR WiFi 1010. Available online: https://store.arduino.cc/mkr-wifi-1010 (accessed on 22 June 2019).

66. Node MCU Arduino. Available online: https://www.nodemcu.com (accessed on 22 June 2019).

67. Interplus Industry Co. Ltd. Available online: http://www.interplus-industry.fr/index.php?option=com_content&view=article&id=52&Itemid=173&lang=en (accessed on 22 June 2019).

68. STC013 Dechang Electronics Co. Ltd. Available online: http://en.yhdc.com/product/SCT013-401.html. (accessed on 22 June 2019).

69. Texas Instruments. Available online: http://www.ti.com/lit/ds/symlink/ads1114.pdf (accessed on 22 June 2019).

70. Arduino Software. Available online: https://www.arduino.cc/en/Main/Software (accessed on 22 June 2019).

71. Sánchez, H.; Gonzalez-Contreras, C.; Agudo, J.E.; Macías, M. IoT and ITV for interconnection, monitoring, and automation of common areas of residents. *Appl. Sci.* **2017**, *7*, 696. [CrossRef]

72. Sridharana, M.; Devi, R.; Dharshini, C.S.; Bhavadarani, M. IoT based performance monitoring and control in counter flow double pipe heat exchanger. *Internet Things* **2019**, *5*, 34–40. [CrossRef]

73. Ramírez-Gil, J.G.; Giraldo Martínez, G.O.; Morales Osorio, J.G. Design of electronic devices for monitoring climatic variables and development of an early warning system for the avocado wilt complex disease. *Comput. Electron. Agric.* **2018**, *153*, 134–143. [CrossRef]

74. Radmannia, S.; Naderzad, M. IoT-based electrosynthesis ecosystem. *Internet Things* **2018**, *3*, 46–51. [CrossRef]

75. Da Cruz, M.A.; Rodrigues, J.J.; Lorenz, P.; Solic, P.; Al-Muhtadi, J.; Albuquerque, V.H.C. A proposal for bridging application layer protocols to HTTP on IoT solutions. *Future Gener. Comput. Syst.* **2019**, *97*, 145–152. [CrossRef]

76. Al-Ali, A.R.; Zualkernan, I.A.; Rashid, M.; Gupta, R.; AliKarar, M. Smart home energy management system using IoT and big data analytics approach. *IEEE Trans. Consum. Electron.* **2017**, *63*, 4. [CrossRef]

77. Moghimi, M.; Liu, J.; Jamborsalamati, P.; Rafi, F.; Rahman, S.; Hossain, J.; Stegen, S.; Lu, J. Internet of things platform for energy management in multi-microgrid system to improve neutral current compensation. *Energies* **2018**, *11*, 3102. [CrossRef]

78. IEC. *IEC Standard 60051-1. Direct Acting Indicating Analogue Electrical Measuring Instruments and their Accessories: Definitions and General Requirements Common to All Parts*; International Electrotechnical Commission: Geneva, Swizerland, 2016.

79. IEC. *IEC Standard 60050-311. International Electrotechnical Vocabulary: Electrical and Electronic Measurements and Measuring Instruments: General Terms Relating to Measurements*; International Electrotechnical Commission: Geneva, Swizerland, 2001.

80. IEC. *IEC Standard 60051-2. Direct Acting Indicating Analogue Electrical Measuring Instruments and their Accessories: Special Requirements for Ammeters and Voltmeters*; International Electrotechnical Commission: Geneva, Swizerland, 2018.

81. IEC. *IEC Standard 60051-3. Direct Acting Indicating Analogue Electrical Measuring Instruments and their Accessories: Special Requirements for Wattmeters and Varmeters*; International Electrotechnical Commission: Geneva, Swizerland, 2018.

82. IEC. *IEC Standard 60051-5. Direct Acting Indicating Analogue Electrical Measuring Instruments and Their Accessories: Special Requirements for Phase Meters, Power Factor Meters and Synchroscopes*; International Electrotechnical Commission: Geneva, Swizerland, 2017.

83. IEC. *IEC Standard 60051-9. Direct acting indicating analogue electrical measuring instruments and their accessories: Recommended test methods*; International Electrotechnical Commission: Geneva, Swizerland, 2019.

84. JCGM/WG 1. *Evaluation of Measurement Data—Guide to the Expression of Uncertainty in Measurement, GUM 50*; JCGM: Sèvres, France, 2008; 134p. [CrossRef]

85. Bell, S. *Measurement Good Practice Guide No. 11 (Issue 2)—A Beginner's Guide to Uncertainty of Measurement*; National Physical Laboratory: Teddington, UK, 2001.

Article

Intermittent Renewable Energy Sources: The Role of Energy Storage in the European Power System of 2040

Henrik Zsiborács [1], Nóra Hegedűsné Baranyai [1,*], András Vincze [2], László Zentkó [3], Zoltán Birkner [4], Kinga Máté [5] and Gábor Pintér [1]

[1]   Department of Economic Methodology, University of Pannonia, Georgikon Faculty, 8360 Keszthely, Hungary
[2]   Department of Foreign Languages, Georgikon Faculty, University of Pannonia, 8360 Keszthely, Hungary
[3]   PANNON Pro Innovations Ltd., 1021 Budapest, Hungary
[4]   Nagykanizsa Campus, University of Pannonia, 8800 Nagykanizsa, Hungary
[5]   Wolf Theiss Faludi Erős Law Firm, 1085 Budapest, Hungary
*   Correspondence: baranyai@georgikon.hu; Tel.: +36-30-373-8550

Received: 21 May 2019; Accepted: 18 June 2019; Published: 26 June 2019

Abstract: Global electricity demand is constantly growing, making the utilization of solar and wind energy sources, which also reduces negative environmental effects, more and more important. These variable energy sources have an increasing role in the global energy mix, including generating capacity. Therefore, the need for energy storage in electricity networks is becoming increasingly important. This paper presents the challenges of European variable renewable energy integration in terms of the power capacity and energy capacity of stationary storage technologies. In this research, the sustainable transition, distributed generation, and global climate action scenarios of the European Network of Transmission System Operators for 2040 were examined. The article introduces and explains the feasibility of the European variable renewable energy electricity generation targets and the theoretical maximum related to the 2040 scenarios. It also explains the determination of the storage fractions and power capacity in a new context. The aim is to clarify whether it is possible to achieve the European variable renewable energy integration targets considering the technology-specific storage aspects. According to the results, energy storage market developments and regulations which motivate the increased use of stationary energy storage systems are of great importance for a successful European solar and wind energy integration. The paper also proves that not only the energy capacity but also the power capacity of storage systems is a key factor for the effective integration of variable renewable energy sources.

Keywords: solar energy; wind energy; energy storage; renewable energy integration; Europe

---

1. Introduction

1.1. Changes in the Spread of Photovoltaic and Wind Energy Technologies in the World

Today's boost in energy demand and shift towards a low-carbon economy brings about an increased need for the deployment of cutting-edge technologies and services in the energy sector [1,2]. Addressing climate change and the excessive greenhouse gas (GHG) emissions are among the top urging issues at a global level. On the pathway towards a low-carbon future, the use of renewable energies will undoubtedly have a key role [3,4]. Variable renewable energy (VRE) sources, such as photovoltaic (PV) energy, may serve as a remedy in order to mitigate the adverse effect of the above factors, given their sustainable, clean, and ecofriendly nature [5–7]. Current ambitions targeting the reduction of GHG-emissions attribute a growing importance to the electricity sector alongside a more distributed generation (DG). When it comes to tackling climate change, PV and wind energy technologies will be key drivers in paving the way towards sustainability and energy conservation.

However, today the integration of VRE sources poses a challenge to be addressed for the successful decentralization of the electricity network. From the point of view of power quality, PV and wind energy have some disadvantages. The intermittent nature of VRE sources and distributed generation remain a challenge to grid operators when scheduling power generation. On the other hand, distributed energy generation may enhance the further spread of smart grids and micro grids and, therefore, ensure a greater share of clean energy in the energy mix [8–14].

PV and wind technologies play a key role in the shift towards green growth, a low-carbon economy, and a greater share of renewables in the energy mix [15]. In the last decade, support schemes such as the feed-in-tariff system, the declining initial capital expenditure due to the boost in innovation, and technology have proved to be essential factors that underpin this phenomenon [16–18]. Statistics show a considerable growth of PV and wind energy globally; 7.5% of the total 26.5% share of renewables in electricity generation was produced by VRE installations in 2017. In the same year, the global built-in PV and wind capacity amounted to 941 GW (Figure 1). The key players of the PV electricity market were China (131.1 GW), followed by the EU (108 GW), the USA (51 GW), and Japan (49 GW). China (188.4), the EU (168.7 GW), and the USA (89 GW) were also leaders in terms of wind capacity, followed by India with a share of 32.8 GW [17,19].

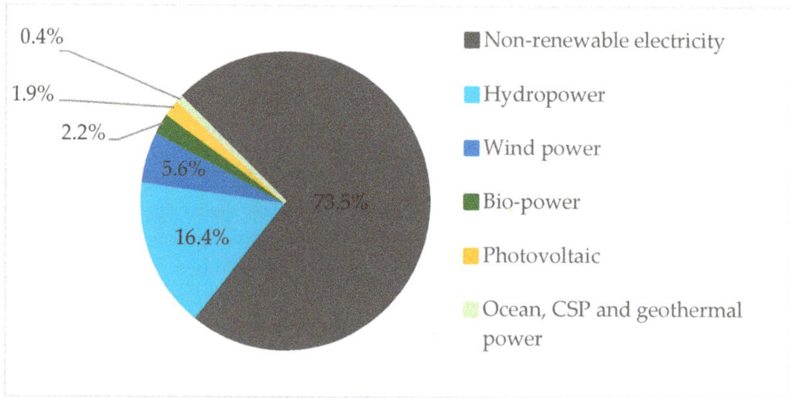

Figure 1. Estimated renewable energy share of global electricity production, 2017, based on [17].

1.2. Energy Challenges with the Spread of Variable Renewable Energy Sources

Today, the integration of VRE sources into the electricity grid is one of the crucial issues to be addressed at an international and national level. For example, the European Union (EU) has set the ambitious goal to cut its overall GHG emission by more than 80% by 2050, as well as to become the global leader in the usage of renewable energy sources (RES). To achieve this goal, member states shall endeavor in the coming years to significantly increase the share of intermittent renewable energy sources in their energy mix. By integrating more VRE sources into the European grid system, it will be essential to tackle the need for a more flexible electricity grid. Subsequently, cost competitive energy storage technologies will be drivers in creating the necessary secure balance between distributed and centralized electricity generation and the integration of a higher share of viable renewables such as solar and wind energy [20]. However, due to their variable power generation nature, the integration of PV and wind power into the electricity grid is a challenge, since the existing grids and their capacities were established to comply with less or non-variable energy sources, dispatchable power generation, and predictable load peaks. In general, today's electricity grids are able to handle a low increase in load as a result of newly built-in VRE capacities, but a massive load increase can cause discrepancies in the macro energy system. In order to mitigate and successfully tackle regional differences arising from the variable solar and wind potentials, the electricity system of the new era should not only be flexible but also possess a sufficient backup capacity. The flexibility of the grid is an essential factor in handling network constraints caused by VRE generation during the peak hours of demand. On the other hand, storage capacity may be beneficial when there is an incline in sunshine hours and wind speed [21–24].

1.3. The Importance of Energy Storage Systems and their Future Role

According to the European vision, the energy system will rely significantly on renewables by 2040, more specifically on non-dispatchable and VRE power, which at the same time will bring about the partial decentralization of the energy system [25–27]. The optimal share of VRE sources in the energy mix depends on various factors. The flexibility of the grid, the back-up capacity, the quality and capacity of the transmission system [28–31], as well as load performance characteristics [32–34] and the actual local weather patterns may determine the volume of VREs that can be safely fed into the system [35–37]. A potential solution to compensate for the uncertainty arising from the variable nature of VREs is to upgrade and enhance the overall flexibility of the electricity grid. By adding storage capacity to the energy system, greater flexibility can be achieved through the provision of a back-up potential for shaving of peak loads or filling valleys [35,36,38].

Today, there exist multiple storage technologies and solutions that are able to compensate for the intermittent nature of VRE sources (Figure 2), namely electro-chemical energy storage, electro-mechanical energy storage, electrical energy storage, thermal storage, and chemical energy storage. The key solutions for large-scale energy storage include compressed air storage, pumped hydro storage (PHS), molten salt thermal storage, or flow batteries (Figure 3). The details of the specific features of these energy storage technologies, however, are not included in this manuscript. The abbreviations for all technologies are listed in the abbreviations section. Overall, the global energy storage capacity including both stationary and grid-connected capacities amounted to approximately 159 GW, of which 153 GW was PHS in 2017 [17,39].

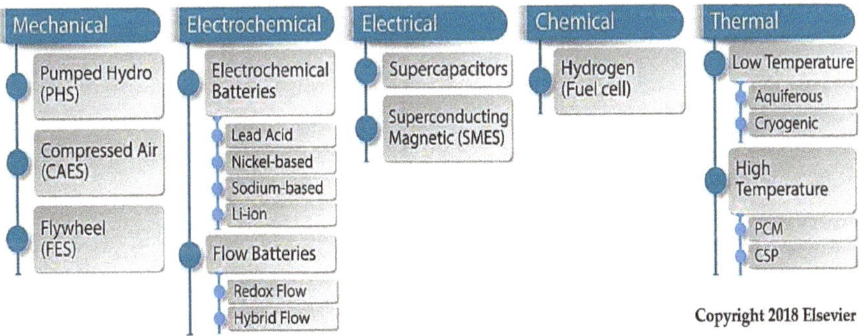

Figure 2. Classification of energy storage technologies by the form of stored energy [39].

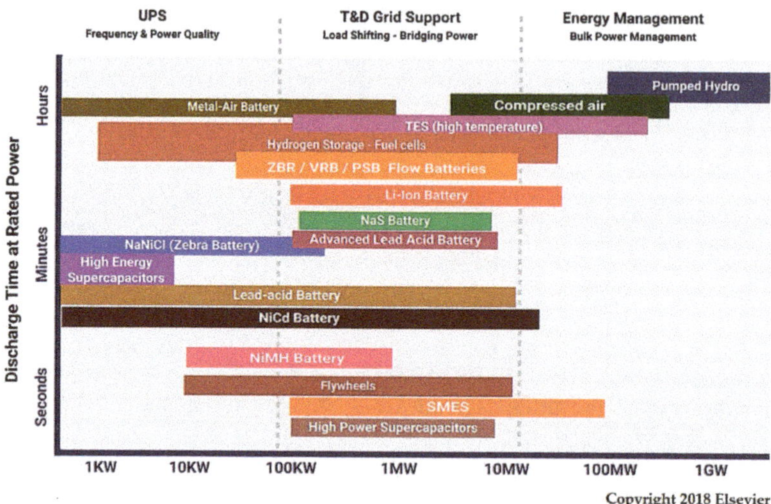

Figure 3. Energy storage technologies by discharge time and power capacity [39].

At the beginning of 2017, the global, overall, new advanced energy storage capacity amounted to around 5.9 GW. In this year, the energy these energy storage technologies put into operation accounted for approximately 0.5 GW of the final total. The share of electrochemical storage solutions (battery) had increased considerably, by 0.4 GW, at the beginning of 2017 up to a total of 2.3 GW [17]. Due to its user-friendly, economical nature and rather simple deployment, battery storage is one of the most popular options when considering energy storage solutions both at a domestic and industrial level. The use cases of energy storage are shown in Figure 4, but an explanation of these features is not provided in the present paper [9,14,40].

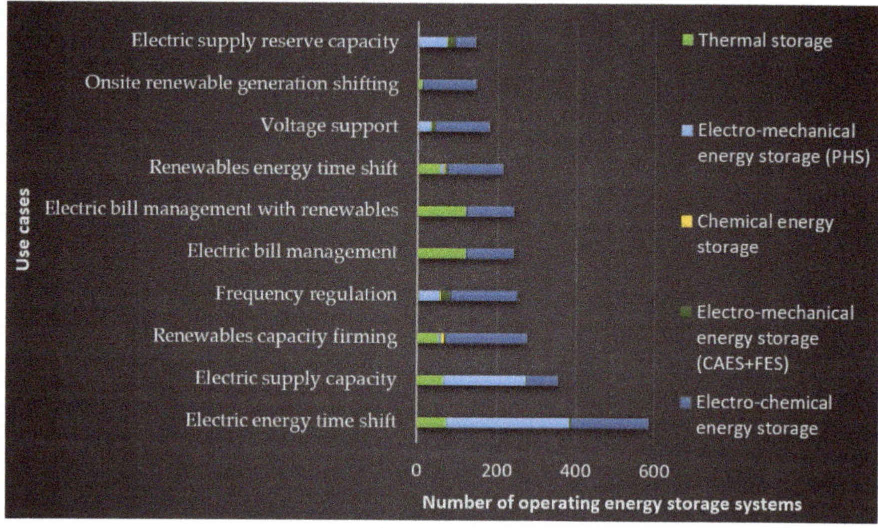

Figure 4. The 10 most common use cases of energy storage systems until August 16, 2016, based on [9] (Projects can have multiple use cases).

Compared to the approximately 4.67 TWh of 2017, if the current trends persist and the share of VREs in the global energy mix doubles, there will be a considerable growth in the overall energy storage capacity by 2030 up to about 6.62–15.89 TWh [8]. However, this increase is slightly unpredictable at present. According to the relevant forecasts, the share of PHS technologies will fall to approximately 90% of the overall installed storage capacity by 2030. In the meantime, the declining production cost of storage appliances will bring about an explosive development of cutting-edge battery technologies and the diversification of their possible uses, inter alia both at grid and self-consumption level (e.g., rooftop solar PV). Based on current projections and future scenarios [10], non-pumped hydroelectricity storage will grow from an estimated 162 GWh in 2017 to 5.8–8.4 TWh by 2030. Key drivers behind the boost in the energy storage market will be behind-the-meter and utility-scale solutions. The battery capacity of stationary applications is expected to increase from the estimated 11 GWh in 2017 up to about 100–421 GWh by 2030. Using battery electricity storage integrated into small-scale PV systems will be one of the most common and marketable ways of battery deployment in the period until 2030. In the near future, the economic viability of stationary battery electricity storage solutions is expected to drastically advance across Europe and beyond due to beneficial factors, such as increased residential and commercial electricity rates, better support schemes (e.g., relatively low feed-in-tariffs), and competitive cost structures. Moreover, alongside with the spread of favorable support schemes for VRE's, new markets for additional products and services are also expected to appear [8–14].

1.4. European Electricity Consumption and Energy Storage Aspects

According to data from the European Network of Transmission System Operators for Electricity (ENTSO-E), in 2017 the European total electricity consumption amounted to 3,329 TWh, showing a moderate increase (+0.2%) compared to the previous year's data. In the same year, the peak demand on the electricity grid was measured on 18 January and amounted to 542 GW (4 GW less than in 2016). With regard to the net generating capacity (NGC), the figures show a slight decline for nuclear (−2.3%) and fossil fuels (−3.1%) from 2016 to 2017. On the other hand, the net generating capacity for solar and wind energy grew by 6.1% and 9.8% in the same period. The thirty-six member countries of ENTSO-E are Austria, Albania, Bosnia and Herzegovina, Belgium, Bulgaria, Switzerland, Cyprus, Czech Republic, Germany, Denmark, Estonia, Spain, Finland, France, United Kingdom, Greece, Croatia, Hungary, Ireland, Iceland, Italy, Lithuania, Luxembourg, Latvia, Montenegro, Macedonia, Netherlands, Norway, Poland, Portugal, Romania, Serbia, Sweden, Slovenia, Slovakia, and Turkey. It should, however, be noted that Albania (member since March 2017) and Turkey (observer member) are not included in the statistics. The summary data are shown in Tables 1 and 2 [41].

Table 1. European electricity consumption and maximum peak loads between 2013 and 2017 [41].

Year	2013	2014	2015	2016	2017
Electricity consumption [tWh]	3293	3241	3301	3322	3329
Maximum peak load [GW]	516	511	520	546	542

Table 2. Evolution of electricity consumption between 2016 and 2017 in Europe [41].

Country	Change of Electricity Consumption between 2016 and 2017 in Europe [%]
United Kingdom	−3.0
Germany	−1.8
Austria	−1.2
Denmark	−1.1
Latvia	−0.6
France	−0.3
Luxembourg	0.0
Norway	0.2
Sweden	0.2
Netherlands	0.4
Switzerland	0.4
Finland	0.6
Portugal	0.7
Belgium	0.8
Ireland	1.0
Macedonia	1.1
Greece	1.2
Spain	1.2
Estonia	1.4
Cyprus	1.6
Bulgaria	1.9
Serbia	2.1
Bosnia and Herzegovina	2.2
Czech Republic	2.5
Romania	2.5
Hungary	2.6
Lithuania	2.6
Poland	2.6
Slovenia	2.8
Croatia	3.2
Iceland	3.2
Slovakia	3.2
Italy	3.9
Montenegro	5.6
Albania	no information
Turkey	no information

The overall net generating capacity from renewable electricity sources (RES) (without hydro energy) has a share of approximately 30% of the total NGC. Meanwhile, electricity produced from hydro power showed a considerable decline caused by decreased water discharge (9.3%). On the regional level, the energy demand shows disparities. While the energy demand is extensively growing in Eastern Europe and shows a moderate growth in the Hispanic Peninsula, there is a slight decline in electricity consumption in some European countries, such as Germany, Austria, and Great Britain. However, the Central European countries (except for Germany and Austria) remain stable in their demand [41]. In Europe the installed PHS capacity in 2017 reached 50.5 GW (approx. 1.9 TWh energy capacity based on [11,40,42–44]), of which a capacity of more than 59% was found in 5 countries, namely Italy, France, Germany, Austria, and Spain (Figure 5). Based on the ENTSO-E scenarios, the PHS capacity increase is expected to be in the range of 58–76 GW by 2040 [27,45]. At the beginning of 2017 other storage technologies represented about 1.3% of the total storage capacity based on thirty ENTSO-E countries [9,46].

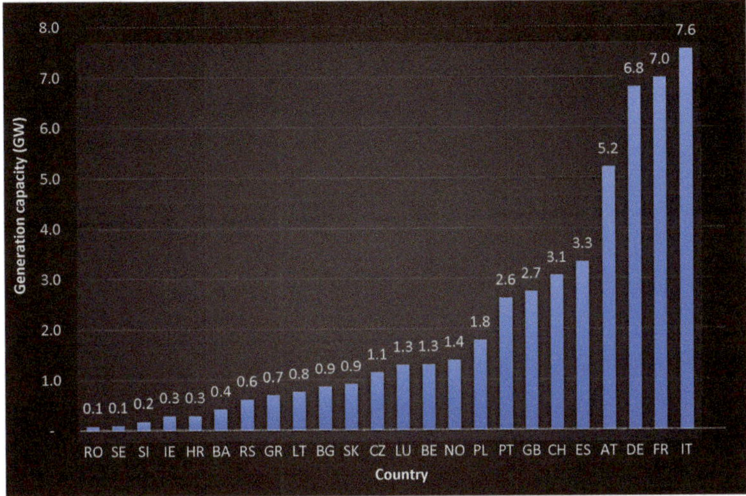

Figure 5. European installed pumped hydro storage (PHS) generation capacity in 2017 based on [45].

2. Material and Methods

In the research project, the 2040 scenarios titled Sustainable Transition (ST), Distributed Generation (DG), and Global Climate Action (GCA) developed by the European Network of Transmission System Operators were examined in relation to the VRE integration targets and the theoretical maximum considering the technology-specific storage aspects. In the modelling we combined only the main findings of the manuscripts that examined the European level and the general, global conclusion from Blanco-Faaij (2018) [47]. These manuscripts also analyzed the power capacity and/or the energy capacity of stationary storage technologies for secure European grid balancing. Thus, the first common point of the analyzed manuscripts is the provision of the balance of the European grid system by stationary storage technologies. Other key components for analysis at the European level were annual demand, VRE penetration, and energy storage capacity, which can be well defined. This approach was published by Blanco-Faaij in 2018 [47]. VRE gross electricity generation is a percentage of the total electricity demand and this value can also be easily determined. According to the authors, the carefully selected articles complement each other's results, and the conclusions of the manuscripts were applied to the European grid sector.

2.1. Description of the ENTSO-E Scenarios

The ST scenario seeks economical, quick, and sustainable CO_2 reduction by replacing lignite and coal by gas in the power sector. In this case energy generation by gas is popular due to the relatively cheap global gas prices and the strong growth of bio-methane. A regulatory framework in place decreases the use of coal power plants. Gas-based energy generation largely provides the necessary flexibility to balance renewables in the power system. In this storyline, climate action is achieved with a mixture of emission trading, national regulation, subsidies, and schemes [27].

In the DG scenario, significant leaps in the innovation of commercial/residential storage technologies and small-scale generation are a key driver in climate action. This case represents a more decentralized development with focus on end-user technologies. Smart technologies, PV systems, electric vehicles, and dual-fuel appliances allow consumers to switch energy depending on market conditions. Biomethane growth is strong as connections to distribution systems grow, utilizing local feedstocks. In this storyline, the electricity demand flexibility is substantially increased, both in industrial and residential solutions, helping electric power adequacy. Wintertime, however, with low solar availability and high heating needs remains a challenge, since batteries cannot be used for seasonal storage [27].

The GCA scenario represents a global effort towards full speed decarbonization. The emphasis is on renewables and nuclear energy in the power sector. Commercial and residential heat becomes more electrified, leading to a steady decline in demand for gas in this sector. The decarbonization of transportation is achieved through gas and electric vehicle growth and the power-to-gas production sees its strongest development within this scenario. Gas power plants provide the flexibility needed within the power market, helping facilitate intermittent renewable technologies within it [27]. The European electricity consumption and maximum peak load features, production capacities and generation in 2040 based on the examined scenarios are shown in Table 3 and Figure 6. These input data were important for modeling.

Table 3. European electricity consumption and maximum peak load features in 2040 based on the examined scenarios [27].

Year	2040, ST	2040, DG	2040, GCA
Electricity demand [TWh]	4030	4450	4100
Maximum peak load [GW]	650	730	690
VRE annual electricity generation compared to the demand [TWh]	1600	2250	2430
VRE annual electricity generation target compared to the demand [%]	39.7	50.6	59.3

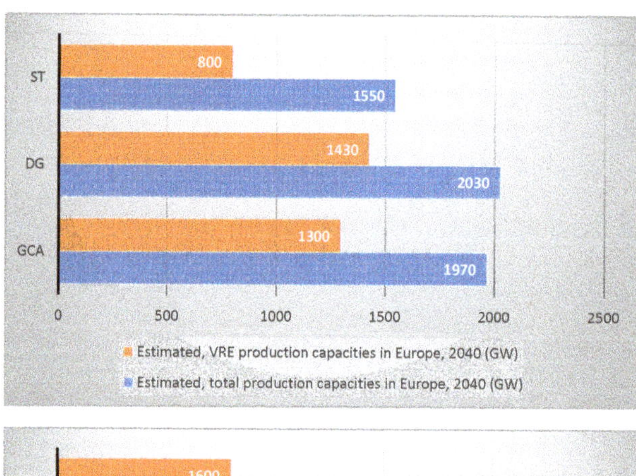

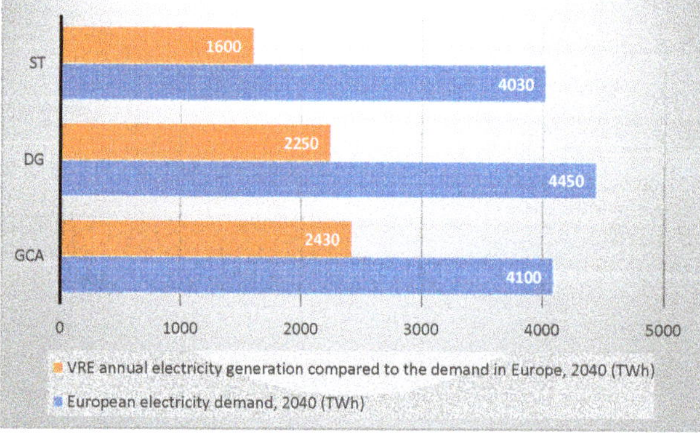

Figure 6. Installed production capacities and generation in Europe based on the 2040 scenarios [48].

The ENTSO-E scenarios take into account the impact of EV penetration in the electricity demand. However, it is stationary storage technologies that ensure that the potential uncertainty of the power supply resulting from VRE penetration is eliminated [27].

2.2. European Energy Storage Case Studies for VRE Integration

An important variable that defines the energy storage capacity requirement is the energy production from the VRE sources in the energy mix [47,49]. Several studies were reviewed to estimate the European stationary storage size as a fraction of VRE penetration and annual demand. With the solution, it is possible to determine the average energy storage fraction requirements expressed in energy storage capacity (TWh). This refers to the amount of energy that can be stored at the same time and not energy delivered throughout a year. The energy storage fraction and the energy storage capacity are relative numbers to compare across studies. In this manuscript a polynomial regression model was developed in MATLAB by combining the logic of 7 studies [13,22,47,50–53]. These sources are meta-analyses, in which hundreds of manuscripts were analyzed and evaluated. The model calculates the average energy storage fraction in the context of VRE gross electricity generation, expressed as a percentage of the total electricity demand. The issue of the energy storage fraction has been analyzed in many studies in the context of VRE energy production. Blanco and Faaij [47] summarize the factors that determine this fraction based on 79 sources and conclude that there are significant differences between countries (Tables 4 and 5). For this reason, the model proposes that the storage requirement should be examined at the European level, defining an average value. Tables 4 and 5 show the summarized VRE penetrations of 10–100%. A general conclusion by Blanco and Faaij [47] is that even for high VRE penetrations of 90–95%, the storage fraction is at most 1.5% of the annual demand, while that for a VRE penetration of 100% this share is highly uncertain. However, it should be noted that according to [13,51–53], there is no need for a high degree of storage flexibility for a VRE percentage of 40–50%. The figures presented in Table 4 are listed in an increasing order of 'VRE penetration', while Table 5 displays the data in the increasing order of the 'energy storage fraction'.

Table 4. Recommended annual storage features in Europe, with less than 100% VRE penetration, based on [47].

Country	Annual Demand [TWh]	VRE Penetration [%]	Energy Storage Capacity [TWh]	Energy Storage Fraction [%]	Ref.
Europe	3746	16	1.15	0.0308	[22,47]
Spain	375	25	0.66	0.18	[54]
Netherlands	123	28.3	0.05	0.04	[55]
West Europe	4647	30	2.4	0.05	[56]
UK	~700	30	0.06	0.01	[57]
Ireland	32.7	34.5	0.07	0.21	[58]
Germany	478	38.6	0.06	0.01	[59]
Europe	4670	48	2.08	0.0445	[22,47]
Germany	562	50	3.5	0.62	[13]
Germany (Region)	53	20–50	0.15	0.28	[60]
Greece	88.3	50	0.4–1.4	1.02	[61]
Austria	83	55	0.2	0.24	[62]
UK	300	60	0.1	0.03	[63]
Spain	420	60	0.6–2.2	0.33	[54]
Germany	2030	66	18	0.89	[31]
Belgium	268	80	1.3	0.32	[64]
Denmark	41	80	0.66	1.61	[65]
Germany	413	80	0.9–1.3	0.27	[59]
Germany	~600	80	7–8	1.25	[66]
Germany	586	80	0.5	0.09	[67]
Germany (Region)	22.7	80	0.184	0.81	[68]
Ireland	45	80	2.8	6.00	[58]
Europe	4900	80	50	1.02	[50]
General, global conclusion	-	90–95	-	1.5	[47]

Table 5. Recommended annual storage features in Europe, with a renewable energy sources (RES) penetration of 100% based on [47].

Region	Year	Annual Demand [TWh]	Wind/Solar [%]	Energy Storage Capacity [TWh]	Energy Storage, Power Capacity [GW]	Energy Storage Fraction [%]	Ref.
Ireland	2050	125	13/2	0.24	10	0.19	[69]
Europe	2050	4200	73/21	13.5	–	0.32	[70]
France	2050	425	40/17	3	3	0.71	[71]
Europe	–	3240	55/45	25	360	0.77	[37]
Germany	2050	475	60/40	9.1	–	1.92	[72]
Germany (Region)	2030	19.9	55/40	0.53	1.5	2.66	[68]
UK	2030	900	55/6	27	35	3.00	[73]
Greece	2050	55.7	100/0	2	0.2–0.3	3.59	[74]
Europe	–	3400	55/45	216	65	6.35	[75]
Europe	2007	3240	55/45	400–480	400	13.58	[35]
Europe	-	-	-	-	-	20–40	[76]

3. Results

3.1. Determination of the European Storage Fractions

The issue of storage fractions has been analyzed in many studies in the context of gross VRE electricity generation. In this manuscript, a polynomial regression model was developed in MATLAB by combining the logic of seven studies [13,22,47,50–53]. Blanco and Faaij [47] summarize the factors that determine the mentioned fractions based on 79 sources and conclude that there are significant differences between countries. Other sources [22,47,50] were suitable for investigating the storage fractions as a function of gross VRE electricity generation at the European level. These baseline values have been applied to modelling:

- European storage fraction: 0.0308–0.0372%; VRE penetration: 16–45% [22].
- European storage fraction: 1.02%; VRE penetration: 80% [50].
- European storage fraction: 1.5%; VRE penetration: 95% [47].

From the reviewed studies it became evident that the 40–50% energy production share from VRE in the European power grid sector is a critical value [13,51–53]. Above this level the need for energy storage dramatically increases (Figure 7) as evidenced by most studies. Below 40–50% of VRE share the storage fraction increases basically linearly, but most studies gave diverging storage fraction values for the 50–100% VRE share. With the relationship created, fraction values up to a VRE penetration of 95% were analyzed (Figure 7). Based on the reviewed studies, the figures given for the recommended storage capacities at an all European level in the case of generating 100% of the annual demand by using RES show far too great a variation to be reliable (Table 5) [35,37,47,49,70,75,76]; therefore, a VRE penetration of 100% was not examined in this research. With the help of a polynomial regression model, an equation that describes the average European storage fraction related to the percentage of gross VRE electricity generation was developed. To build the equation (Equation (1)) that best models the storage fraction as a function of VRE share in consumption, the equation takes into consideration the joint slopes of source [22] for VRE shares below 45% and [47,50] above 45%, resulting in the final figures of storage fractions as shown in the figure below. With Equation number 1 (Table 6), it is possible to determine the energy storage capacity volumes of electricity (at stationary storage systems) at the European level, assuming appropriate demand-side management, smart market regulations, advanced weather forecasting systems, and continuous, ideal European network development to maintain secure electricity supplies through balancing. The R-square and adjusted R-square values of the MATLAB model derived from the input data are 1.

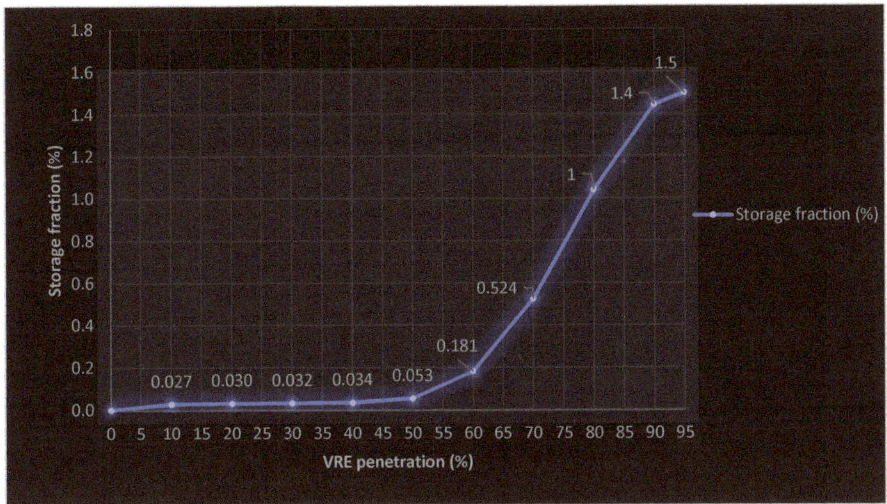

Figure 7. Result of the average European storage fraction analysis.

Table 6. Result of the European storage fraction analysis.

Description	Equation
Equation (1), storage fraction [%]	$Storage\ fraction - p_1 VRE^8 \mid p_2 VRE^7 + p_3 VRE^6 + p_4 VRE^5 + p_5 VRE^4 + p_6 VRE^3 + p_7 VRE^2 + p_8 VRE + p_9$
p_i parameter values	$p_1 = -3.758^{-14}$; $p_2 = -1.327^{-11}$; $p_3 = 1.818^{-09}$; $p_4 = -1.234^{-07}$; $p_5 = 4.443^{-06}$; $p_6 = -8.163^{-05}$; $p_7 = 5.844^{-04}$; $p_8 = 1.646^{-03}$; $p_8 = 3.687^{-04}$

3.2. Determination of the European Storage Power Capacity

It should be noted that there is no unified solution for determining the necessary future storage power capacity size requirements in Europe, but there are some well-defined ranges. Due to the lack of data, it is the studies [22,47,50,77] that can provide adequate information on the necessary future storage capacity requirements needed for balancing the European grid. In these sources, the 16–48% and the 80% VRE penetration ranges can be determined [22,47,50,77]. In the case of a European annual demand of 4900 TWh and 80% VRE integration, a 125 GW storage power capacity is recommended for grid balancing. In the model, the value of 125 GW was adjusted proportionally to the annual demand values of the 2040 ENTSO-E scenarios [22,47,50,77]. The starting point of the modeling was the power capacity context of [22] up to 45% of VRE penetration, then due to the lack of data, polynomial regression models were applied to the GW values of VRE gross electricity generation of 45–80%. The relationship is confirmed by the work of Cebulla et al. [49] and the results approximate the values of 'balanced' and 'wind+' of [49] (the PV and wind ratios between installed capacities for the first category were 2:1–1:1.5, while the second category value was 1:1.5). Based on this, the power capacity requirements for the VRE objectives of the ENTSO-E scenarios were determined (Figure 8). In addition, the theoretical maximum VRE integration potentials of the summarized European power capacity of all storage technologies were also determined.

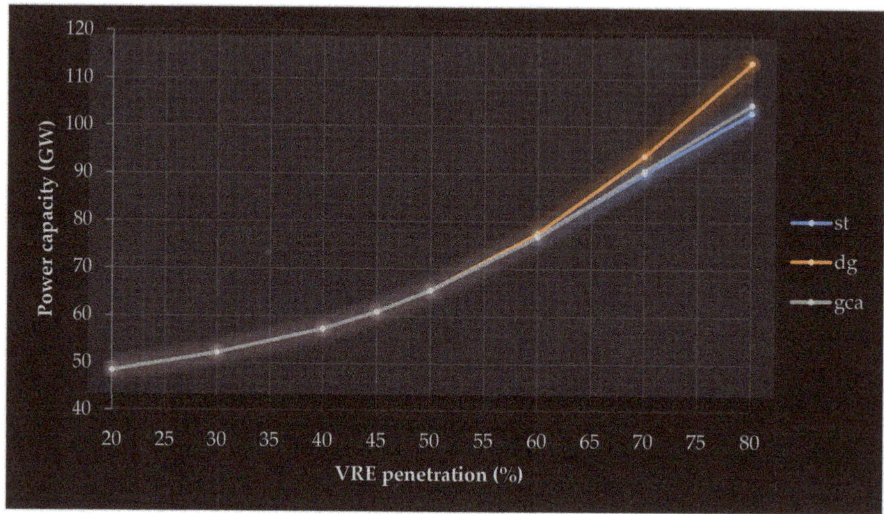

Figure 8. European power capacity requirement analysis results based on different VRE penetration levels of the 2040 ENTSO-E scenarios.

3.3. European Variable Renewable Energy Integration Possibilities

The power capacity of PHS was calculated on the basis of the ENTSO-E ST, DG, and GCA scenarios. From the energy capacity input data of [11,40,42–44] the sources, the PHS energy capacity was estimated by linear change due to lack of data (Tables 7 and 8). The International Renewable Energy Agency (IRENA) published a comprehensive study [10] on the future of energy storage trends, cost, and markets. This report breaks down the electricity storage energy capacity growth by storage technology according to four scenarios in 2030. However, the scenarios differ significantly and IRENA'S reference data were taken into account for the calculations. Considering the IRENA, the Global Energy Storage Database (DOE) studies, and the ENTSO-E Ten Years Network Development Plan 2018 Storage project database [9,10,78], it was assumed that the estimated European power capacity of other storage technologies will be 5% (scenario 1) and 25% (scenario 2) compared to the PHS values in 2040, and the average charge/discharge period was assumed as (other storage technologies) 8/8 h (scenario 1) and 12/12 h (scenario 2) (Tables 7 and 8).

Table 7. Power capacity and energy storage capacity results of the European energy storage systems in 2040, based on [9–11,27,40,42–44,78,79], scenario 1.

Year	2040 ST	2040, DG	2040, GCA
Power capacity of PHS [GW]	58		76
Estimated energy storage capacity, PHS [GWh]	3500		3900
Estimated European power capacity of other storage technologies, scenario 1 [GW]	2.9		3.8
Estimated energy storage capacity, other storage technologies, scenario 1 [GWh]	23,2		30.4
Summarized European power capacity of all storage technologies, scenario 1 [GW]	60.9		79.8
Summarized energy storage capacity of all storage technologies, scenario 1 [GWh]	3523.2		3930.4

Table 8. Power capacity and energy storage capacity results of the European energy storage systems in 2040, based on [9–11,27,40,42–44,78,79], scenario 2.

Year	2040 ST	2040, DG	2040, GCA
Power capacity of PHS [GW]	58		76
Estimated energy storage capacity, PHS [GWh]	3500		3900
Estimated European power capacity of other storage technologies, scenario 2 [GW]	14.5		19
Estimated energy storage capacity, other storage technologies, scenario 2 [GWh]	174		228
Summarized European power capacity of all storage technologies, scenario 2 [GW]	72.5		95
Summarized energy storage capacity of all storage technologies, scenario 2 [GWh]	3674		4128

From the electricity demand and the VRE penetration in the ENTSO-E ST, DG, and GCA scenarios, the energy storage capacity requirements and the storage fraction requirements were calculated by using Equation (1) (Table 9). Based on the analyzed scenarios, the fraction values were between 0.033% and 0.166%, which would mean 1.35–6.82 TWh energy storage capacities. In addition, the power capacity requirements of the energy storage systems of the three scenarios for the VRE integration would be in the range of 57–76 GW.

Table 9. Results related to the European storage power capacity and energy storage capacity requirements in 2040.

Year	2040, ST	2040, DG	2040, GCA
Electricity demand [TWh]	4030	4450	4100
ENTSO-E, maximum peak load [GW]	650	730	690
Annual VRE gross electricity generation compared to the demand [%]	39.7	50.6	59.3
Required storage fraction for the scenarios, based on equation 1 [%]	0.033	0.057	0.166
Required energy storage capacity for the scenarios, based on equation 1 [GWh]	1348	2518	6819
Required storage power capacity to the scenarios, based on the logic of Figure 8 [GW]	57	66	76

It was examined whether the VRE penetration targets of the ENTSO-E scenarios would be feasible considering the estimated storage power capacity and the energy capacity of stationary storage systems based on the approaches of scenario 1 and 2. It was also determined whether the power capacity or the energy storage capacity is the limiting factor in terms of successful VRE integration. Based on the results, we came to the conclusion that due to the need for a secure electricity supply both factors are equally important for successful VRE integration (Figure 9). The results showed that achieving a minimum of approximately 45–50% VRE penetration integration could be a realistic target in the European power grid sector until 2040. The ENTSO-E ST and DG scenarios appear to be rational goals. For the GCA scenario, the 55% VRE penetration rate seems to be feasible compared to the 59% target. According to the results, energy storage market developments and regulations that motivate the increased use of energy storage systems are of great importance for a successful European solar and wind energy integration.

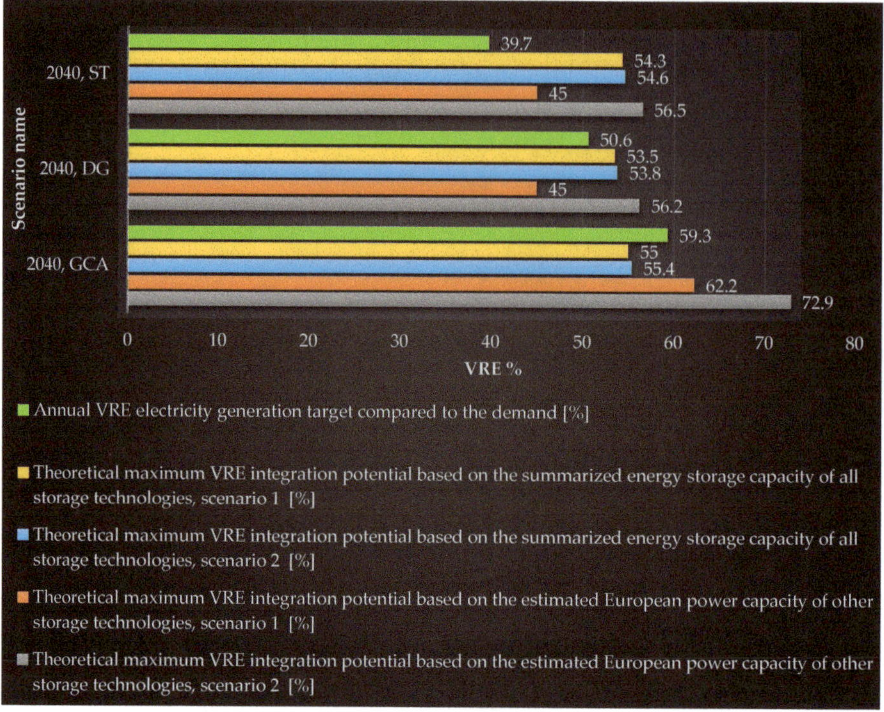

Figure 9. The feasibility of the European VRE integration target based on the various scenarios.

4. Conclusions

This study examined the European variable renewable energy integration challenges related to the power capacity and energy capacity of stationary storage technologies. It also analyzed and presented the feasibility of the European VRE electricity generation targets and the theoretical maximum related to the 2040 scenarios. The determination of the storage fractions, the power capacity, and the energy storage capacity were modelled in a new context. Based on the results we came to the conclusion that due to the requirement of a secure electricity supply, all factors are equally important for successful VRE integration. The results showed that achieving a minimum of approximately 45–50% VRE penetration unitl 2040 could be a realistic target in the European energy grid sector. The ENTSO-E ST and DG scenarios appear to be rational goals. For the GCA scenario, a 55% VRE penetration rate seems feasible compared to the 59% target. For the success of European VRE integration, energy storage market developments and regulations that motivate the increased use of energy storage systems are crucial.

Author Contributions: H.Z. was mainly responsible for the technical and modelling aspects, and conceived and designed the manuscript. All authors contributed equally in the analysis of the data and the writing and revision of the manuscript.

Acknowledgments: We acknowledge the financial support of Széchenyi 2020 under the EFOP-3.6.1-16-2016-00015.

Conflicts of Interest: The authors declare no conflict of interest.

Abbreviations

The following abbreviations are used in this manuscript:

CAES	Compressed air energy storage (-)
CF	Capacity Factor (%)
DG	Distributed Generation (-)
EU	European Union (-)
FES	Flywheel energy storage (-)
JRC-IDEES	Integrated Database of the European Energy Sector (-)
GCA	Global Climate Action (-)
GHG	Greenhouse gas (-)
NGC	Net generating capacity (-)
PHS	Pumped hydro storage (-)
PV	Photovoltaic (-)
RES	Renewable energy sources (-)
ST	Sustainable Transition (-)
SMES	Superconducting magnetic energy storage (-)
T&D	Transmission and distribution (-)
TES	Thermal energy storage (-)
RES	Renewable electricity sources (-)
UPS	Uninterruptable power supply
VRE	Variable renewable energy (-)

References

1. Zafar, R.; Mahmood, A.; Razzaq, S.; Ali, W.; Naeem, U.; Shehzad, K. Prosumer based energy management and sharing in smart grid. *Renew. Sustain. Energy Rev.* **2018**, *82*, 1675–1684. [CrossRef]
2. Zame, K.K.; Brehm, C.A.; Nitica, A.T.; Richard, C.L.; Schweitzer, G.D., III. Smart grid and energy storage: Policy recommendations. *Renew. Sustain. Energy Rev.* **2018**, *82*, 1646–1654. [CrossRef]
3. Kordmahaleh, A.A.; Naghashzadegan, M.; Javaherdeh, K.; Khoshgoftar, M. Design of a 25 MWe Solar Thermal Power Plant in Iran with Using Parabolic Trough Collectors and a Two-Tank Molten Salt Storage System. *Int. J. Photoenergy* **2017**, *2017*, 4210184. [CrossRef]
4. Noman, A.M.; Addoweesh, K.E.; Alolah, A.I. Simulation and Practical Implementation of ANFIS-Based MPPT Method for PV Applications Using Isolated Ćuk Converter. *Int. J. Photoenergy* **2017**, *2017*, 3106734. [CrossRef]
5. Daliento, S.; Chouder, A.; Guerriero, P.; Pavan, A.M.; Mellit, A.; Moeini, R.; Tricoli, P. Monitoring, Diagnosis, and Power Forecasting for Photovoltaic Fields: A Review. *Int. J. Photoenergy* **2017**, *2017*, 1356851. [CrossRef]
6. Cucchiella, F.; D'Adamo, I. A Multicriteria Analysis of Photovoltaic Systems: Energetic, Environmental, and Economic Assessments. *Int. J. Photoenergy* **2015**, *2015*, 627454. [CrossRef]
7. D'Adamo, I.; Miliacca, M.; Rosa, P. Economic Feasibility for Recycling of Waste Crystalline Silicon Photovoltaic Modules. *Int. J. Photoenergy* **2017**, *2017*, 4184676. [CrossRef]
8. Fathima, A.H.; Palanisamy, K. Energy Storage Systems for Energy Management of Renewables in Distributed Generation Systems. In *Energy Management of Distributed Generation Systems*; InTech: London, UK, 2016.
9. Sandia National Laboratories. DOE Global Energy Storage Database. Available online: https://www.energystorageexchange.org/projects/data_visualization (accessed on 2 March 2019).
10. International Renewable Energy Agency. *Electricity Storage and Renewables: Costs and Markets to 2030*; International Renewable Energy Agency: Abu Dhabi, UAE, 2017.
11. Gimeno-Gutiérrez, M.; Lacal-Arántegui, R. Assessment of the European potential for pumped hydropower energy storage based on two existing reservoirs. *Renew. Energy* **2015**, *75*, 856–868. [CrossRef]
12. The Conversation Trust (UK) Limited. As Nuclear Power Plants Close, States Need to Bet Big on Energy Storage. Available online: https://theconversation.com/as-nuclear-power-plants-close-states-need-to-bet-big-on-energy-storage-62032 (accessed on 26 September 2018).
13. Schill, W.-P. Residual load, renewable surplus generation and storage requirements in Germany. *Energy Policy* **2014**, *73*, 65–79. [CrossRef]

14. Luo, X.; Wang, J.; Dooner, M.; Clarke, J. Overview of current development in electrical energy storage technologies and the application potential in power system operation. *Appl. Energy* **2015**. [CrossRef]

15. Fraunhofer Institute for Solar Energy Systems ISE. *Annual Report 2016/2017*; Fraunhofer institute for solar energy systems ISE: Freiburg, Germany, 2017; pp. 1–88.

16. Zsiborács, H.; Pályi, B.; Pintér, G.; Popp, J.; Balogh, P.; Gabnai, Z.; Pető, K.; Farkas, I.; Baranyai, N.H.; Bai, A. Technical-economic study of cooled crystalline solar modules. *Sol. Energy* **2016**, *140*. [CrossRef]

17. *Renewables 2018 Global Status Report—REN21*; REN21: Paris, France, 2018.

18. Enjavi-Arsanjani, M.; Hirbodi, K.; Yaghoubi, M. Solar Energy Potential and Performance Assessment of CSP Plants in Different Areas of Iran. *Energy Procedia* **2015**, *69*, 2039–2048. [CrossRef]

19. Statista, I. Cumulative Solar Photovoltaic Capacity Globally as of 2017, by Select Country (in Gigawatts). Available online: https://www.statista.com/statistics/264629/existing-solar-pv-capacity-worldwide/ (accessed on 12 September 2018).

20. European Commission. *Energy Storage—The Role of Electricity*; European Commission: Brussels, Belgium, 2017.

21. Rodríguez, R.A.; Becker, S.; Andresen, G.B.; Heide, D.; Greiner, M. Transmission needs across a fully renewable European power system. *Renew. Energy* **2014**, *63*, 467–476. [CrossRef]

22. Bertsch, J.; Growitsch, C.; Lorenczik, S.; Nagl, S. Flexibility in Europe's power sector—An additional requirement or an automatic complement? *Energy Econ.* **2016**, *53*, 118–131. [CrossRef]

23. Cho, A. Energy's tricky tradeoffs. *Science* **2010**, *329*, 786–787. [CrossRef] [PubMed]

24. Jacobson, M.Z.; Delucchi, M.A. Providing all global energy with wind, water, and solar power, Part I: Technologies, energy resources, quantities and areas of infrastructure, and materials. *Energy Policy* **2011**, *39*, 1154–1169. [CrossRef]

25. Delucchi, M.A.; Jacobson, M.Z. Providing all global energy with wind, water, and solar power, Part II: Reliability, system and transmission costs, and policies. *Energy Policy* **2011**, *39*, 1170–1190. [CrossRef]

26. Czisch, G. Szenarien zur Zukünftigen Stromversorgung, Kostenoptimierte Variationen zur Versorgung Europas und Seiner Nachbarn mit Strom aus Erneuerbaren Energien. Ph.D. Thesis, Universitat Kassel, Kassel, Germany, 2005.

27. ENTSO-E. TYNDP 2018—Scenario Report. Available online: https://tyndp.entsoe.eu/tyndp2018/scenario-report/ (accessed on 12 April 2019).

28. Czisch, G.; Giebel, G. *Realisable Scenarios for a Future Electricity Supply Based 100% on Renewable Energies in: Energy Solutions for Sustainable Development: Proceedings of the Risø International Energy Conference 2007; [...took place 22–24 May 2007]*; Risø National Laboratory: Roskilde, Denmark, 2007; ISBN 9788755036031.

29. Kempton, W.; Pimenta, F.M.; Veron, D.E.; Colle, B.A. Electric power from offshore wind via synoptic-scale interconnection. *Proc. Natl. Acad. Sci. USA* **2010**, *107*, 7240–7245. [CrossRef]

30. Schaber, K.; Steinke, F.; Hamacher, T. Transmission grid extensions for the integration of variable renewable energies in Europe: Who benefits where? *Energy Policy* **2012**, *43*, 123–135. [CrossRef]

31. Schaber, K.; Steinke, F.; Mühlich, P.; Hamacher, T. Parametric study of variable renewable energy integration in Europe: Advantages and costs of transmission grid extensions. *Energy Policy* **2012**, *42*, 498–508. [CrossRef]

32. Widen, J. Correlations Between Large-Scale Solar and Wind Power in a Future Scenario for Sweden. *IEEE Trans. Sustain. Energy* **2011**, *2*, 177–184. [CrossRef]

33. Aboumahboub, T.; Schaber, K.; Tzscheutschler, P.; Hamacher, T. Optimal Configuration of a Renewable-based Electricity Supply Sector. *WSEAS Trans. POWER Syst.* **2010**, *5*, 120–129.

34. Yao, R.; Steemers, K. A method of formulating energy load profile for domestic buildings in the UK. *Energy Build.* **2005**, *37*, 663–671. [CrossRef]

35. Heide, D.; von Bremen, L.; Greiner, M.; Hoffmann, C.; Speckmann, M.; Bofinger, S. Seasonal optimal mix of wind and solar power in a future, highly renewable Europe. *Renew. Energy* **2010**, *35*, 2483–2489. [CrossRef]

36. Heide, D.; Greiner, M.; von Bremen, L.; Hoffmann, C. Reduced storage and balancing needs in a fully renewable European power system with excess wind and solar power generation. *Renew. Energy* **2011**, *36*, 2515–2523. [CrossRef]

37. Rasmussen, M.G.; Andresen, G.B.; Greiner, M. Storage and balancing synergies in a fully or highly renewable pan-European power system. *Energy Policy* **2012**, *51*, 642–651. [CrossRef]

38. Hedegaard, K.; Meibom, P. Wind power impacts and electricity storage—A time scale perspective. *Renew. Energy* **2012**, *37*, 318–324. [CrossRef]

39. Argyrou, M.C.; Christodoulides, P.; Kalogirou, S.A. Energy storage for electricity generation and related processes: Technologies appraisal and grid scale applications. *Renew. Sustain. Energy Rev.* **2018**, *94*, 804–821. [CrossRef]

40. PANNON Pro Innovations Ltd. Practical Experiences of PV and Storage Systems. Available online: https://klimainnovacio.hu/en/pannon-pro-innovations (accessed on 2 February 2018).

41. Entso-E. Statistics and Data—Electricity in Europe. Available online: https://www.entsoe.eu/publications/statistics-and-data/ (accessed on 11 April 2019).

42. Union of the Electricity Industry–EURELECTRIC. *Hydro in Europe: Powering Renewables Full Report*; EURELECTRIC: Brussels, Belgium, 2011.

43. Mantzos, L.; Matei, N.A.; Mulholland, E.; Rózsai, M.; Tamba, M.; Wiesenthal, T. Joint Research Centre Data Catalogue, JRC-IDEES 2015. Available online: http://data.jrc.ec.europa.eu/dataset/jrc-10110-10001 (accessed on 11 March 2019).

44. International Hydropower Association. Pumped Storage Tracking Tool. Available online: https://www.hydropower.org/hydropower-pumped-storage-tool (accessed on 19 December 2018).

45. International Hydropower Association. *Hydropower Status Report, 2018*; International Hydropower Association: London, UK, 2018.

46. European Academies Science Advisory Council. *ea sac Science Advice for the Benefit of Europe Valuing Dedicated Storage in Electricity Grids*; European Academies Science Advisory Council: Halle (Saale), Germany, 2017.

47. Blanco, H.; Faaij, A. A review at the role of storage in energy systems with a focus on Power to Gas and long-term storage. *Renew. Sustain. Energy Rev.* **2018**, *81*, 1049–1086. [CrossRef]

48. ENTSO-E. Europe Power System 2040: Completing the Map & Assessing the Cost of Non-Grid. Available online: https://tyndp.entsoe.eu/tyndp2018/power-system-2040/ (accessed on 12 April 2019).

49. Cebulla, F.; Haas, J.; Eichman, J.; Nowak, W.; Mancarella, P. How much electrical energy storage do we need? A synthesis for the U.S., Europe, and Germany. *J. Clean. Prod.* **2018**, *181*, 449–459. [CrossRef]

50. European Commission. *Roadmap 2050: A practical Guide to a Prosperous, Low- Carbon Europe*; European Climate Foundation (ECF): AM Den Haag, The Netherlands, 2010.

51. Kondziella, H.; Bruckner, T. Flexibility requirements of renewable energy based electricity systems—A review of research results and methodologies. *Renew. Sustain. Energy Rev.* **2016**, *53*, 10–22. [CrossRef]

52. Agora Energiewende. *Stromspeicher in der Energiewende*; Agora Energiewende: Berlin, Germany, 2014.

53. Fraunhofer Institute for Solar Energy Systems. *Roadmap Speicher—Speicherbedarf für Erneuerbare Energien Speicheralternativen—Speicheranreiz—Überwindung rechtlicher Hemmnisse*; Endbericht: Freiburg, Germany, 2014.

54. *D5.2—SPAIN, Overview of Current Status and Future Development Scenarios of the Electricity System, and Assessment of the Energy Storage Needs*; stoRE: Munich, Germany, 2013.

55. de Boer, H.S.; Grond, L.; Moll, H.; Benders, R. The application of power-to-gas, pumped hydro storage and compressed air energy storage in an electricity system at different wind power penetration levels. *Energy* **2014**, *72*, 360–370. [CrossRef]

56. Inage, S. *Prospects for Large-Scale Energy Storage in Decarbonised Power Grids*; International Energy Agency: Paris, France, 2009.

57. Strbac, G.; Aunedi, M.; Pudjianto, D.; Djapic, P.; Teng, F.; Sturt, A.; Jackravut, D.; Sansom, R.; Yufit, V.; Brandon, N. *Strategic Assessment of the Role and Value of Energy Storage Systems in the UK Low Carbon Energy Future*; Imperial College London: London, UK, 2012.

58. *Power System Overview and RES Integration, Ireland*; stoRE: Munich, Germany, 2013.

59. *Overview of the Electricity Supply System and an Estimation of Future Energy Storage Needs, Germany*; stoRE: Munich, Germany, 2013.

60. Ludig, S.; Haller, M.; Schmid, E.; Bauer, N. Fluctuating renewables in a long-term climate change mitigation strategy. *Energy* **2011**, *36*, 6674–6685. [CrossRef]

61. *D5.1—GREECE, Overview of the Electricity System Status and its Future Development Scenarios—Assessment of the Energy Storage Infrastructure Needs*; stoRE: Munich, Germany, 2013.

62. *Assessment of the Future Energy Storage Needs of Austria for Integration of Variable RES-E Generation, Deliverable 5.1—Austria*; stoRE: Munich, Germany, 2013.

63. Pfenninger, S.; Keirstead, J. Renewables, nuclear, or fossil fuels? Scenarios for Great Britain's power system considering costs, emissions and energy security. *Appl. Energy* **2015**, *152*, 83–93. [CrossRef]

64. Belderbos, A.; Delarue, E.; D'haeseleer, W. Possible role of power-to-gas in future energy systems. In Proceedings of the 2015 12th International Conference on the European Energy Market (EEM), Lisbon, Portugal, 19–22 May 2015; pp. 1–5.

65. *Overview of Current Status and Future Development Scenarios of the Electricity System in Denmark—Allowing Integration of Large Quantities of Wind Power*; stoRE: Munich, Germany, 2013.

66. Doetsch, C.; Droste-Franke, B.; Perrin, M. *Electric Energy Storage, Future Energy Storage Demand*; International Energy Agency: Paris, France, 2015.

67. Bertuccioli, L.; Chan, A.; Hart, D.; Lehner, F.; Madden, B.; Standen, E. *Development of Water Electrolysis in the European Union, Final Report*; Fuel cells and hydrogen Joint Undertaking: Lausann, Switzerland; Cambridge, UK, 2014.

68. Moeller, C.; Meiss, J.; Mueller, B.; Hlusiak, M.; Breyer, C.; Kastner, M.; Twele, J. Transforming the electricity generation of the Berlin–Brandenburg region, Germany. *Renew. Energy* **2014**, *72*, 39–50. [CrossRef]

69. Connolly, D.; Lund, H.; Mathiesen, B.V.; Leahy, M. The first step towards a 100% renewable energy-system for Ireland. *Appl. Energy* **2011**, *88*, 502–507. [CrossRef]

70. Aboumahboub, T.; Schaber, K.; Tzscheutschler, P.; Hamacher, T. Optimization of the utilization of renewable energy sources in the electricity sector. In Proceedings of the EE'10 Proceedings of the 5th IASME/WSEAS International Conference on Energy & Environment, Cambridge, UK, 23–25 February 2010; pp. 196–204.

71. Krakowski, V.; Assoumou, E.; Mazauric, V.; Maïzi, N. Feasible path toward 40–100% renewable energy shares for power supply in France by 2050: A prospective analysis. *Appl. Energy* **2016**, *171*, 501–522. [CrossRef]

72. Weitemeyer, S.; Kleinhans, D.; Vogt, T.; Agert, C. Integration of Renewable Energy Sources in future power systems: The role of storage. *Renew. Energy* **2015**, *75*, 14–20. [CrossRef]

73. Centre for Alternative Technology. Zero Carbon: Rethinking the Future—Centre for Alternative Technology. Available online: https://www.cat.org.uk/info-resources/zero-carbon-britain/research-reports/zero-carbon-rethinking-the-future/ (accessed on 12 April 2019).

74. Anagnostopoulos, J.S.; Papantonis, D.E. Study of pumped storage schemes to support high RES penetration in the electric power system of Greece. *Energy* **2012**, *45*, 416–423. [CrossRef]

75. Sauer, D.U.; Moser, A.; Leuthold, M.; Cai, Z.; Chen, H.; Alvarez, R.; Moos, M.; Thien, T. Storage Demand in a 100%-RES-Power System for Europe with Different PV-Installation Levels. In *28th European Photovoltaic Solar Energy Conference and Exhibition*; WIP: Paris, France, 2013; pp. 4261–4265.

76. Steinke, F.; Wolfrum, P.; Hoffmann, C. Grid vs. storage in a 100% renewable Europe. *Renew. Energy* **2013**, *50*, 826–832. [CrossRef]

77. European Climate Foundation. Roadmap 2050—Technical & Economic Analysis—Full Report. Available online: http://www.roadmap2050.eu/project/roadmap-2050 (accessed on 27 September 2018).

78. ENTSO-E. TYNDP 2018 Storage Project Sheets. Available online: https://tyndp.entsoe.eu/tyndp2018/projects/storage_projects (accessed on 1 May 2019).

79. International Renewable Energy Agency (IRENA). *Renewable Energy Capacity Statistics 2017*; IRENA: Abu Dhabi, UAE, 2017.

MDPI
St. Alban-Anlage 66
4052 Basel
Switzerland
Tel. +41 61 683 77 34
Fax +41 61 302 89 18
www.mdpi.com

Electronics Editorial Office
E-mail: electronics@mdpi.com
www.mdpi.com/journal/electronics